0588

CONCEPTS OF MODERN PHYSICS

CONCEPTS OF MODERN PHYSICS
Second Edition

Arthur Beiser

McGRAW-HILL BOOK COMPANY

New York   St. Louis   San Francisco   Düsseldorf   Johannesburg
Kuala Lumpur   London   Mexico   Montreal   New Delhi
Panama   Rio de Janeiro   Singapore   Sydney   Toronto

Library of Congress Cataloging in Publication Data

Beiser, Arthur.
    Concepts of modern physics.

    (McGraw-Hill series in fundamentals of physics)
    1. Matter-Constitution. 2. Quantum theory.
I. Title.
QC173.B413 1973   530.1   72-7089
ISBN 0-07-004363-9

CONCEPTS OF MODERN PHYSICS

        6 7 8 9 0 FGRFGR 7 9 8

This book was set in Laurel by York Graphic Services, Inc. The editors were Jack L. Farnsworth and Andrea Stryker-Rodda; the designer was Rafael Hernandez; and the production supervisor was Joe Campanella.
Fairfield Graphics was printer and binder.

CONTENTS

PART FOUR THE NUCLEUS

PREFACE

This book is intended for use with one-semester courses in modern physics that have elementary classical physics and calculus as prerequisites. Relativity and quantum theory are considered first to provide a framework for understanding the physics of atoms and nuclei. The theory of the atom is then developed with emphasis on quantum-mechanical notions, and is followed by a discussion of the properties of aggregates of atoms. Finally atomic nuclei and elementary particles are examined.

The balance here deliberately leans more toward ideas than toward experimental methods and practical applications, because I believe that the beginning student is better served in his introduction to modern physics by a conceptual framework than by a mass of individual details. However, all physical theories live or die by the sword of experiment, and a number of extended derivations are included in order to demonstrate exactly how an abstract concept can be related to actual measurements. Many instructors will prefer not to hold their students responsible for the more complicated (though not necessarily mathematically difficult) discussions, and I have indicated with asterisks sections that can be passed over lightly without loss of continuity; problems based on the contents of these sections are also marked with asterisks. Other omissions are also possible, of course. Relativity, for example, may well have been covered earlier, and Part 3 in its entirety may be skipped when its contents will be the subject of later work. Thus there is scope for an instructor to fashion the type of course he wishes, whether a general survey or a deeper inquiry into selected subjects, and to choose the level of treatment appropriate to his audience.

An expanded version of this book requiring no higher degree of mathematical preparation is my *Perspectives of Modern Physics,* an Upper Division Text in this series; other Upper Division Texts carry forward specific aspects of modern physics in detail.

In preparing this edition of *Concepts of Modern Physics* much of the original text has been reorganized and rewritten, the coverage of a number of topics has been broadened, and some material of peripheral interest has been dropped. I am grateful to Y. Beers and T. Satoh for their helpful suggestions in this regard.

Arthur Beiser

LIST OF ABBREVIATIONS

Å	angstrom
amp	ampere
atm	atmosphere
b	barn
C	coulomb
Ci	curie
d	day
eV	electron volt
F	farad
fm	fermi
g	gram
h	hour
Hz	hertz
J	joule
K	degree Kelvin
m	meter
mi	mile
min	minute
mol	mole
N	newton
s	second
T	tesla
u	atomic mass unit
V	volt
W	watt
yr	year

BASIC CONCEPTS 1

SPECIAL RELATIVITY 1

Our study of modern physics begins with a consideration of the special theory of relativity. This is a logical starting point, since all physics is ultimately concerned with measurement and relativity involves an analysis of how measurements depend upon the observer as well as upon what is observed. From relativity emerges a new mechanics in which there are intimate relationships between space and time, mass and energy. Without these relationships it would be impossible to understand the microscopic world within the atom whose elucidation is the central problem of modern physics.

1.1 THE MICHELSON-MORLEY EXPERIMENT

The wave theory of light was devised and perfected several decades before the electromagnetic nature of the waves became known. It was reasonable for the pioneers in optics to regard light waves as undulations in an all-pervading elastic medium called the *ether,* and their successful explanation of diffraction and interference phenomena in terms of ether waves made the notion of the ether so familiar that its existence was accepted without question. Maxwell's development of the electromagnetic theory of light in 1864 and Hertz's experimental confirmation of it in 1887 deprived the ether of most of its properties, but nobody at the time seemed willing to discard the fundamental idea represented by the ether: that light propagates relative to some sort of universal frame of reference. Let us consider an example of what this idea implies with the help of a simple analogy.

Figure 1-1 is a sketch of a river of width D which flows with the speed v. Two boats start out from one bank of the river with the same speed V. Boat A crosses the river to a point on the other bank directly opposite the starting point and then returns, while boat B heads downstream for the distance D and then returns to the starting point. Let us calculate the time required for each round trip.

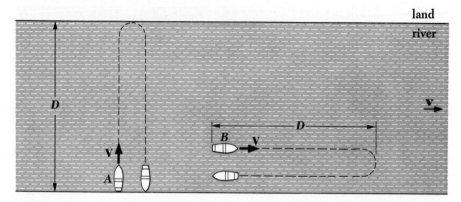

FIGURE 1-1   Boat A goes directly across the river and returns to its starting point, while boat B heads downstream for an identical distance and then returns.

We begin by considering boat A. If A heads perpendicularly across the river, the current will carry it downstream from its goal on the opposite bank (Fig. 1-2). It must therefore head somewhat upstream in order to compensate for the current. In order to accomplish this, its upstream component of velocity should be exactly $-v$ in order to cancel out the river current v, leaving the

FIGURE 1-2   Boat A must head upstream in order to compensate for the river current.

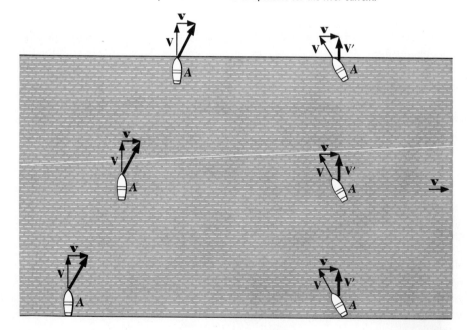

component $\mathbf{V}'$ as its net speed across the river. From Fig. 1-2 we see that these speeds are related by the formula

$$V^2 = V'^2 + v^2$$

so that the actual speed with which boat A crosses the river is

$$V' = \sqrt{V^2 - v^2}$$
$$= V\sqrt{1 - v^2/V^2}$$

Hence the time for the initial crossing is the distance D divided by the speed V'. Since the reverse crossing involves exactly the same amount of time, the total round-trip time t_A is twice D/V', or

1.1 $$t_A = \frac{2D/V}{\sqrt{1 - v^2/V^2}}$$

The case of boat B is somewhat different. As it heads downstream, its speed relative to the shore is its own speed V *plus* the speed v of the river (Fig. 1-3), and it travels the distance D downstream in the time $D/(V + v)$. On its return trip, however, B's speed relative to the shore is its own speed V *minus* the speed v of the river. It therefore requires the longer time $D/(V - v)$ to travel upstream the distance D to its starting point. The total round-trip time t_B is the sum of these times, namely,

$$t_B = \frac{D}{V + v} + \frac{D}{V - v}$$

Using the common denominator $(V + v)(V - v)$ for both terms,

$$t_B = \frac{D(V - v) + D(V + v)}{(V + v)(V - v)}$$
$$= \frac{2DV}{V^2 - v^2}$$

1.2 $$= \frac{2D/V}{1 - v^2/V^2}$$

which is greater than t_A, the corresponding round-trip time for the other boat. The ratio between the times t_A and t_B is

1.3 $$\frac{t_A}{t_B} = \sqrt{1 - v^2/V^2}$$

If we know the common speed V of the two boats and measure the ratio t_A/t_B, we can determine the speed v of the river.

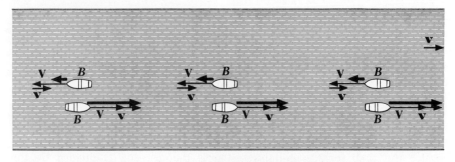

FIGURE 1-3 The speed of boat B downstream relative to the shore is increased by the speed of the river current while its speed upstream is reduced by the same amount.

The reasoning used in this problem may be transferred to the analogous problem of the passage of light waves through the ether. If there is an ether pervading space, we move through it with at least the 3×10^4 m/s (18.5 mi/s) speed of the earth's orbital motion about the sun; if the sun is also in motion, our speed through the ether is even greater (Fig. 1-4). From the point of view of an observer on the earth, the ether is moving past the earth. To detect this motion, we can use the pair of light beams formed by a half-silvered mirror instead of a pair of boats (Fig. 1-5). One of these light beams is directed to a

FIGURE 1-4 Motions of the earth through a hypothetical ether.

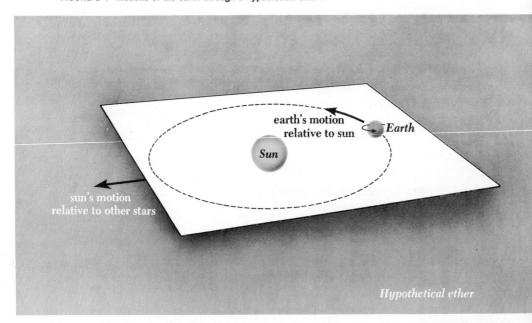

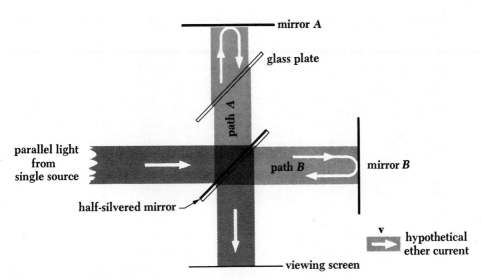

mirror A

glass plate

path A

parallel light
from
single source

path B

mirror B

half-silvered mirror

v

hypothetical
ether current

viewing screen

FIGURE 1-5   The Michelson-Morley experiment.

mirror along a path perpendicular to the ether current, while the other goes to a mirror along a path parallel to the ether current. The optical arrangement is such that both beams return to the same viewing screen. The purpose of the clear glass plate is to ensure that both beams pass through the same thicknesses of air and glass.

If the path lengths of the two beams are *exactly* the same, they will arrive at the screen in phase and will interfere constructively to yield a bright field of view. The presence of an ether current in the direction shown, however, would cause the beams to have different transit times in going from the half-silvered mirror to the screen, so that they would no longer arrive at the screen in phase but would interfere destructively. In essence this is the famous experiment performed in 1887 by the American physicists Michelson and Morley.

In the actual experiment the two mirrors are not perfectly perpendicular, with the result that the viewing screen appears crossed with a series of bright and dark interference fringes due to differences in path length between adjacent light waves (Fig. 1-6). If either of the optical paths in the apparatus is varied in length, the fringes appear to move across the screen as reinforcement and cancellation of the waves succeed one another at each point. The stationary apparatus, then, can tell us nothing about any time difference between the two paths. When the apparatus is rotated by 90°, however, the two paths change their orientations relative to the hypothetical ether stream, so that the beam formerly requiring the time t_A for the round trip now requires t_B and vice versa. If these times are different, the fringes will move across the screen during the rotation.

FIGURE 1-6 Fringe pattern observed in Michelson-Morley experiment.

Let us calculate the fringe shift expected on the basis of the ether theory. From Eqs. 1.1 and 1.2 the time difference between the two paths owing to the ether drift is

$$\Delta t = t_B - \dot{t}_A$$
$$= \frac{2D/V}{1 - v^2/V^2} - \frac{2D/V}{\sqrt{1 - v^2/V^2}}$$

Here v is the ether speed, which we shall take as the earth's orbital speed of 3×10^4 m/s, and V is the speed of light c, where $c = 3 \times 10^8$ m/s. Hence

$$\frac{v^2}{V^2} = \frac{v^2}{c^2}$$
$$= 10^{-8}$$

which is much smaller than 1. According to the binomial theorem, when x is extremely small compared with 1,

$$(1 \pm x)^n \approx 1 \pm nx$$

We may therefore express Δt to a good approximation as

$$\Delta t = \frac{2D}{c}\left[\left(1 + \frac{v^2}{c^2}\right) - \left(1 + \frac{1}{2}\frac{v^2}{c^2}\right)\right]$$
$$= \left(\frac{D}{c}\right)\left(\frac{v^2}{c^2}\right)$$

Here D is the distance between the half-silvered mirror and each of the other mirrors. The path difference d corresponding to a time difference Δt is

$$d = c\,\Delta t$$

If d corresponds to the shifting of n fringes,

$$d = n\lambda$$

where λ is the wavelength of the light used. Equating these two formulas for d, we find that

$$n = \frac{c\,\Delta t}{\lambda}$$

$$= \frac{Dv^2}{\lambda c^2}$$

In the actual experiment Michelson and Morley were able to make D about 10 m in effective length through the use of multiple reflections, and the wavelength of the light they used was about 5,000 Å $(1\text{ Å} = 10^{-10}\text{ m})$. The expected fringe shift in each path when the apparatus is rotated by 90° is therefore

$$n = \frac{Dv^2}{\lambda c^2}$$

$$= \frac{10\text{ m} \times (3 \times 10^4\text{ m/s})^2}{5 \times 10^{-7}\text{ m} \times (3 \times 10^8\text{ m/s})^2}$$

$$= 0.2\text{ fringe}$$

Since both paths experience this fringe shift, the total shift should amount to $2n$ or 0.4 fringe. A shift of this magnitude is readily observable, and therefore Michelson and Morley looked forward to establishing directly the existence of the ether.

To everybody's surprise, *no fringe shift whatever* was found. When the experiment was performed at different seasons of the year and in different locations, and when experiments of other kinds were tried for the same purpose, the conslusions were always identical: no motion through the ether was detected.

The negative result of the Michelson-Morley experiment had two consequences. First, it rendered untenable the hypothesis of the ether by demonstrating that the ether has no measurable properties—an ignominious end for what had once been a respected idea. Second, it suggested a new physical principle: the speed of light in free space is the same everywhere, regardless of any motion of source or observer.

1.2 THE SPECIAL THEORY OF RELATIVITY

We mentioned earlier the role of the ether as a universal frame of reference with respect to which light waves were supposed to propagate. Whenever we speak of "motion," of course, we really mean "motion relative to a frame of reference." The frame of reference may be a road, the earth's surface, the sun, the center of our galaxy; but in every case we must specify it. Stones dropped in Bermuda and in Perth, Australia, both fall "down," and yet the two move in exactly opposite directions relative to the earth's center. Which is the correct location of the frame of reference in this situation, the earth's surface or its center?

The answer is that *all* frames of reference are equally correct, although one may be more convenient to use in a specific case. *If* there were an ether pervading all space, we could refer all motion to it, and the inhabitants of Bermuda and Perth would escape from their quandary. The absence of an ether, then, implies that there is no universal frame of reference, since light (or, in general, electromagnetic waves) is the only means whereby information can be transmitted through empty space. All motion exists solely relative to the person or instrument observing it. If we are in a free balloon above a uniform cloud bank and see another free balloon change its position relative to us, we have no way of knowing which balloon is "really" moving. Should we be isolated in the universe, there would be no way in which we could determine whether we were in motion or not, because without a frame of reference the concept of motion has no meaning.

The theory of relativity resulted from an analysis of the physical consequences implied by the absence of a universal frame of reference. The special theory of relativity, developed by Albert Einstein in 1905, treats problems involving *inertial frames of reference*, which are frames of reference moving at constant velocity with respect to one another. The general theory of relativity, proposed by Einstein a decade later, treats problems involving frames of reference accelerated with respect to one another. An observer in an isolated laboratory *can* detect accelerations. Anybody who has been in an elevator or on a merry-go-round can verify this statement from his own experience. The special theory has had a profound influence on all physics, and we shall concentrate on it with only a brief glance at the general theory.

The special theory of relativity is based upon two postulates. The first states that **the laws of physics may be expressed in equations having the same form in all frames of reference moving at constant velocity with respect to one another.** This postulate expresses the absence of a universal frame of reference. If the laws of physics had different forms for different observers in relative motion, it could be determined from these differences which objects are "stationary" in space and which are "moving." But because there is no universal frame of reference, this distinction does not exist in nature; hence the above postulate.

The second postulate of special relativity states that **the speed of light in free space has the same value for all observers, regardless of their state of motion.**

This postulate follows directly from the results of the Michelson-Morley experiment and many others.

At first sight these postulates hardly seem radical. Actually they subvert almost all the intuitive concepts of time and space we form on the basis of our daily experience. A simple example will illustrate this statement. In Fig. 1-7 we have the two boats *A* and *B* once more, with boat *A* at rest in the water while boat *B* drifts at the constant velocity v. There is a low-lying fog present, and so on neither boat does the observer have any idea which is the moving one. At the instant that *B* is abreast of *A*, a flare is fired. The light from the flare travels uniformly in all directions, according to the second postulate of special relativity. An observer on either boat must find a sphere of light expanding with *himself* at its center, according to the first postulate of special relativity, even though one of them is changing his position with respect to the point where the flare went off. The observers cannot detect which of them is undergoing such a change in position since the fog eliminates any frame of reference other than each boat itself, and so, since the speed of light is the same for both of them, they must both see the identical phenomenon.

Why is the situation of Fig. 1-7 unusual? Let us consider a more familiar analog. The boats are at sea on a clear day and somebody on one of them drops a stone into the water when they are abreast of each other. A circular pattern

FIGURE 1-7 Relativistic phenomena differ from everyday experience.

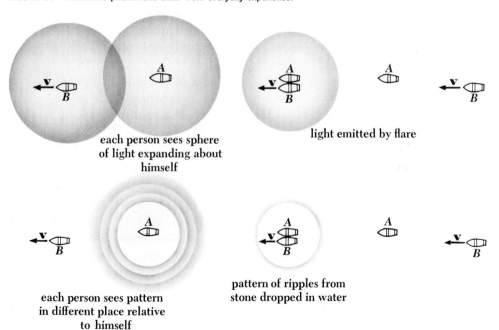

each person sees sphere
of light expanding about
himself

light emitted by flare

each person sees pattern
in different place relative
to himself

pattern of ripples from
stone dropped in water

of ripples spreads out, as at the bottom of Fig. 1-7, *which appears different* to observers on each boat. Merely by observing whether or not he is at the center of the pattern of ripples, each observer can tell whether he is moving relative to the water or not. Water is in itself a frame of reference, and an observer on a boat moving through it measures ripple speeds with respect to himself that are different in different directions, in contrast to the uniform ripple speed measured by an observer on a stationary boat. It is important to recognize that motion and waves *in water* are entirely different from motion and waves *in space;* water is in itself a frame of reference while space is not, and wave speeds in water vary with the observer's motion while wave speeds of light in space do not.

The only way of interpreting the fact that observers in the two boats in our example perceive identical expanding spheres of light is to regard the coordinate system of each observer, from the point of view of the other, as being affected by their relative motion. When this idea is developed, using only accepted laws of physics and Einstein's postulates, we shall see that many peculiar effects are predicted. One of the triumphs of modern physics is the experimental confirmation of these effects.

1.3 TIME DILATION

We shall first use the postulates of special relativity to investigate how relative motion affects measurements of time intervals and lengths.

A clock moving with respect to an observer appears to tick less rapidly than it does when at rest with respect to him. That is, if someone in a spacecraft finds that the time interval between two events in the spacecraft is t_0, we on the ground would find that the same interval has the longer duration t. The quantity t_0, which is determined by events that occur *at the same place* in an observer's frame of reference, is called the *proper time* of the interval between the events. When witnessed from the ground, the events that mark the beginning and end of the time interval occur at different places, and in consequence the duration of the interval appears longer than the proper time. This effect is called *time dilation.*

To see how time dilation comes about, let us examine the operation of the particularly simple clock shown in Fig. 1-8 and inquire how relative motion affects what we measure. This clock consists of a stick L_0 long with a mirror at each end. A pulse of light is reflected up and down between the mirrors, and an appropriate device is attached to one of the mirrors to give a "tick" of some kind each time the pulse of light strikes it. (Such a device might be

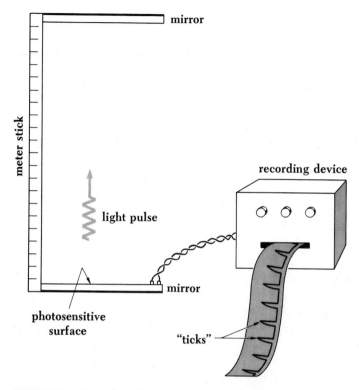

mirror

meter stick

light pulse

recording device

photosensitive
surface

mirror

"ticks"

FIGURE 1-8   A simple clock. Each "tick" corresponds to a round trip of the light
pulse from the lower mirror to the upper one and back.

a photosensitive surface on the mirror which can be arranged to give an electric
signal when the light pulse arrives.)   The proper time t_0 between ticks is

1.4 $$t_0 = \frac{2L_0}{c}$$

If the stick is 1 m long,

$$t_0 = \frac{2 \text{ m}}{3 \times 10^8 \text{ m/s}} = 0.67 \times 10^{-8} \text{ s}$$

and there are 1.5×10^8 ticks/s. Two identical clocks of this kind are built, and
one is attached to a spaceship mounted perpendicular to the direction of motion
while the other remains at rest on the earth's surface.

Now we ask how much time t elapses between ticks in the moving clock as
measured by an observer on the ground with an identical clock that is stationary
with respect to him. Each tick involves the passage of a pulse of light at speed

c from the lower mirror to the upper one and back. During this round-trip passage, the entire clock in the spaceship is in motion, which means that the pulse of light, as seen from the ground, actually follows a zigzag path (Fig. 1-9). On its way from the lower mirror to the upper one in the time $t/2$, the pulse of light travels a horizontal distance of $vt/2$ and a total distance of $ct/2$. Since L_0 is the vertical distance between the mirrors.

$$\left(\frac{ct}{2}\right)^2 = L_0{}^2 + \left(\frac{vt}{2}\right)^2$$

$$\frac{t^2}{4}(c^2 - v^2) = L_0{}^2$$

$$t^2 = \frac{4L_0{}^2}{c^2 - v^2} = \frac{(2L_0)^2}{c^2(1 - v^2/c^2)}$$

and

1.5 $$t = \frac{2L_0/c}{\sqrt{1 - v^2/c^2}}$$

But $2L_0/c$ is the time interval t_0 between ticks on the clock on the ground, as in Eq. 1.4, and so

1.6 $$t = \frac{t_0}{\sqrt{1 - v^2/c^2}}$$ **Time dilation**

FIGURE 1-9 A light clock in a spacecraft as seen by an observer at rest on the ground. The mirrors are parallel to the direction of motion of the spacecraft.

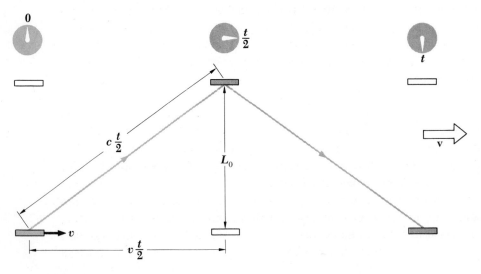

The moving clock in the spaceship appears to tick at a slower rate than the stationary one on the ground, as seen by an observer on the ground.

Exactly the same analysis holds for measurements of the clock on the ground by the pilot of the spaceship. To him, the light pulse of the ground clock follows a zigzag path which requires a total time t per round trip, while his own clock, at rest in the spaceship, ticks at intervals of t_0. He too finds that

$$t = \frac{t_0}{\sqrt{1 - v^2/c^2}}$$

so the effect is reciprocal: *every* observer finds that clocks in motion relative to him tick more slowly than when they are at rest.

Our discussion has been based on a somewhat unusual clock that employs a light pulse bouncing back and forth between two mirrors. Do the same conclusions apply to more conventional clocks that use machinery—spring-controlled escapements, tuning forks, or whatever—to produce ticks at constant time intervals? The answer must be yes, since if a mirror clock and a conventional clock in the spaceship agree with each other on the ground but not when in flight, the disagreement between them could be used to determine the speed of the spaceship without reference to any other object—which contradicts the principle that all motion is relative. Detailed calculations of what happens to conventional clocks in motion—as seen from the ground—confirm this answer. For example, as we shall learn in Sec. 1.10, the mass of an object is greater when it is in motion, so that the period of an oscillating object must be greater in the moving spaceship. Therefore *all* clocks at rest relative to one another behave the same to all observers, regardless of any motion at constant velocity of either the group of clocks or the observers.

The relative character of time has many implications. For example, events that seem to take place simultaneously to one observer may not be simultaneous to another observer in relative motion, and vice versa. Who is right? The question is, of course, meaningless: *both* observers are "right," since each simply measures what he sees.

Because simultaneity is a relative notion and not an absolute one, physical theories which require simultaneity in events at different locations must be discarded. The principle of conservation of energy in its elementary form states that the total energy content of the universe is constant, but it does not rule out a process in which a certain amount of energy ΔE vanishes at one point while an equal amount of energy ΔE spontaneously comes into being somewhere else with no actual transport of energy from one place to the other. Because simultaneity is relative, some observers of the process will find energy not being conserved. To rescue conservation of energy in the light of special relativity, then, it is necessary to say that, when energy disappears somewhere and appears elsewhere, it has actually *flowed* from the first location to the second. (There are many ways in which a flow of energy can occur, of course.) Thus energy is conserved *locally* in any arbitrary region of space at any time, not merely when the universe as a whole is considered—a much stronger statement of this principle.

Although time is a relative quantity, not all the notions of time formed by everyday experience are incorrect. Time does not run backward to *any* observer, for instance: a sequence of events that occur somewhere at t_1, t_2, t_3, ... will appear in the same order to all observers everywhere, though not necessarily with the same time intervals $t_2 - t_1$, $t_3 - t_2$, ... between each pair of events. Similarly, no distant observer, regardless of his state of motion, can see an event before it happens—more precisely, before a nearby observer sees it—since the speed of light is finite and signals require the minimum period of time L/c to travel a distance L. There is no way to peer into the future, although temporal (and, as we shall see, spatial) perspectives of past events may appear different to different observers.

1.4 THE TWIN PARADOX

We are now in a position to understand the famous relativistic phenomenon known as the twin paradox. This paradox involves two identical clocks, one of which remains on the earth while the other goes on a voyage into space at the speed v and returns a time t later. It is customary to replace actual clocks with a pair of identical male twins named A and B; this substitution is perfectly acceptable, because the processes of life—heartbeats, respiration, and so on—constitute biological clocks of reasonable regularity.

Twin A takes off when he is 20 yr old and travels at a speed of $0.99c$. To his brother B on the earth, A seems to be living more slowly, in fact at a rate only

$$\sqrt{1 - v^2/c^2} = \sqrt{1 - (0.99c)^2/c^2} = 0.14 = 14 \text{ percent}$$

as fast as B goes. For every breath that A takes, B takes 7; for every meal that A eats, B eats 7; for every thought that A thinks, B thinks 7. Finally, after 70 yr have elapsed by B's reckoning, A returns home, a man of only 30 while B is then 90 yr old.

Where is the paradox? If we examine the situation from the point of view of twin A in the spaceship, B on the earth is in motion at $0.99c$. Therefore we might expect B to be 30 yr old upon the return of the spaceship while A is 90 at this time—the precise opposite of what was concluded in the preceding paragraph.

The resolution of the paradox depends upon the fact that the spaceship is accelerated at various times in its journey: when it takes off, when it turns around, and when it finally comes to a stop. During each of these accelerations A was not in an inertial frame of reference, and the inertial frames corresponding to the outward and return trips were different. The earthbound twin B, on the other hand, was not accelerated and stayed in the same inertial frame all the time. What B measured may therefore be interpreted on the basis of special relativity, and his conclusion—that A is younger when he comes back—is correct. Of course, A's life-span has not been extended *to A himself*, since however long his 10 yr on the spacecraft may have seemed to his brother B, it has been only 10 yr as far as he is concerned. What has happened is that A's accelerations affected his life processes, and by applying the conclusions of general relativity for accelerated clocks we find that A is younger than B on his return by the exact amount expected on the basis of B's analysis using the formula for time dilation.

1.5 LENGTH CONTRACTION

Measurements of lengths as well as of time intervals are affected by relative motion. The length L of an object in motion with respect to an observer always appears to the observer to be shorter than its length L_0 when it is at rest with respect to him, a phenomenon known as the *Lorentz-FitzGerald contraction*. This contraction occurs only in the direction of the relative motion. The length L_0 of an object in its rest frame is called its *proper length*.

We can use the light clock of the previous section to investigate the Lorentz contraction. For this purpose we imagine the clock oriented so that the light pulse travels back and forth parallel to the direction in which the clock is moving relative to the observer (Fig. 1-10). At $t = 0$ the light pulse starts from the rear mirror, and it arrives at the front mirror at t_1. The pulse has traveled the distance ct_1 to reach the front mirror, where from the diagram

$$ct_1 = L + vt_1$$

Hence

1.7
$$t_1 = \frac{L}{c - v}$$

where L is the distance between the mirrors as measured by the observer at rest.

The pulse is then reflected by the front mirror and returns to the rear mirror at t after traveling the distance $c(t - t_1)$, where

$$c(t - t_1) = L - v(t - t_1)$$

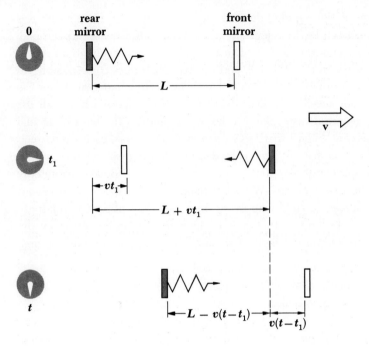

FIGURE 1-10 A light clock in a spacecraft as seen by an observer on the ground. The mirrors are perpendicular to the direction of motion of the spacecraft.

Hence the entire time interval t, as determined from the ground, is

1.8 $$t = \frac{L}{c+v} + t_1$$

We eliminate t_1 with the help of Eq. 1.7 to find that

$$t = \frac{L}{c+v} + \frac{L}{c-v}$$

$$= \frac{2Lc}{(c+v)(c-v)}$$

$$= \frac{2Lc}{c^2 - v^2}$$

1.9 $$= \frac{2L/c}{1 - v^2/c^2}$$

Equation 1.9 gives the time interval t between ticks of the moving clock as measured by an observer on the ground.

We earlier found another expression for t,

$$(1.5) \qquad t = \frac{2L_0/c}{\sqrt{1 - v^2/c^2}}$$

which is in terms of L_0, the proper distance between the mirrors, instead of in terms of L, the distance as measured by an observer in relative motion. The two formulas must be equivalent, and hence we have

$$\frac{2L/c}{1 - v^2/c^2} = \frac{2L_0/c}{\sqrt{1 - v^2/c^2}}$$

1.10 $\qquad L = L_0\sqrt{1 - v^2/c^2}$ <u>Lorentz contraction</u>

Because the relative velocity appears only as v^2 in Eq. 1.10, the Lorentz contraction is a reciprocal effect. To a man in a spacecraft, objects on the earth appear shorter than they did when he was on the ground by the same factor, $\sqrt{1 - v^2/c^2}$, that the spacecraft appears shorter to somebody at rest. The proper length of an object is the maximum length any observer will find.

The relativistic length contraction is negligible for ordinary speeds, but it is an important effect at speeds close to the speed of light. A speed of 1,000 mi/s seems enormous to us, and yet it results in a shortening in the direction of motion to only

$$\frac{L}{L_0} = \sqrt{1 - \frac{v^2}{c^2}}$$
$$= \sqrt{1 - \frac{(1,000 \text{ mi/s})^2}{(186,000 \text{ mi/s})^2}}$$
$$= 0.999985$$
$$= 99.9985 \text{ percent}$$

of the length at rest. On the other hand, a body traveling at 0.9 the speed of light is shortened to

$$\frac{L}{L_0} = \sqrt{1 - \frac{(0.9c)^2}{c^2}}$$
$$= 0.436$$
$$= 43.6 \text{ percent}$$

of the length at rest, a significant change.

The ratio between L and L_0 in Eq. 1.10 is the same as that in Eq. 1.3 when it is applied to the times of travel of the two light beams, so that we might be tempted to consider the Michelson-Morley result solely as evidence for the contraction of the length of their apparatus in the direction of the earth's motion.

This interpretation was tested by Kennedy and Thorndike in a similar experiment using an interferometer with arms of unequal length. They also found no fringe shift, which means that these experiments must be considered evidence for the absence of an ether with all this implies and not only for contractions of the apparatus.

An actual photograph of an object in very rapid relative motion would reveal a somewhat different distortion, depending upon the direction from which the object is viewed and the ratio v/c. The reason for this effect is that light reaching the camera (or eye, for that matter) from the more distant parts of the object was emitted earlier than that coming from the nearer parts; the camera "sees" a picture that is actually a composite, since the object was at different locations when the various elements of the single image that reaches the film left it. This effect supplements the Lorentz contraction by extending the apparent length of a moving object in the direction of motion. As a result, a three-dimensional body, such as a cube, may be seen as rotated in orientation as well as changed in shape, again depending upon the position of the observer and the value of v/c. This result must be distinguished from the Lorentz contraction itself, which is a physical phenomenon. If there were no Lorentz contraction, the appearance of a moving body would be also different from what it is at rest, but in another way.

It is interesting to note that the above approach to the visual appearance of rapidly moving objects was not made until 1959, 54 years after the publication of the special theory of relativity.

1.6 MESON DECAY

A striking illustration of both time dilation and length contraction occurs in the decay of unstable particles called μ *mesons,* whose properties we shall discuss in greater detail later. For the moment our interest lies in the fact that a μ meson decays into an electron an average of 2×10^{-6} s after it comes into being. Now μ mesons are created high in the atmosphere by fast cosmic-ray particles arriving at the earth from space, and reach sea level in profusion. Such mesons have a typical speed of 2.994×10^8 m/s, which is 0.998 of the velocity of light c. But in $t_0 = 2 \times 10^{-6}$ s, the meson's mean lifetime, they can travel a distance of only

$$
\begin{aligned}
y &= vt_0 \\
&= 2.994 \times 10^8 \text{ m/s} \times 2 \times 10^{-6} \text{ s} \\
&= 600 \text{ m}
\end{aligned}
$$

while they are actually created at altitudes more than 10 times greater than this.

We can resolve the meson paradox by using the results of the special theory of relativity. Let us examine the problem from the frame of reference of the meson, in which its lifetime is 2×10^{-6} s. In this case the distance from the meson to the ground appears shortened by the factor

$$\frac{y}{y_0} = \sqrt{1 - v^2/c^2}$$

That is, while we, on the ground, measure the altitude at which the meson is produced as y_0, the meson "sees" it as y. If we let y be 600 m, the maximum distance the meson can go *in its own frame of reference* at the speed 0.998c before decaying, we find that the corresponding distance y_0 in *our reference frame* is

$$y_0 = \frac{y}{\sqrt{1 - v^2/c^2}}$$

$$= \frac{600}{\sqrt{1 - \dfrac{(0.998c)^2}{c^2}}} \text{ m}$$

$$= \frac{600}{\sqrt{1 - 0.996}} \text{ m}$$

$$= \frac{600}{0.063} \text{ m}$$

$$= 9{,}500 \text{ m}$$

Hence, despite their brief life-spans, it is possible for the mesons to reach the ground from the considerable altitudes at which they are actually formed.

Now let us examine the problem from the frame of reference of an observer on the ground. From the ground the altitude at which the meson is produced is y_0, but its lifetime in *our reference frame* has been extended, owing to the relative motion, to the value

$$t = \frac{t_0}{\sqrt{1 - v^2/c^2}}$$

$$= \frac{2 \times 10^{-6}}{\sqrt{1 - \dfrac{(0.998c)^2}{c^2}}} \text{ s}$$

$$= \frac{2 \times 10^{-6}}{0.063} \text{ s}$$

$$= 31.7 \times 10^{-6} \text{ s}$$

almost 16 times greater than when it is at rest with respect to us. In 31.7×10^{-6} s a meson whose speed is $0.998c$ can travel a distance

$$y_0 = vt$$
$$= 2.994 \times 10^8 \text{ m/s} \times 31.7 \times 10^{-6} \text{ s}$$
$$= 9,500 \text{ m}$$

the same distance obtained before. The two points of view give identical results.

*1.7 THE LORENTZ TRANSFORMATION

Let us suppose that we are in a frame of reference S and find that the coordinates of some event that occurs at the time t are x, y, z. An observer located in a different frame of reference S' which is moving with respect to S at the constant velocity v will find that the same event occurs at the time t' and has the coordinates x', y', z'. (In order to simplify our work, we shall assume that $\mathbf{v}$ is in the $+x$ direction, as in Fig. 1-11.) How are the measurements x, y, z, t related to x', y', z', t'?

If we are unaware of special relativity, the answer seems obvious enough. If time in both systems is measured from the instant when the origins of S and

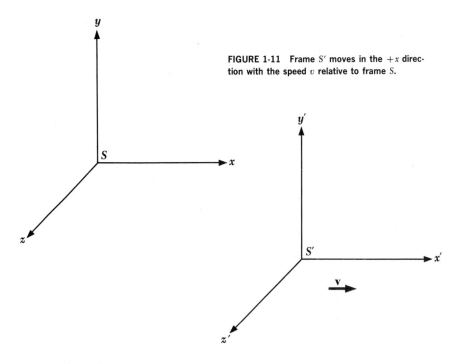

FIGURE 1-11 Frame S' moves in the $+x$ direction with the speed v relative to frame S.

S' coincided, measurements in the x direction made in S will exceed those made in S' by the amount vt, which represents the distance that S' has moved in the x direction. That is

1.11 $\qquad x' = x - vt$

There is no relative motion in the y and z directions, and so

1.12 $\qquad y' = y$

1.13 $\qquad z' = z$

In the absence of any indication to the contrary in our everyday experience, we further assume that

1.14 $\qquad t' = t$

The set of Eqs. 1.11 to 1.14 is known as the *Galilean transformation*.

To convert velocity components measured in the S frame to their equivalents in the S' frame according to the Galilean transformation, we simply differentiate x', y', and z' with respect to time:

1.15 $\qquad v'_x = \dfrac{dx'}{dt'} = v_x - v$

1.16 $\qquad v'_y = \dfrac{dy'}{dt'} = v_y$

1.17 $\qquad v'_z = \dfrac{dz'}{dt'} = v_z$

While the Galilean transformation and the velocity transformation it leads to are both in accord with our intuitive expectations, they violate both of the postulates of special relativity. The first postulate calls for identical equations of physics in both the S and S' frames of reference, but the fundamental equations of electricity and magnetism assume very different forms when the Galilean transformation is used to convert quantities measured in one frame into their equivalents in the other. The second postulate calls for the same value of the speed of light c whether determined in S or S'. If we measure the speed of light in the x direction in the S system to be c, however, in the S' system it will be

$$c' = c - v$$

according to Eq. 1.15. Clearly a different transformation is required if the postulates of special relativity are to be satisfied. We would expect both time dilation and length contraction to follow naturally from this new transformation.

A reasonable guess as to the nature of the correct relationship between x and x' is

1.18 $x' = k(x - vt)$

where k is a factor of proportionality that does not depend upon either x or t but may be a function of v. The choice of Eq. 1.18 follows from several considerations:

1. It is linear in x and x', so that a single event in frame S corresponds to a single event in frame S', as it must.
2. It is simple, and a simple solution to a problem should always be explored first.
3. It has the possibility of reducing to Eq. 1.11, which we know to be correct in ordinary mechanics.

Because the equations of physics must have the same form in both S and S', we need only change the sign of v (in order to take into account the difference in the direction of relative motion) to write the corresponding equation for x in terms of x' and t':

1.19 $x = k(x' + vt')$

The factor k must be the same in both frames of reference since there is no difference between S and S' other than in the sign of v.

As in the case of the Galilean transformation, there is nothing to indicate that there might be differences between the corresponding coordinates y, y' and z, z' which are normal to the direction of v. Hence we again take

1.20 $y' = y$

1.21 $z' = z$

The time coordinates t and t', however, are *not* equal. We can see this by substituting the value of x' given by Eq. 1.18 into Eq. 1.19. We obtain

$$x = k^2(x - vt) + kvt'$$

from which we find that

1.22 $t' = kt + \left(\dfrac{1 - k^2}{kv}\right)x \; \approx \; \dfrac{x}{v}\left(\dfrac{1}{k} - k\right) - T\,K$

Equations 1.18 and 1.20 to 1.22 constitute a coordinate transformation that satisfies the first postulate of special relativity.

The second postulate of relativity enables us to evaluate k. At the instant $t = 0$, the origins of the two frames of reference S and S' are in the same place, according to our initial conditions, and $t' = 0$ then also. Suppose that a flare is set off at the common origin of S and S' at $t = t' = 0$, and the observers in each system proceed to measure the speed with which the light from it spreads out. Both observers must find the same speed c (Fig. 1-7), which means that in the S frame

1.23 $$x = ct$$

while in the S' frame

1.24 $$x' = ct'$$

Substituting for x' and t' in Eq. 1.24 with the help of Eqs. 1.18 and 1.22,

$$k(x - vt) = ckt + \left(\frac{1 - k^2}{kv}\right)cx$$

and solving for x,

$$x = \frac{ckt + vkt}{k - \left(\dfrac{1 - k^2}{kv}\right)c}$$

$$= ct\left[\frac{k + \dfrac{v}{c}k}{k - \left(\dfrac{1 - k^2}{kv}\right)c}\right]$$

$$= ct\left[\frac{1 + \dfrac{v}{c}}{1 - \left(\dfrac{1}{k^2} - 1\right)\dfrac{c}{v}}\right]$$

This expression for x will be the same as that given by Eq. 1.23, namely $x = ct$, provided that the quantity in the brackets equals 1. Therefore

$$\frac{1 + \dfrac{v}{c}}{1 - \left(\dfrac{1}{k^2} - 1\right)\dfrac{c}{v}} = 1$$

and

1.25 $$k = \frac{1}{\sqrt{1 - v^2/c^2}}$$

Inserting the above value of k in Eqs. 1.18 and 1.22, we have, for the complete transformation of measurements of an event made in S to the corresponding measurements made in S', the equations

1.26
$$x' = \frac{x - vt}{\sqrt{1 - v^2/c^2}}$$

1.27
$$y' = y$$

1.28
$$z' = z$$ **Lorentz transformation**

1.29
$$t' = \frac{t - \dfrac{vx}{c^2}}{\sqrt{1 - v^2/c^2}}$$

These equations comprise the *Lorentz transformation*. They were first obtained by the Dutch physicist H. A. Lorentz, who showed that the basic formulas of electromagnetism are the same in all frames of reference in uniform relative motion only when these transformation equations are used. It was not until a number of years later that Einstein discovered their full significance. It is obvious that the Lorentz transformation reduces to the Galilean transformation when the relative velocity v is small compared with the velocity of light c.

The relativistic length contraction follows directly from the Lorentz transformation. Let us consider a rod lying along the x' axis in the moving frame S'. An observer in this frame determines the coordinates of its ends to be x_1' and x_2', and so the proper length of the rod is

$$L_0 = x_2' - x_1'$$

In order to find $L = x_2 - x_1$, the length of the rod as measured in the stationary frame S at the time t, we make use of Eq. 1.26. We have

$$x_1' = \frac{x_1 - vt}{\sqrt{1 - v^2/c^2}}$$

$$x_2' = \frac{x_2 - vt}{\sqrt{1 - v^2/c^2}}$$

and so

$$L = x_2 - x_1$$
$$= (x_2' - x_1')\sqrt{1 - v^2/c^2}$$
$$= L_0\sqrt{1 - v^2/c^2}$$

which is the same as Eq. 1.10.

In the previous section the coordinates of the ends of the moving rod were measured in the stationary frame S at the same time t, and it was easy to use Eq. 1.26 to find L in terms of L_0 and v. If we want to examine time dilation, though, Eq. 1.29 is not convenient, because t_1 and t_2, the start and finish of the chosen time interval, must be measured when the moving clock is at the respective *different* positions x_1 and x_2. In situations of this kind it is easier to use the *inverse Lorentz transformation*, which converts measurements made in the moving frame S' to their equivalents in S. To obtain the inverse transformation, primed and unprimed quantities in Eqs. 1.26 to 1.29 are exchanged, and v is replaced by $-v$:

1.30
$$x = \frac{x' + vt'}{\sqrt{1 - v^2/c^2}}$$

1.31
$$y = y'$$

Inverse Lorentz transformation

1.32
$$z = z'$$

1.33
$$t = \frac{t' + \frac{vx'}{c^2}}{\sqrt{1 - v^2/c^2}}$$

Let us consider a clock at the point x' in the moving frame S'. When an observer in S' finds that the time is t_1', an observer in S will find it to be t_1, where, from Eq. 1.33,

$$t_1 = \frac{t_1' + \frac{vx'}{c^2}}{\sqrt{1 - v^2/c^2}}$$

After a time interval of t_0 (to him), the observer in the moving system finds that the time is now t_2' according to his clock. That is,

$$t_0 = t_2' - t_1'$$

The observer in S, however, measures the end of the same time interval to be

$$t_2 = \frac{t_2' + \frac{vx'}{c^2}}{\sqrt{1 - v^2/c^2}}$$

so to him the duration of the interval t is

$$t = t_2 - t_1$$

$$= \frac{t_2' - t_1'}{\sqrt{1 - v^2/c^2}}$$

or

$$t = \frac{t_0}{\sqrt{1 - v^2/c^2}}$$

as we found earlier with the help of a light-pulse clock.

*1.9 VELOCITY ADDITION

One of the postulates of special relativity states that the speed of light c in free space has the same value for all observers, regardless of their relative motion. But "common sense" tells us that, if we throw a ball forward at 50 ft/s from a car moving at 80 ft/s, the ball's speed relative to the ground is 130 ft/s, the sum of the two speeds. Hence we would expect that a ray of light emitted in a frame of reference S' in the direction of its motion at velocity v relative to another frame S will have a speed of $c + v$ as measured in S, contradicting the above postulate. "Common sense" is no more reliable as a guide in science than it is elsewhere, and we must turn to the Lorentz transformation equations for the correct scheme of velocity addition.

Let us consider something moving relative to both S and S'. An observer in S measures its three velocity components to be

$$V_x = \frac{dx}{dt} \qquad V_y = \frac{dy}{dt} \qquad V_z = \frac{dz}{dt}$$

while to an observer in S' they are

$$V_x' = \frac{dx'}{dt'} \qquad V_y' = \frac{dy'}{dt'} \qquad V_z' = \frac{dz'}{dt'}$$

By differentiating the inverse Lorentz transformation equations for x, y, z, and t, we obtain

$$dx = \frac{dx' + v\,dt'}{\sqrt{1 - v^2/c^2}}$$

$$dy = dy'$$

$$dz = dz'$$

$$dt = \frac{dt' + \dfrac{v\,dx'}{c^2}}{\sqrt{1 - v^2/c^2}}$$

and so

$$V_x = \frac{dx}{dt}$$

$$= \frac{dx' + v\,dt'}{dt' + \dfrac{v\,dx'}{c^2}}$$

$$= \frac{\dfrac{dx'}{dt'} + v}{1 + \dfrac{v}{c^2}\dfrac{dx'}{dt'}}$$

1.34 $$V_x = \frac{V'_x + v}{1 + \dfrac{vV'_x}{c^2}}$$ **Relativistic velocity transformation**

Similarly

1.35 $$V_y = \frac{V'_y \sqrt{1 - v^2/c^2}}{1 + \dfrac{vV'_x}{c^2}}$$

1.36 $$V_z = \frac{V'_z \sqrt{1 - v^2/c^2}}{1 + \dfrac{vV'_x}{c^2}}$$

If $V'_x = c$, that is, if a ray of light is emitted in the moving reference frame S' in its direction of motion relative to S, an observer in frame S will measure the velocity

$$V_x = \frac{V'_x + v}{1 + \dfrac{vV'_x}{c^2}}$$

$$= \frac{c + v}{1 + \dfrac{vc}{c^2}}$$

$$= \frac{c(c + v)}{c + v}$$

$$= c$$

Both observers determine the same value for the speed of light, as they must.

The relativistic velocity transformation has other peculiar consequences. For instance, we might imagine wishing to pass a space ship whose speed with respect to the earth in $0.9c$ at a relative speed of $0.5c$. According to conventional mechanics our required speed relative to the earth would have to be $1.4c$, more than the velocity of light. According to Eq. 1.34, however, with $V'_x = 0.5c$ and $v = 0.9c$, the necessary speed is only

$$V_x = \frac{V'_x + v}{1 + \dfrac{vV'_x}{c^2}}$$

$$= \frac{0.5c + 0.9c}{1 + \dfrac{(0.9c)(0.5c)}{c^2}}$$

$$= 0.9655c$$

which is less than c. We need go less than 10 percent faster than a space ship traveling at $0.9c$ in order to pass it at a relative speed of $0.5c$.

1.10 THE RELATIVITY OF MASS

Until now we have been considering only the purely kinematical aspects of special relativity. The dynamical consequences of relativity are at least as remarkable, including as they do the variation of mass with velocity and the equivalence of mass and energy.

We begin by considering an elastic collision (that is, a collision in which kinetic energy is conserved) between two particles A and B, as witnessed by observers in the reference frames S and S' which are in uniform relative motion. The properties of A and B are identical when determined in reference frames in which they are at rest. The frames S and S' are oriented as in Fig. 1-12, with S' moving in the $+x$ direction with respect to S at the velocity v.

Before the collision, particle A had been at rest in frame S and particle B in frame S'. Then, at the same instant, A was thrown in the $+y$ direction at the speed V_A while B was thrown in the $-y'$ direction at the speed V'_B, where

1.37 $\qquad V_A = V'_B$

Hence the behavior of A as seen from S is exactly the same as the behavior of B as seen from S'. When the two particles collide, A rebounds in the $-y$ direction at the speed V_A, while B rebounds in the $+y'$ direction at the speed V'_B. If the particles are thrown from positions Y apart, an observer in S finds that the collision occurs at $y = \frac{1}{2}Y$ and one in S' finds that it occurs at $y' = \frac{1}{2}Y$.

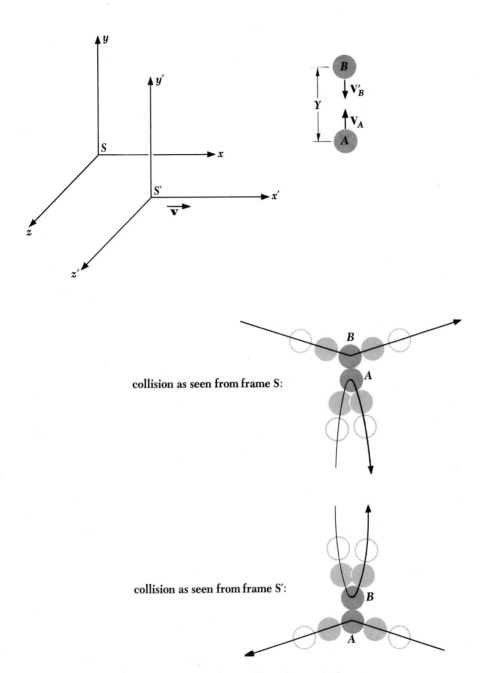

collision as seen from frame S:

collision as seen from frame S':

FIGURE 1-12 An elastic collision as observed in two different frames of reference.

The round-trip time T_0 for A as measured in frame S is therefore

1.38
$$T_0 = \frac{Y}{V_A}$$

and it is the same for B in S',

$$T_0 = \frac{Y}{V'_B}$$

If momentum is conserved in the S frame, it must be true that

1.39
$$m_A V_A = m_B V_B$$

where m_A and m_B are the masses of A and B, and V_A and V_B their velocities *as measured in the S frame.* In S the speed V_B is found from

1.40
$$V_B = \frac{Y}{T}$$

where T is the time required for B to make its round trip *as measured in S.* In S', however, B's trip requires the time T_0, where

1.41
$$T = \frac{T_0}{\sqrt{1 - v^2/c^2}}$$

according to our previous results. Although observers in both frames see the same event, they disagree as to the length of time the particle thrown from the other frame requires to make the collision and return.

Replacing T in Eq. 1.40 with its equivalent in terms of T_0, we have

$$V_B = \frac{Y\sqrt{1 - v^2/c^2}}{T_0}$$

From Eq. 1.38

$$V_A = \frac{Y}{T_0}$$

Inserting these expressions for V_A and V_B in Eq. 1.39, we see that momentum is conserved provided that

1.42
$$m_A = m_B\sqrt{1 - v^2/c^2} \qquad \frac{V_B}{V_A}$$

Our original hypothesis was that A and B are identical when at rest with respect to an observer; the difference between m_A and m_B therefore means that measurements of mass, like those of space and time, depend upon the relative speed between an observer and whatever he is observing.

In the above example both A and B are moving in S. In order to obtain a formula giving the mass m of a body measured while in motion in terms of its mass m_0 when measured at rest, we need only consider a similar example in which V_A and V_B' are very small. In this case an observer in S will see B approach A with the velocity v, make a glancing collision (since $V_B' \ll v$), and then continue on. In S

$$m_A = m_0$$

and

$$m_B = m$$

and so

1.43
$$m = \frac{m_0 \ \text{REST}}{\sqrt{1 - v^2/c^2}}$$
    Relativistic mass

The mass of a body moving at the speed v relative to an observer is larger than its mass when at rest relative to the observer by the factor $1/\sqrt{1 - v^2/c^2}$. This mass increase is reciprocal; to an observer in S'

$$m_A = m$$

and

$$m_B = m_0$$

Measured from the earth, a rocket ship in flight is shorter than its twin still on the ground and its mass is greater. To somebody on the rocket ship in flight the ship on the ground also appears shorter and to have a greater mass. (The effect is, of course, unobservably small for actual rocket speeds.) Equation 1.43 is plotted in Fig. 1-13.

$$KE = mc^2 - m_0 c^2$$

Provided that momentum is defined as

1.44
$$mv = \frac{m_0 v}{\sqrt{1 - v^2/c^2}}$$

conservation of momentum is valid in special relativity just as in classical physics. However, Newton's second law of motion is correct only in the form

$$F = \frac{d}{dt}(mv)$$

1.45
$$= \frac{d}{dt}\left[\frac{m_0 v}{\sqrt{1 - v^2/c^2}} \right]$$

This is *not* equivalent to saying that

$$F = ma$$

$$= m\frac{dv}{dt}$$

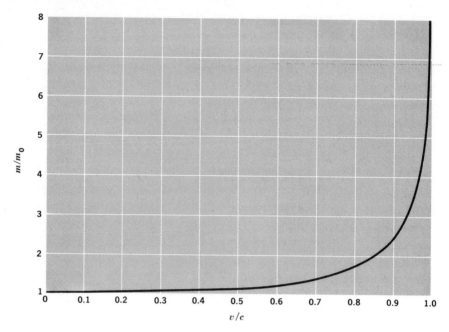

FIGURE 1-13 The relativity of mass.

even with m given by Eq. 1.43, because

$$\frac{d}{dt}(mv) = m\frac{dv}{dt} + v\frac{dm}{dt}$$

and dm/dt does not vanish if the speed of the body varies with time. The resultant force on a body is always equal to the time rate of change of its momentum.

Relativistic mass increases are significant only at speeds approaching that of light. At a speed one-tenth that of light the mass increase amounts to only 0.5 percent, but this increase is over 100 percent at a speed nine-tenths that of light. Only atomic particles such as electrons, protons, mesons, and so on have sufficiently high speeds for relativistic effects to be measurable, and in dealing with these particles the "ordinary" laws of physics cannot be used. Historically, the first confirmation of Eq. 1.43 was the discovery by Bücherer in 1908 that the ratio e/m of the electron's charge to its mass is smaller for fast electrons than for slow ones; this equation, like the others of special relativity, has been verified by so many experiments that it is now recognized as one of the basic formulas of physics.

1.11 MASS AND ENERGY

The most famous relationship Einstein obtained from the postulates of special relativity concerns mass and energy. This relationship can be derived directly from the definition of the kinetic energy T of a moving body as the work done in bringing it from rest to its state of motion. That is,

$$T = \int_0^s F \, ds$$

where F is the component of the applied force in the direction of the displacement ds and s is the distance over which the force acts. Using the relativistic form of the second law of motion

$$F = \frac{d(mv)}{dt}$$

the expression for kinetic energy becomes

$$T = \int_0^s \frac{d(mv)}{dt} \, ds$$

$$= \int_0^{mv} v \, d(mv)$$

$$= \int_0^v v \, d\left(\frac{m_0 v}{\sqrt{1 - v^2/c^2}}\right)$$

Integrating by parts ($\int x \, dy = xy - \int y \, dx$),

$$T = \frac{m_0 v^2}{\sqrt{1 - v^2/c^2}} - m_0 \int_0^v \frac{v \, dv}{\sqrt{1 - v^2/c^2}}$$

$$= \frac{m_0 v^2}{\sqrt{1 - v^2/c^2}} + m_0 c^2 \sqrt{1 - v^2/c^2} \Bigg|_0^v$$

$$= \frac{m_0 c^2}{\sqrt{1 - v^2/c^2}} - m_0 c^2$$

1.46
$$= mc^2 - m_0 c^2$$

Equation 1.46 states that the kinetic energy of a body is equal to the increase in its mass consequent upon its relative motion multiplied by the square of the speed of light.

Equation 1.46 may be rewritten

1.47
$$mc^2 = T + m_0 c^2$$

If we interpret mc^2 as the *total energy* E of the body, it follows that, when the body is at rest and $T = 0$, it nevertheless possesses the energy m_0c^2. Accordingly m_0c^2 is called the *rest energy* E_0 of a body whose mass at rest is m_0. Equation 1.47 therefore becomes

$$E = E_0 + T$$

where

1.48 $\qquad E_0 = m_0c^2$ **Rest energy**

In addition to its kinetic, potential, electromagnetic, thermal, and other familiar guises, then, energy can manifest itself as mass. The conversion factor between the unit of mass (kg) and the unit of energy (J) is c^2, so 1 kg of matter has an energy content of 9×10^{16} J. Even a minute bit of matter represents a vast amount of energy, and, in fact, the conversion of matter into energy is the source of the power liberated in all the exothermic reactions of physics and chemistry.

Since mass and energy are not independent entities, the separate conservation principles of energy and mass are properly a single one, the principle of conservation of mass energy. Mass *can* be created or destroyed, but when this happens an equivalent amount of energy simultaneously vanishes or comes into being, and vice versa. Mass and energy are different aspects of the same thing.

When the relative speed v is small compared with c, the formula for kinetic energy must reduce to the familiar $\frac{1}{2}m_0v^2$, which has been verified by experiment at low speeds. Let us see whether this is true. The binomial theorem of algebra tells us that if some quantity x is much smaller than 1,

$$(1 \pm x)^n \approx 1 \pm nx$$

The relativistic formula for kinetic energy is

$$T = mc^2 - m_0c^2$$
$$= \frac{m_0c^2}{\sqrt{1 - v^2/c^2}} - m_0c^2$$

Expanding the first term of this formula with the help of the binomial theorem, with $v^2/c^2 \ll 1$ since v is much less than c,

$$T = (1 + \tfrac{1}{2}v^2/c^2)m_0c^2 - m_0c^2$$
$$= \tfrac{1}{2}m_0v^2$$

Hence at low speeds the relativistic expression for the kinetic energy of a moving particle reduces to the classical one. The total energy of such a particle is

$$E = m_0c^2 + \tfrac{1}{2}m_0v^2$$

In the foregoing calculation relativity has once again met an important test; it has yielded exactly the same results as those of ordinary mechanics at low speeds, where we know by experience that the latter are perfectly valid. It is nevertheless important to keep in mind that, so far as is known, the correct formulation of mechanics has its basis in relativity, with classical mechanics no more than an approximation correct only under certain circumstances.

It is often convenient to express several of the relativistic formulas obtained above in forms somewhat different from their original ones. The new equations are so easy to derive that we shall simply state them without proof:

1.49
$$E = \sqrt{m_0^2 c^4 + p^2 c^2}$$

1.50
$$p = m_0 c \sqrt{\frac{1}{1 - v^2/c^2} - 1}$$

1.51
$$T = m_0 c^2 \left(\frac{1}{\sqrt{1 - v^2/c^2}} - 1 \right)$$

1.52
$$\frac{v}{c} = \sqrt{1 - \frac{1}{[1 + (T/m_0 c^2)]^2}}$$

1.53
$$\frac{1}{\sqrt{1 - v^2/c^2}} = \sqrt{1 + \frac{p^2}{m_0^2 c^2}}$$

1.54
$$= 1 + \frac{T}{m_0 c^2}$$

$$p = m v$$

The symbol p is used for the magnitude of the linear momentum mv.

These formulas are particularly useful in nuclear and elementary-particle physics, where the kinetic energies of moving particles are customarily specified, rather than their velocities. Equation 1.52, for instance, permits us to find v/c directly from $T/m_0 c^2$, the ratio between the kinetic and rest energies of a particle.

1.12 MASS AND ENERGY: ALTERNATIVE DERIVATION

The equivalence of mass and energy can be demonstrated in a number of different ways. An interesting derivation that is somewhat different from the one given above, but also suggested by Einstein, makes use of the basic notion that the center of mass of an isolated system (one that does not interact with its surroundings) cannot be changed by any process occurring within the system. In this derivation we imagine a closed box from one end of which a burst of electromagnetic radiation is emitted, as in Fig. 1-14. This radiation carries energy and momentum, and when the emission occurs, the box recoils in order that the

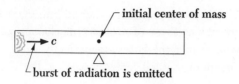

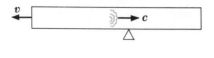

initial center of mass

burst of radiation is emitted

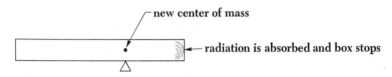

new center of mass

radiation is absorbed and box stops

FIGURE 1-14   Radiant energy possesses inertial mass.

total momentum of the system remain constant. When the radiation is absorbed at the opposite end of the box, its momentum cancels the momentum of the box, which then comes to rest. During the time in which the radiation was in transit, the box has moved a distance s. If the center of mass of the system is still to be in the same location in space, the radiation must have transferred mass from the end at which it was emitted to the end at which it was absorbed. We shall compute the amount of mass that must be transferred if the center of mass of the system is to remain unchanged.

For simplicity we shall consider the sides of the box to be massless and its ends to have the mass $\frac{1}{2}M$ each. The center of mass is therefore at the center of the box, a distance $\frac{1}{2}L$ from each end. A burst of electromagnetic radiation that has the energy E carries the momentum E/c according to electromagnetic theory, and, by hypothesis, has associated with it an amount of mass m. When the radiation is emitted, the box, whose mass is now $M - m$, recoils with the velocity v. From the principle of conservation of momentum,

$$p_{\text{box}} = p_{\text{radiation}}$$
$$(M - m)v = \frac{E}{c}$$

and so the recoil velocity of the box is

$$v = \frac{E}{(M - m)c} \approx \frac{E}{Mc}$$

since m is much smaller than M. The time t during which the box moves is equal to the time required by the radiation to reach the opposite end of the box, a distance L away; this means that $t = L/c$ (assuming that $v \ll c$, which is true when $m \ll M$). During the time t the box is displaced to the left by $s = vt = EL/Mc^2$.

After the box has stopped, the mass of its left-hand end is $\frac{1}{2}M - m$ and the mass of its right-hand end is $\frac{1}{2}M + m$ owing to the transfer of the mass m associated with the energy E of the radiation. If the center of mass is to be in the same place it was originally,

$$(\tfrac{1}{2}M - m)(\tfrac{1}{2}L + s) = (\tfrac{1}{2}M + m)(\tfrac{1}{2}L - s)$$

or

$$m = \frac{Ms}{L}$$

Inserting the value of the displacement s,

$$m = \frac{E}{c^2}$$

The mass associated with an amount of energy E is equal to E/c^2.

In the above derivation we assumed that the box is a perfectly rigid body: that the entire box starts to move when the radiation is emitted and the entire box comes to a stop when the radiation is absorbed. Actually, of course, there is no such thing as a rigid body that meets this specification; for example, the radiation, which travels with the speed of light, will arrive at the right-hand end of the box *before* that end begins to move! When the finite speed of elastic waves in the box is taken into account in a more elaborate calculation, however, the same result that $m = E/c^2$ is obtained.

Problems

1. A certain particle has a lifetime of 10^{-7} s when measured at rest. How far does it go before decaying if its speed is $0.99c$ when it is created?

2. An airplane is flying at 300 m/s (672 mi/h). How much time must elapse before a clock in the airplane and one on the ground differ by 1 s?

3. How fast must a spacecraft travel relative to the earth for each day on the spacecraft to correspond to 2 d on the earth?

4. A rocket ship leaves the earth at a speed of $0.98c$. How much time does it take for the minute hand of a clock in the ship to make a complete revolution as measured by an observer on the earth? 5.025 hr.

5. An astronaut whose height on the earth is exactly 6 ft is lying parallel to the axis of a spacecraft moving at 0.9c relative to the earth. What is his height as measured by an observer in the same spacecraft? By an observer on the earth?

6. A meter stick is projected into space at so great a speed that its length appears contracted to only 50 cm. How fast is it going in miles per second?

7. A rocket ship is 100 m long on the ground. When it is in flight, its length is 99 m to an observer on the ground. What is its speed?

° 8. An observer moving in the $+x$ direction at a speed (in the laboratory system) of 2.9×10^8 m/s finds the speed of an object moving in the $-x$ direction to be 2.998×10^8 m/s. What is the speed of the object in the laboratory system?

° 9. A man on the moon sees two spacecraft, A and B, coming toward him from opposite directions at the respective speeds of 0.8c and 0.9c. (a) What does a man on A measure for the speed with which he is approaching the moon? For the speed with which he is approaching B? (b) What does a man on B measure for the speed with which he is approaching the moon? For the speed with which he is approaching A?

10. It is possible for the electron beam in a television picture tube to move across the screen at a speed faster than the speed of light. Why does this not contradict special relativity?

11. A man has a mass of 100 kg on the ground. When he is in a rocket ship in flight, his mass is 101 kg as determined by an observer on the ground. What is the speed of the rocket ship?

12. How fast must an electron move in order that its mass equal the rest mass of the proton?

13. Find the speed of a 0.1-MeV electron according to classical and relativistic mechanics.

14. How much mass does a proton gain when it is accelerated to a kinetic energy of 500 MeV?

15. How much mass does an electron gain when it is accelerated to a kinetic energy of 500 MeV?

16. The total energy of a particle is exactly twice its rest energy. Find its speed.

17. How much work must be done in order to increase the speed of an electron from 1.2×10^8 m/s to 2.4×10^8 m/s?

18. (a) The density of a substance is ρ in the S frame in which it is at rest.

Find the density ρ' that an observer in the S' frame moving at a speed relative to S of v would determine. (b) Gold has a density of 19.3 g/cm³ when the sample is at rest relative to the observer. What is its density when the relative velocity is $0.9c$.

19. A certain quantity of ice at $0°C$ melts into water at $0°C$ and in so doing gains 1 kg of mass. What was its initial mass?

20. Dynamite liberates about 5.4×10^6 J/kg when it explodes. What fraction of its total energy content is this?

21. Solar energy reaches the earth at the rate of about 1,400 W/m² of surface perpendicular to the direction of the sun. By how much does the mass of the sun decrease in each second? (The mean radius of the earth's orbit is 1.5×10^{11} m.)

22. Prove that $\frac{1}{2}mv^2$, where $m = m_0/\sqrt{1 - v^2/c^2}$, does *not* equal the kinetic energy of a particle moving at relativistic speeds.

23. Express the relativistic form of the second law of motion, $F = d\,(mv)/dt$, in terms of m_0, v, c, and dv/dt.

24. A man leaves the earth in a rocket ship that makes a round trip to the nearest star, 4 light-years distant, at a speed of $0.9c$. How much younger is he upon his return than his twin brother who remained behind? (A light-year is the distance light travels in a year. It is equal to 9.46×10^{15} m.)

25. Light of frequency ν is emitted by a source. An observer moving away from the source at the speed v measures a frequency of ν'. By considering the source as a clock that ticks ν times per second and gives off a pulse of light with each tick, show that

$$\nu' = \nu \sqrt{\frac{1 - v/c}{1 + v/c}}$$

This constitutes the *longitudinal doppler effect* in light. (If the observer is moving *toward* the source at the relative speed v, the $+$ and $-$ signs in the radical of the above formula are interchanged.) Why does this result differ from the corresponding one for sound waves in air?

26. The *transverse doppler effect*, which has no nonrelativistic counterpart, applies to measurements of light waves made by an observer in relative motion perpendicular to the direction of propagation of the waves. (In the preceding problem the observer moves parallel to the direction of propagation.) Show that in the transverse doppler effect

$$\nu' = \nu \sqrt{1 - v^2/c^2}$$

27. Twin A makes a round trip at a speed of $0.8c$ to a star 4 light-years away, while twin B stays behind on the earth. Each twin sends the other a signal once a year by his own reckoning. (a) How many signals does A send during the trip? How many does B send? (b) Use the doppler effect formula of Prob. 25 to analyze this situation. How many signals does A receive during the trip? How many does B receive? Are these results consistent with those of part (a)?

PARTICLE PROPERTIES OF WAVES 2

In our everyday experience there is nothing mysterious or ambiguous about the concepts of *particle* and *wave*. A stone dropped into a lake and the ripples that spread out from its point of impact apparently have in common only the ability to carry energy and momentum from one place to another. Classical physics, which mirrors the "physical reality" of our sense impressions, treats particles and waves as separate components of that reality. The mechanics of particles and the optics of waves are traditionally independent disciplines, each with its own chain of experiments and hypotheses.

The physical reality we perceive arises from phenomena that occur in the microscopic world of atoms and molecules, electrons and nuclei, but in this world there are neither particles nor waves in our sense of these terms. We regard electrons as particles because they possess charge and mass and behave according to the laws of particle mechanics in such familiar devices as television picture tubes. We shall see, however, that there is as much evidence in favor of interpreting a moving electron as a wave manifestation as there is in favor of interpreting it as a particle manifestation. We regard electromagnetic waves as waves because under suitable circumstances they exhibit diffraction, interference, and polarization. Similarly, we shall see that under other circumstances electromagnetic waves behave as though they consist of streams of particles. Together with special relativity, the wave-particle duality is central to an understanding of modern physics, and in this book there are few arguments that do not draw upon one or the other of these fundamental principles.

2.1 THE PHOTOELECTRIC EFFECT

Late in the nineteenth century a series of experiments revealed that electrons are emitted from a metal surface when light of sufficiently high frequency (ultraviolet light is required for all but the alkali metals) falls upon it. This phenomenon is known as the *photoelectric effect*. Figure 2-1 illustrates the type

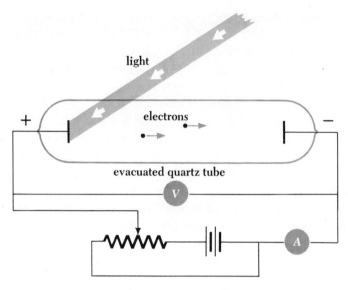

light

electrons

+

−

evacuated quartz tube

V

A

FIGURE 2-1 Experimental observation of the photoelectric effect.

of apparatus that was employed in the more precise of these experiments. An evacuated tube contains two electrodes connected to an external circuit like that shown schematically, with the metal plate whose surface is to be irradiated as the anode. Some of the photoelectrons that emerge from the irradiated surface have sufficient energy to reach the cathode despite its negative polarity, and they constitute the current that is measured by the ammeter in the circuit. As the retarding potential V is increased, fewer and fewer electrons get to the cathode and the current drops. Ultimately, when V equals or exceeds a certain value V_0, of the order of a few volts, no further electrons strike the cathode and the current ceases.

The existence of the photoelectric effect ought not to be surprising; after all, light waves carry energy, and some of the energy absorbed by the metal may somehow concentrate on individual electrons and reappear as kinetic energy. When we look more closely at the data, however, we find that the photoelectric effect can hardly be interpreted so simply.

One of the features of the photoelectric effect that particularly puzzled its discoverers is that the energy distribution in the emitted electrons (called *photoelectrons*) is independent of the intensity of the light. A strong light beam yields more photoelectrons than a weak one of the same frequency, but the average electron energy is the same (Fig. 2-2). Also, within the limits of experimental accuracy (about 10^{-9} s), there is no time lag between the arrival of light at a metal surface and the emission of photoelectrons. These observations cannot be understood from the electromagnetic theory of light.

Let us consider violet light falling on a sodium surface, in an apparatus like that of Fig. 2-1. There will be a detectable photoelectric current when 10^{-6} W/m^2 of electromagnetic energy is absorbed by the surface (a more intense beam than this is required, of course, since sodium is a good reflector of light). Now there are about 10^{19} atoms in a layer of sodium 1 atom thick and 1 m^2 in area, so that, if we assume that the incident light is absorbed in the 10 uppermost layers of sodium atoms, the 10^{-6} W/m^2 is distributed among 10^{20} atoms. Hence each atom receives energy at the average rate of 10^{-26} W, which is less than 10^{-7} eV/s. It should therefore take more than 10^7 s, or almost a year, for any single electron to accumulate the 1 eV or so of energy that the photo-electrons are found to have! In the maximum possible time of 10^{-9} s, an average electron, according to electromagnetic theory, will have gained only 10^{-16} eV. Even if we call upon some kind of resonance process to explain why some

FIGURE 2-2 Photoelectron current is proportional to light intensity for all retarding voltages. The extinction voltage V_0 is the same for all intensities of light of a given frequency ν.

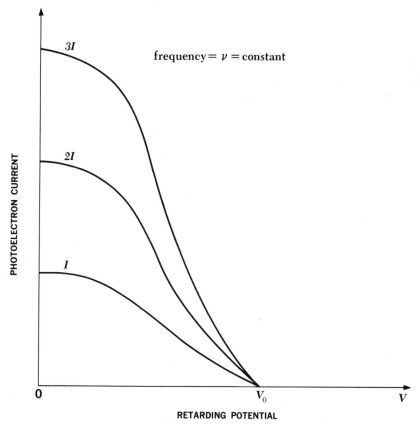

electrons acquire more energy than others, the fortunate electrons could hardly have more than 10^{-10} of the observed energy.

Equally odd from the point of view of the wave theory is the fact that the photoelectron energy depends upon the *frequency* of the light employed (Fig. 2-3). At frequencies below a certain critical frequency characteristic of each particular metal, no electrons whatever are emitted. Above this threshold frequency the photoelectrons have a range of energies from 0 to a certain maximum value, and this maximum energy increases linearly with increasing frequency. High frequencies result in high-maximum photoelectron energies, low frequencies in low-maximum photoelectron energies. Thus a faint blue light produces electrons with more energy than those produced by a bright red light, although the latter yields a greater number of them.

Figure 2-4 is a plot of maximum photoelectron energy T_{max} versus the frequency ν of the incident light in a particular experiment that employed a sodium surface. It is clear that the relationship between T_{max} and the frequency ν involves a proportionality, which we can express in the form

$$\begin{aligned} T_{max} &= h(\nu - \nu_0) \\ &= h\nu - h\nu_0 \end{aligned}$$

2.1

FIGURE 2-3   The extinction voltage V_0 depends upon the frequency ν of the light. When the retarding potential is $V = 0$, the photoelectric current is the same for light of a given intensity regardless of its frequency.

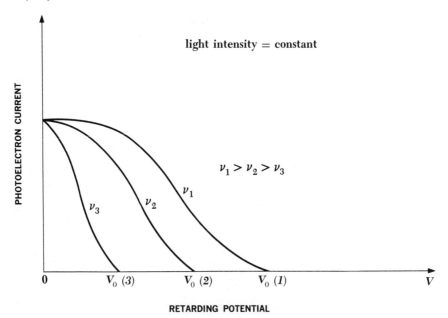

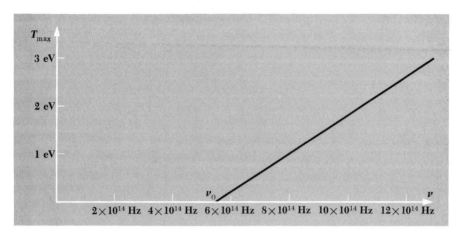

FIGURE 2-4 Maximum photoelectron energy as a function of the frequency of the incident light for a sodium surface.

where ν_0 is the threshold frequency below which no photoemission occurs and h is a constant. Significantly, the value of h,

$$h = 6.626 \times 10^{-34} \text{ J-s}$$

is *always the same*, although ν_0 varies with the particular metal being illuminated.

2.2 THE QUANTUM THEORY OF LIGHT

The electromagnetic theory of light accounts so well for such a variety of phenomena that it must contain some measure of truth. Yet this well-founded theory is completely at odds with the photoelectric effect. In 1905 Albert Einstein found that the paradox presented by the photoelectric effect could be understood only by taking seriously a notion proposed five years earlier by the German theoretical physicist Max Planck. Planck was seeking to explain the characteristics of the radiation emitted by bodies hot enough to be luminous, a problem notorious at the time for its resistance to solution. Planck was able to derive a formula for the spectrum of this radiation (that is, the relative brightness of the various colors present) as a function of the temperature of the body that was in agreement with experiment provided he assumed that the radiation is emitted *discontinuously* as little bursts of energy. These bursts of energy are called *quanta*. Planck found that the quanta associated with a particular frequency ν of light all have the same energy and that this energy E is directly proportional to ν. That is, where h, today known as *Planck's constant,* has the value

$$h = 6.626 \times 10^{-34} \text{ J-s}$$

We shall examine some of the details of this interesting problem and its solution in Chap. 9.

While he had to assume that the electromagnetic energy radiated by a hot object emerges intermittently, Planck did not doubt that it propagates continuously through space as electromagnetic waves. Einstein proposed that light not only is emitted a quantum at a time, but also propagates as individual quanta. In terms of this hypothesis the photoelectric effect can be readily explained. The empirical formula Eq. 2.1 may be rewritten

2.3 $h\nu = T_{\max} + h\nu_0$ **Photoelectric effect**

Einstein's proposal means that the three terms of Eq. 2.3 are to be interpreted as follows:

$h\nu$ = the energy content of each quantum of the incident light

$T_{\max}$ = the maximum photoelectron energy

$h\nu_0$ = the minimum energy needed to dislodge an electron from the metal surface being illuminated

There must be a minimum energy required by an electron in order to escape from a metal surface, or else electrons would pour out even in the absence of light. The energy $h\nu_0$ characteristic of a particular surface is called its *work function.* Hence Eq. 2.3 states that

$$\frac{\text{Quantum}}{\text{energy}} = \frac{\text{maximum electron}}{\text{energy}} + \frac{\text{work function}}{\text{of surface}}$$

It is easy to see why not all photoelectrons have the same energy, but emerge with all energies up to $T_{\max}$: $h\nu_0$ is the work that must be done to take an electron through the metal surface from just beneath it, and more work is required when the electron originates deeper in the metal.

The validity of this interpretation of the photoelectric effect is confirmed by studies of *thermionic emission.* Long ago it became known that the presence of a very hot object increases the electrical conductivity of the surrounding air, and late in the nineteenth century the reason for this phenomenon was found to be the emission of electrons from such an object. Thermionic emission makes possible the operation of such devices as television picture tubes, in which metal filaments or specially coated cathodes at high temperature supply dense streams of electrons. The emitted electrons evidently obtain their energy from the thermal agitation of the particles constituting the metal, and we should expect that the electrons must acquire a certain minimum energy in order to escape. This minimum energy can be determined for many surfaces, and it is always close to the photoelectric work function for the same surfaces. In photoelectric emission, photons of light provide the energy required by an electron to escape, while in thermionic emission heat does so: in both cases the physical processes involved in the emergence of an electron from a metal surface are the same.

Let us apply Eq. 2.3 to a specific situation. The work function of potassium is 2.2 eV. When ultraviolet light of wavelength 3,500 Å (1 Å = 1 angstrom unit = 10^{-10} m) falls on a potassium surface, what is the maximum energy in electron volts of the photoelectrons? From Eq. 2.3,

$$T_{max} = h\nu - h\nu_0$$

Since $h\nu_0$ is already expressed in electron volts, we need only compute the quantum energy $h\nu$ of 3,500-Å light. This is

$$h\nu = \frac{hc}{\lambda}$$

$$= \frac{6.63 \times 10^{-34}\,\text{J-s} \times 3 \times 10^8\,\text{m/s} \times 10^{10}\,\text{Å/m}}{3,500\,\text{Å}}$$

$$= 5.7 \times 10^{-19}\,\text{J}$$

To convert this energy from joules to electron volts, we recall that

$$1\,\text{eV} = 1.6 \times 10^{-19}\,\text{J}$$

and so

$$h\nu = \frac{5.7 \times 10^{-19}\,\text{J}}{1.6 \times 10^{-19}\,\text{J}}$$

$$= 3.6\,\text{eV}$$

Hence the maximum photoelectron energy is

$$T_{max} = h\nu - h\nu_0$$
$$= 3.6\,\text{eV} - 2.2\,\text{eV}$$
$$= 1.4\,\text{eV}$$

The view that light propagates as a series of little packets of energy (usually called photons) is directly opposed to the wave theory of light. The latter, which provides the sole means of explaining a host of optical effects—notably diffraction and interference—is one of the most securely established of physical theories. Planck's suggestion that a hot object emits light in separate quanta was not incompatible with the propagation of light as a wave. Einstein's suggestion in 1905 that light travels through space in the form of distinct photons, however, elicited incredulity from his contemporaries. According to the wave theory, light waves spread out from a source in the way ripples spread out on the surface of a lake when a stone falls into it. The energy carried by the light, in this analogy, is distributed continuously throughout the wave pattern. According to the quantum theory, on the other hand, light spreads out from a source as a series of localized concentrations of energy, each sufficiently small to be capable of absorption by a single electron. Curiously, the quantum theory of light, which treats it strictly as a particle phenomenon, explicitly involves the light frequency v, strictly a wave concept.

The quantum theory of light is strikingly successful in explaining the photoelectric effect. It predicts correctly that the maximum photoelectron energy should depend upon the frequency of the incident light and not upon its intensity, contrary to what the wave theory suggests, and it is able to explain why even the feeblest light can lead to the immediate emission of photoelectrons, again contrary to the wave theory. The wave theory can give no reason why there should be a threshold frequency such that, when light of lower frequency is employed, no photoelectrons are observed no matter how strong the light beam, something that follows naturally from the quantum theory.

Which theory are we to believe? A great many physical hypotheses have had to be altered or discarded when they were found to disagree with experiment, but never before have we had to devise two totally different theories to account for a single physical phenomenon. The situation here is fundamentally different from what it is, say, in the case of relativistic versus Newtonian mechanics, where the latter turns out to be an approximation of the former. There is no way of deriving the quantum theory of light from the wave theory of light or vice versa.

In a specific event light exhibits *either* a wave or a particle nature, never both simultaneously. The same light beam that is diffracted by a grating can cause the emission of photoelectrons from a suitable surface, but these processes occur independently. **The wave theory of light and the quantum theory of light complement each other.** Electromagnetic waves account for the observed manner in which light propagates, while photons account for the observed manner in which energy is transferred between light and matter. We have no alternative to regarding light as something that manifests itself as a stream of discrete photons on occasion and as a wave train the rest of the time. The "true nature" of light is no longer something that can be visualized in terms of everyday experience, and we must accept both wave and quantum theories, contradictions and all, as the closest we can get to a complete description of light.

2.3 X RAYS

The photoelectric effect provides convincing evidence that photons of light can transfer energy to electrons. Is the inverse process also possible? That is, can part or all of the kinetic energy of a moving electron be converted into a photon? As it happens, the inverse photoelectric effect not only does occur, but had been discovered (though not at all understood) prior to the theoretical work of Planck and Einstein.

In 1895 Wilhelm Roentgen made the classic observation that a highly penetrating radiation of unknown nature is produced when fast electrons impinge on matter. These *X rays* were soon found to travel in straight lines, even through electric and magnetic fields, to pass readily through opaque materials, to cause phosphorescent substances to glow, and to expose photographic plates. The faster the original electrons, the more penetrating the resulting X rays, and the greater the number of electrons, the greater the intensity of the X-ray beam.

Not long after this discovery it began to be suspected that X rays are electromagnetic waves. After all, electromagnetic theory predicts that an accelerated electric charge will radiate electromagnetic waves, and a rapidly moving electron suddenly brought to rest is certainly accelerated. Radiation produced under these circumstances is given the German name *bremsstrahlung* ("braking radiation"). The absence of any perceptible X-ray refraction in the early work could be attributed to very short wavelengths, below those in the ultraviolet range, since the refractive index of a substance decreases to unity (corresponding to straight-line propagation) with decreasing wavelength.

The wave nature of X rays was first established in 1906 by Barkla, who was able to exhibit their polarization. Barkla's experimental arrangement is sketched in Fig. 2-5. We shall analyze this classic experiment under the assumption that X rays are electromagnetic waves. At the left a beam of unpolarized X rays heading in the $-z$ direction impinges on a small block of carbon. These X rays are scattered by the carbon; this means that electrons in the carbon atoms are set in vibration by the electric vectors of the X rays and then reradiate. Because the electric vector in an electromagnetic wave is perpendicular to its direction of propagation, the initial beam of X rays contains electric vectors that lie in the xy plane only. The target electrons therefore are induced to vibrate in the xy plane. A scattered X ray that proceeds in the $+x$ direction can have an electric vector in the y direction only, and so it is plane-polarized. To demonstrate this polarization, another carbon block is placed in the path of the ray, as at the right. The electrons in this block are restricted to vibrate in the y direction and therefore reradiate X rays that propagate in the xz plane exclusively, and not at all in the y direction. The observed absence of scattered X rays outside the xz plane confirms the wave character of X rays.

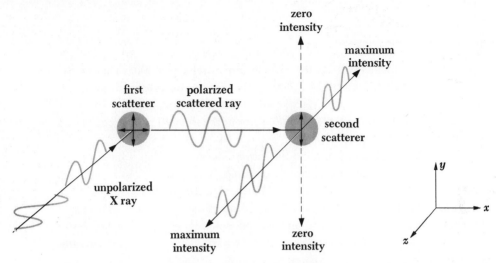

FIGURE 2-5 Barkla's experiment to demonstrate X-ray polarization.

ASSUMES   X-RAYS   ARE   ELECTROMAGNETIC   WAVES

In 1912 a method was devised for measuring the wavelengths of X rays. A diffraction experiment had been recognized as ideal, but, as we recall from physical optics, the spacing between adjacent lines on a diffraction grating must be of the same order of magnitude as the wavelength of the light for satisfactory results, and gratings cannot be ruled with the minute spacing required by X rays. In 1912, however, Max von Laue recognized that the wavelengths hypothesized for X rays were about the same order of magnitude as the spacing between adjacent atoms in crystals, which is about 1 Å. He therefore proposed that crystals be used to diffract X rays, with their regular lattices acting as a kind of three-dimensional grating. Suitable experiments were performed in the next year, and the wave nature of X rays was successfully demonstrated. In these experiments wavelengths from 1.3×10^{-11} to 4.8×10^{-11} m were found, 0.13 to 0.48 Å, 10^{-4} of those in visible light and hence having quanta 10^4 times as energetic. We shall consider X-ray diffraction in Sec. 2.4.

For purposes of classification, electromagnetic radiations with wavelengths in the approximate interval from 10^{-11} to 10^{-8} m (0.1 to 100 Å) are today considered as X rays.

Figure 2-6 is a diagram of an X-ray tube. A cathode, heated by an adjacent filament through which an electric current is passed, supplies electrons copiously by thermionic emission. The high potential difference V maintained between the cathode and a metallic target accelerates the electrons toward the latter. The face of the target is at an angle relative to the electron beam, and the X rays that emerge from the target pass through the side of the tube. The tube is evacuated to permit the electrons to get to the target unimpeded.

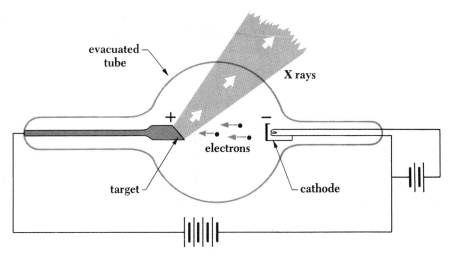

evacuated tube

X rays

+

electrons

−

target

cathode

FIGURE 2-6 An X-ray tube.

As was said earlier, classical electromagnetic theory predicts the production of bremsstrahlung when electrons are accelerated, thereby apparently accounting for the X rays emitted when fast electrons are stopped by the target of an X-ray tube. However, the agreement between the classical theory and the experimental data is not satisfactory in certain important respects. Figures 2-7 and 2-8 show the X-ray spectra that result when tungsten and molybdenum targets are bombarded by electrons at several different accelerating potentials. The curves exhibit two distinctive features not explainable in terms of electromagnetic theory:

1. In the case of molybdenum there are pronounced intensity peaks at certain wavelengths, indicating the enhanced production of X rays. These peaks occur at various specific wavelengths for each target material and originate in rearrangements of the electron structures of the target atoms after having been disturbed by the bombarding electrons. The important thing to note at this point is the production of X rays of specific wavelengths, a decidedly nonclassical effect, in addition to the production of a continuous X-ray spectrum.

2. The X rays produced at a given accelerating potential V vary in wavelength, but none has a wavelength shorter than a certain value λ_{min}. Increasing V decreases λ_{min}. At a particular V, λ_{min} is the *same* for both the tungsten and molybdenum targets. Duane and Hunt have found that λ_{min} is inversely proportional to V; their precise relationship is

2.4
$$\lambda_{min} = \frac{1.24 \times 10^{-6} \text{ V-m}}{V}$$

X-ray production

PARTICLE PROPERTIES OF WAVES   53

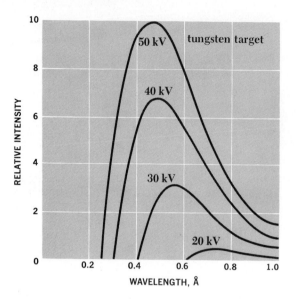

FIGURE 2-7 X-ray spectra of tungsten at various accelerating potentials.

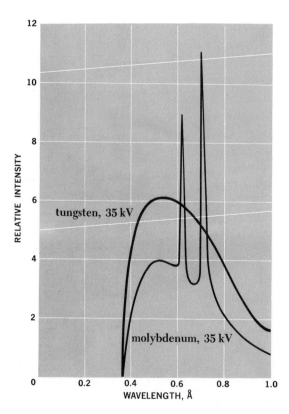

FIGURE 2-8 X-ray spectra of tungsten and molybdenum at 35 kV accelerating potential.

The second observation is readily understood in terms of the quantum theory of radiation. Most of the electrons incident upon the target lose their kinetic energy gradually in numerous collisions, their energy going simply into heat. (This is the reason that the targets in X-ray tubes are normally of high-melting-point metals, and an efficient means of cooling the target is often employed.) A few electrons, though, lose most or all of their energy in single collisions with target atoms; this is the energy that is evolved as X rays. X-ray production, then, except for the peaks mentioned in observation 1 above, represents an inverse photoelectric effect. Instead of photon energy being transformed into electron kinetic energy, electron kinetic energy is being transformed into photon energy. A short wavelength means a high frequency, and a high frequency means a high photon energy $h\nu$. It is therefore logical to interpret the short wavelength limit λ_{min} of Eq. 2.4 as corresponding to a maximum photon energy $h\nu_{max}$, where

2.5 $$h\nu_{max} = \frac{hc}{\lambda_{min}} = eV = E_{NERGY}$$

Since work functions are only several electron volts while the accelerating potentials in X-ray tubes are tens or hundreds of thousands of volts, we may assume that the kinetic energy T of the bombarding electrons is

2.6 $$T = eV$$

When the entire kinetic energy of an electron is lost to create a single photon, then

2.7 $$h\nu_{max} = T$$

Substituting Eqs. 2.5 and 2.6 into 2.7, we see that

$$h\nu_{max} = T$$

$$\frac{hc}{\lambda_{min}} = eV$$

$$\lambda_{min} = \frac{hc}{eV}$$

$$= \frac{6.63 \times 10^{-34} \text{ J-s} \times 3 \times 10^8 \text{ m/s}}{1.6 \times 10^{-19} \text{ C} \times V}$$

$$= \frac{1.24 \times 10^{-6}}{V} \text{ V-m}$$

which is just the experimental relationship of Eq. 2.4. It is therefore correct to regard X-ray production as the inverse of the photoelectric effect.

A conventional X-ray machine might have an accelerating potential of 50,000 V. To find the shortest wavelength present in its radiation, we use

Eq. 2.4, with the result that

$$\lambda_{min} = \frac{1.24 \times 10^{-6} \text{ V-m}}{5 \times 10^4 \text{ V}}$$

$$= 2.5 \times 10^{-11} \text{ m}$$

$$= 0.25 \text{ Å}$$

This wavelength corresponds to the frequency

$$\nu_{max} = \frac{c}{\lambda_{min}}$$

$$= \frac{3 \times 10^8 \text{ m/s}}{2.5 \times 10^{-11} \text{ m}}$$

$$= 1.2 \times 10^{19} \text{ Hz}$$

2.4 X-RAY DIFFRACTION

Let us now return to the question of how X rays may be demonstrated to consist of electromagnetic waves. A crystal consists of a regular array of atoms, each of which is able to scatter any electromagnetic waves that happen to strike it. The mechanism of scattering is straightforward. An atom in a constant electric field becomes polarized since its negatively charged electrons and positively charged nucleus experience forces in opposite directions; these forces are small compared with the forces holding the atom together, and so the result is a distorted charge distribution equivalent to an electric dipole. In the presence of the alternating electric field of an electromagnetic wave of frequency ν, the polarization changes back and forth with the same frequency ν. An oscillating electric dipole is thus created at the expense of some of the energy of the incoming wave, whose amplitude is accordingly decreased. The oscillating dipole in turn radiates electromagnetic waves of frequency ν, and these secondary waves proceed in all directions except along the dipole axis. In an assembly of atoms exposed to unpolarized radiation, the secondary radiation is isotropic since the contributions of the individual atoms are random. In wave terminology, the secondary waves have spherical wavefronts in place of the plane wavefronts of the incoming waves (Fig. 2-9). The scattering process, then, involves an atom absorbing incident plane waves and reemitting spherical waves of the same frequency.

A monochromatic beam of X rays that falls upon a crystal will be scattered in all directions within it, but, owing to the regular arrangement of the atoms, in certain directions the scattered waves will constructively interfere with one

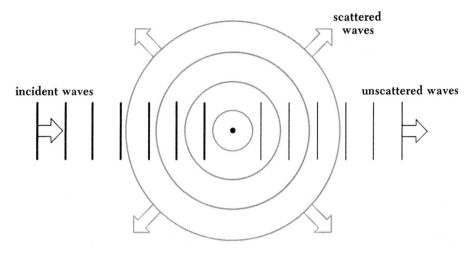

FIGURE 2-9 The scattering of electromagnetic radiation by a group of atoms. Incident plane waves are reemitted as spherical waves.

another while in others they will destructively interfere. The atoms in a crystal may be thought of as defining families of parallel planes, as in Fig. 2-10, with each family having a characteristic separation between its component planes. This analysis was suggested in 1913 by W. L. Bragg, in honor of whom the above planes are called *Bragg planes*. The conditions that must be fulfilled for radiation scattered by crystal atoms to undergo constructive interference may be obtained

FIGURE 2-10 Two sets of Bragg planes in a NaCl crystal.

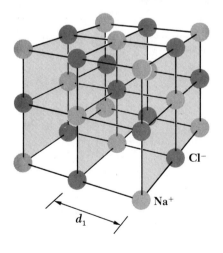

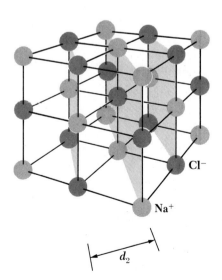

from a diagram like that in Fig. 2-11. A beam containing X rays of wavelength λ is incident upon a crystal at an angle θ with a family of Bragg planes whose spacing is d. The beam goes past atom A in the first plane and atom B in the next, and each of them scatters part of the beam in random directions. Constructive interference takes place only between those scattered rays that are parallel and whose paths differ by exactly λ, 2λ, 3λ, and so on. That is, the path difference must be nλ, where n is an integer. The only rays scattered by A and B for which this is true are those labeled I and II in Fig. 2-11. The first condition upon I and II is that their common scattering angle be equal to the angle of incidence θ of the original beam. The second condition is that

2.8 $$2d \sin \theta = n\lambda \qquad n = 1, 2, 3, \ldots$$

since ray II must travel the distance $2d \sin \theta$ farther than ray I. The integer n is the *order* of the scattered beam.

The schematic design of an X-ray spectrometer based upon Bragg's analysis is shown in Fig. 2-12. A collimated beam of X rays falls upon a crystal at an angle θ, and a detector is placed so that it records those rays whose scattering angle is also θ. Any X rays reaching the detector therefore obey the first Bragg condition. As θ is varied, the detector will record intensity peaks corresponding to the orders predicted by Eq. 2.8. If the spacing d between adjacent Bragg planes in the crystal is known, the X-ray wavelength λ may be calculated.

How can we determine the value of d? This is a simple task in the case of crystals whose atoms are arranged in cubic lattices similar to that of rock salt (NaCl), illustrated in Fig. 2-10. As an example, let us compute the separation of adjacent atoms in a crystal of NaCl. The molecular weight of NaCl is 58.5, which means there are 58.5 kg of NaCl per kilomole (kmol). Since there are

FIGURE 2-11   X-ray scattering from cubic crystal.

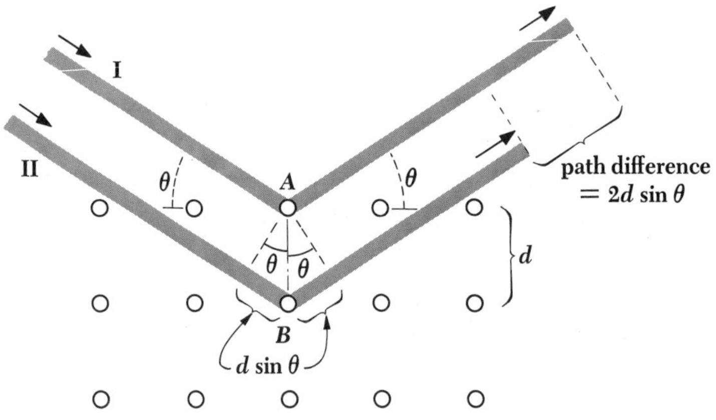

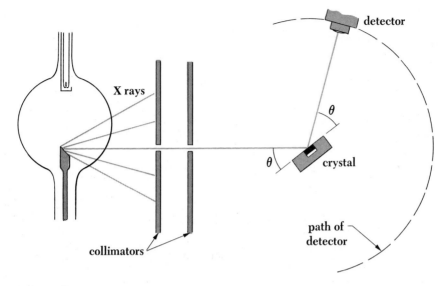

FIGURE 2-12 X-ray spectrometer.

$N_0 = 6.02 \times 10^{26}$ molecules in a kmol of any substance (N_0 is Avogadro's number), the mass of each NaCl "molecule"—that is, of each Na + Cl pair of atoms—is given by

$$m_{\text{NaCl}} = 58.5 \frac{\text{kg}}{\text{kmol}} \times \frac{1}{6.02 \times 10^{26} \text{ molecules/kmol}}$$
$$= 9.72 \times 10^{-26} \text{ kg/molecule}$$

Crystalline NaCl has a density of 2.16×10^3 kg/m³, and so, taking into account the presence of two atoms in each NaCl "molecule," the number of atoms in 1 m³ of NaCl is

$$n = 2 \frac{\text{atoms}}{\text{molecule}} \times 2.16 \times 10^3 \frac{\text{kg}}{\text{m}^3} \times \frac{1}{9.72 \times 10^{-26} \text{ kg/molecule}}$$
$$= 4.45 \times 10^{28} \text{ atoms/m}^3$$

If d is the distance between adjacent atoms in a crystal, there are d^{-1} atoms/m along any of the crystal axes and d^{-3} atoms/m³ in the entire crystal. Hence

$$d^{-3} = n$$

and

$$d = n^{-\frac{1}{3}} = (4.45 \times 10^{28})^{-\frac{1}{3}} \text{ m}$$
$$= 2.82 \times 10^{-10} \text{ m}$$
$$= 2.82 \text{ Å}$$

2.5 THE COMPTON EFFECT

The quantum theory of light postulates that photons behave like particles except for the absence of any rest mass. If this is true, then it should be possible for us to treat collisions between photons and, say, electrons in the same manner as billiard-ball collisions are treated in elementary mechanics.

Figure 2-13 shows how such a collision might be represented, with an X-ray photon striking an electron (assumed to be initially at rest in the laboratory coordinate system) and being scattered away from its original direction of motion while the electron receives an impulse and begins to move. In the collision the photon may be regarded as having lost an amount of energy that is the same as the kinetic energy T gained by the electron, though actually separate photons are involved. If the initial photon has the frequency ν associated with it, the scattered photon has the lower frequency ν', where

$$\text{Loss in photon energy} = \text{gain in electron energy}$$

2.9
$$h\nu - h\nu' = T$$

From the previous chapter we recall that

$$E = \sqrt{m_0{}^2 c^4 + p^2 c^2}$$

so that, since the photon has no rest mass, its total energy is

$$E = pc$$

Since

$$E = h\nu$$

FIGURE 2-13   The Compton effect.

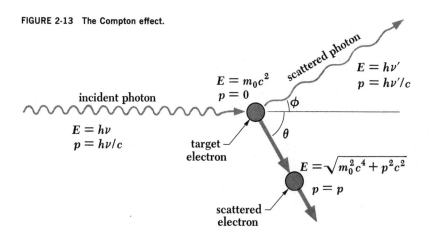

for a photon, its momentum is

$$p = \frac{E}{c}$$

2.10
$$= \frac{h\nu}{c}$$

Momentum, unlike energy, is a vector quantity, incorporating direction as well as magnitude, and in the collision momentum must be conserved in each of two mutually perpendicular directions. (When more than two bodies participate in a collision, of course, momentum must be conserved in each of three mutually perpendicular directions.) The directions we choose here are that of the original photon and one perpendicular to it in the plane containing the electron and the scattered photon (Fig. 2-13). The initial photon momentum is $h\nu/c$, the scattered photon momentum is $h\nu'/c$, and the initial and final electron momenta are respectively 0 and p. In the original photon direction

Initial momentum = final momentum

2.11
$$\frac{h\nu}{c} + 0 = \frac{h\nu'}{c} \cos \phi + p \cos \theta$$

and perpendicular to this direction

Initial momentum = final momentum

2.12
$$0 = \frac{h\nu'}{c} \sin \phi - p \sin \theta$$

The angle ϕ is that between the directions of the initial and scattered photons, and θ is that between the directions of the initial photon and the recoil electron. From Eqs. 2.9, 2.11, and 2.12 we shall now obtain a formula relating the wavelength difference between initial and scattered photons with the angle ϕ between their directions, both of which are readily measurable quantities.

The first step is to multiply Eqs. 2.11 and 2.12 by c and rewrite them as

$$pc \cos \theta = h\nu - h\nu' \cos \phi$$
$$pc \sin \theta = h\nu' \sin \phi$$

By squaring each of these equations and adding the new ones together, the angle θ is eliminated, leaving

2.13
$$p^2c^2 = (h\nu)^2 - 2(h\nu)(h\nu') \cos \phi + (h\nu')^2$$

Next we equate the two expressions for the total energy of a particle

$$E = T + m_0c^2$$
$$E = \sqrt{m_0^2c^4 + p^2c^2}$$

from the previous chapter to give

$$(T + m_0 c^2)^2 = m_0{}^2 c^4 + p^2 c^2$$
$$p^2 c^2 = T^2 + 2 m_0 c^2 T$$

Since

$$T = h\nu - h\nu',$$

we have

2.14 $\qquad p^2 c^2 = (h\nu)^2 - 2(h\nu)(h\nu') + (h\nu')^2 + 2m_0 c^2 (h\nu - h\nu')$

Substituting this value of $p^2 c^2$ in Eq. 2.13, we finally obtain

2.15 $\qquad 2m_0 c^2 (h\nu - h\nu') = 2(h\nu)(h\nu')(1 - \cos\phi)$

This relationship is simpler when expressed in terms of wavelength rather than frequency. Dividing Eq. 2.15 by $2h^2 c^2$,

$$\frac{m_0 c}{h}\left(\frac{\nu}{c} - \frac{\nu'}{c}\right) = \frac{\nu}{c}\frac{\nu'}{c}(1 - \cos\phi)$$

and so, since $\nu/c = 1/\lambda$ and $\nu'/c = 1/\lambda'$,

$$\frac{m_0 c}{h}\left(\frac{1}{\lambda} - \frac{1}{\lambda'}\right) = \frac{(1 - \cos\phi)}{\lambda\lambda'}$$

2.16 $\qquad \lambda' - \lambda = \dfrac{h}{m_0 c}(1 - \cos\phi) \qquad\qquad$ **Compton effect**

Equation 2.16 was derived by Arthur H. Compton in the early 1920s, and the phenomenon it describes, which he was the first to observe, is known as the *Compton effect*. It constitutes very strong evidence in support of the quantum theory of radiation.

Equation 2.16 gives the change in wavelength expected for a photon that is scattered through the angle ϕ by a particle of rest mass m_0; it is independent of the wavelength λ of the incident photon. The quantity $h/m_0 c$ is called the *Compton wavelength* of the scattering particle, which for an electron is 0.024 Å $(2.4 \times 10^{-12}$ m). From Eq. 2.16 we note that the greatest wavelength change that can occur will take place for $\phi = 180°$, when the wavelength change will be twice the Compton wavelength $h/m_0 c$. Because the Compton wavelength of an electron is 0.024 Å, while it is considerably less for other particles owing to their larger rest masses, the maximum wavelength change in the Compton effect is 0.048 Å. Changes of this magnitude or less are readily observable only in X rays, since the shift in wavelength for visible light is less than 0.01 percent of the initial wavelength, while for X rays of $\lambda = 1$ Å it is several percent.

The experimental demonstration of the Compton effect is straightforward. As in Fig. 2-14, a beam of X rays of a single, known wavelength is directed

FIGURE 2-14 Experimental demonstration of the Compton effect.

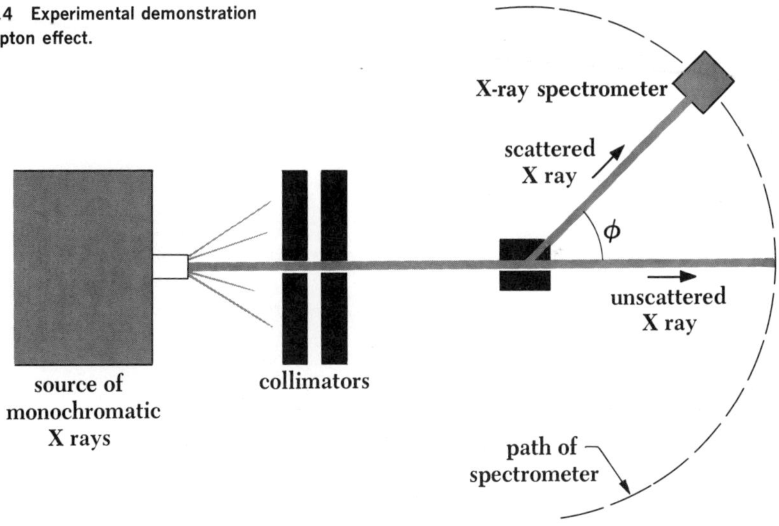

X-ray spectrometer

scattered X ray

ϕ

unscattered X ray

source of monochromatic X rays

collimators

path of spectrometer

at a target, and the wavelengths of the scattered X rays are determined at various angles ϕ. The results, shown in Fig. 2-15, exhibit the wavelength shift predicted by Eq. 2.16, but at each angle the scattered X rays also include a substantial proportion having the initial wavelength. This is not hard to understand. In deriving Eq. 2.16, we assumed that the scattering particle is able to move freely, a reasonable assumption since many of the electrons in matter are only loosely bound to their parent atoms. Other electrons, however, are very tightly bound and, when struck by a photon, the entire atom recoils instead of the single electron. In this event the value of m_0 to use in Eq. 2.16 is that of the entire atom, which is tens of thousands of times greater than that of an electron, and the resulting Compton shift is accordingly so minute as to be undetectable.

2.6 PAIR PRODUCTION

As we have seen, a photon can give up all or part of its energy $h\nu$ to an electron. It is also possible for a photon to materialize into an electron and a positron (positive electron), a process in which electromagnetic energy is converted into rest energy. No conservation principles are violated when an electron-positron pair is created near an atomic nucleus (Fig. 2-16). The sum of the charges of the electron ($q = -e$) and of the positron ($q = +e$) is zero, as is the charge of the photon; the total energy, including mass energy, of the electron and positron equals the photon energy; and linear momentum is conserved with the help of the nucleus, which carries away enough photon momentum for the

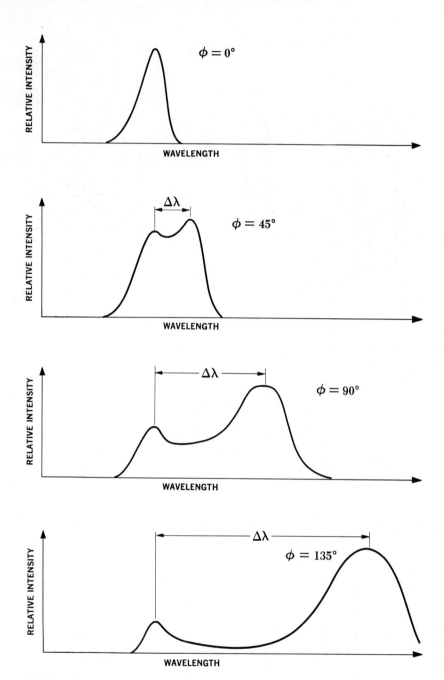

FIGURE 2-15 Compton scattering.

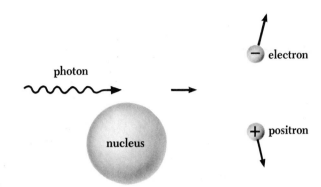

FIGURE 2-16 Pair production.

process to occur but, because of its relatively enormous mass, only a negligible fraction of the photon energy. (Energy and linear momentum could not both be conserved if pair production were to occur in empty space, so it does not occur there.)

$$f = \frac{9.1 \times 10^{-31} (3 \times 10^{8})^{2}}{1.6 \times 10^{-19}} = \frac{mv^{2}}{q}$$

The rest energy $m_0 c^2$ of an electron or positron is 0.51 MeV, and so pair production requires a photon energy of at least 1.02 MeV. Any additional photon energy becomes kinetic energy of the electron and positron. The corresponding maximum photon wavelength is 0.012 Å. Electromagnetic waves with such wavelengths are called *gamma rays*, and are found in nature as one of the emissions from radioactive nuclei and in cosmic rays.

The inverse of pair production occurs when an electron and positron come together and are annihilated to create a pair of photons. The directions of the photons are such as to conserve both energy and linear momentum, and so no nucleus or other particle is required for annihilation to take place.

Three processes in all are therefore responsible for the absorption of X and gamma rays in matter. At low photon energies Compton scattering is the sole mechanism, since there are definite thresholds for both the photoelectric effect (several eV) and electron pair production (1.02 MeV). Both Compton scattering and the photoelectric effect decrease in importance with increasing energy, as shown in Fig. 2-17 for the case of a lead absorber, while the likelihood of pair production increases. At high photon energies the dominant mechanism of energy loss is pair production. The curve representing the total absorption in lead has its minimum at about 2 MeV. The ordinate of the graph is the linear absorption coefficient μ, which is equal to the ratio between the fractional decrease in radiation intensity $-dI/I$ and the absorber thickness dx. That is,

$$\frac{dI}{I} = -\mu \, dx$$

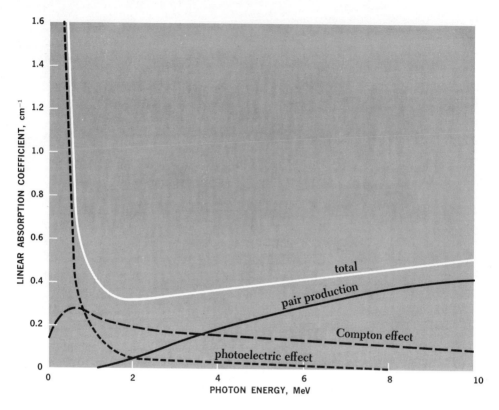

FIGURE 2-17   Linear absorption coefficients for photons in lead. These curves refer to energy absorption, not to the likelihood of interactions in the medium.

whose solution is

$$I = I_0 e^{-\mu x}$$

The intensity of the radiation decreases exponentially with the thickness of the absorber.

*2.7   GRAVITATIONAL RED SHIFT

Although a photon has no rest mass, it nevertheless behaves as though it possesses the inertial mass

2.17 $$m = \frac{h\nu}{c^2}$$ **Photon "mass"**

Does a photon possess gravitational mass as well? Since the inertial and gravitational masses of all material bodies are found experimentally to be equal (this *principle of equivalence* is one of the starting points of Einstein's general theory of relativity), it would seem worth looking into the question of whether photons have the same gravitational behavior as other particles.

Let us consider a photon of frequency ν emitted from the surface of a star of mass M and radius R (Fig. 2-18). The potential energy of a mass m on the star's surface is

2.18
$$V = -\frac{GMm}{R}$$

The potential energy of the photon is accordingly

$$V = -\frac{GMh\nu}{c^2 R}$$

and its total energy E, the sum of V and the quantum energy $h\nu$, is

$$E = h\nu - \frac{GMh\nu}{c^2 R}$$

2.19
$$= h\nu\left(1 - \frac{GM}{c^2 R}\right)$$

At a large distance from the star, for instance at the earth, the photon is beyond the star's gravitational field but its total energy remains the same. The photon's energy is now entirely electromagnetic, and

2.20
$$E = h\nu'$$

FIGURE 2-18 The frequency of a photon emitted from the surface of a star decreases as it moves away from the star.

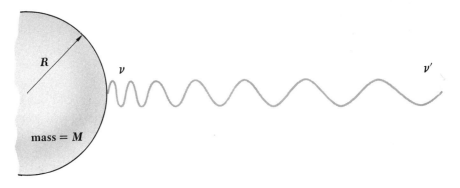

R

ν

ν'

mass $= M$

where ν' is the frequency of the arriving photon. (The potential energy of the photon in the earth's gravitational field is negligible compared with that in the star's field.) Hence

$$h\nu' = h\nu\left(1 - \frac{GM}{c^2R}\right)$$

$$\frac{\nu'}{\nu} = 1 - \frac{GM}{c^2R}$$

2.21 $\qquad \dfrac{\Delta\nu}{\nu} = \dfrac{\nu - \nu'}{\nu} = 1 - \dfrac{\nu'}{\nu} = \dfrac{GM}{c^2R}$

The photon has a *lower* frequency at the earth, corresponding to its loss in energy as it leaves the field of the star. A photon in the visible region of the spectrum is thus shifted toward the red end, and this phenomenon is accordingly known as the *gravitational red shift*. It must be distinguished from the doppler red shift observed in the spectra of distant galaxies due to their apparent recession from the earth, a recession attributed to a general expansion of the universe.

As we shall learn in Chap. 4, the atoms of every element, when suitably excited, emit photons of certain specific frequencies only. The validity of Eq. 2.21 can therefore be checked by comparing the frequencies found in stellar spectra with those in spectra obtained in the laboratory. Since G/c^2 is only

$$\frac{G}{c^2} = \frac{6.67 \times 10^{-11}\,\text{N-m}^2/\text{kg}^2}{(3 \times 10^8\,\text{m/s})^2} = 7.41 \times 10^{-28}\,\text{m/kg}$$

the gravitational red shift can be observed only in radiation from very dense stars. In the case of the sun, a more or less ordinary star, $R = 6.96 \times 10^8$ m and $M = 1.99 \times 10^{30}$ kg, and

$$\frac{\Delta\nu}{\nu} = \frac{GM}{c^2R} = 7.41 \times 10^{-28}\,\frac{\text{m}}{\text{kg}} \times \frac{1.99 \times 10^{30}\,\text{kg}}{6.96 \times 10^8\,\text{m}}$$

$$= 2.12 \times 10^{-6}$$

Since $\Delta\lambda/\lambda = \Delta\nu/\nu$, the gravitational red shift in solar radiation amounts to only about 0.01 Å for green light of wavelength 5,000 Å and is undetectable due to the doppler broadening of the spectral lines.

However, there is a class of stars in the final stages of their evolution called *white dwarfs* that are composed of atoms whose electron structures have "collapsed," and such stars have quite enormous densities—typically $\sim$5 tons/in.^{3} A white dwarf might have a radius of 9×10^6 m, about 0.01 that of the sun, and a mass of 1.2×10^{30} kg, about 0.6 that of the sun, so that

$$\frac{\Delta \nu}{\nu} = \frac{GM}{c^2 R} \approx 7.41 \times 10^{-28} \frac{\text{m}}{\text{kg}} \times \frac{1.2 \times 10^{30} \text{ kg}}{9 \times 10^6 \text{ m}}$$

$$\approx 10^{-4}$$

Here the gravitational red shift would be ≈ 0.5 Å for light of wavelength 5,000 Å, which is measurable under favorable circumstances. In the case of the white dwarf Sirius B (the "companion of Sirius"), the predicted red shift is $\Delta \nu / \nu \approx 5.9 \times 10^{-5}$ and the observed shift is 6.6×10^{-5}; in view of the uncertainty in the M/R ratio for Sirius B, these figures would seem to confirm the attribution of gravitational mass to photons.

If there is a star for which $GM/c^2 R \geq 1$, we see from Eq. 2.19 that no photon can ever leave it. Such a star could not radiate and would be invisible—a "black hole" in space. There seems to be no fundamental reason why black holes should not actually exist, and it should be possible to detect them by virtue of the combination of their light-absorbing ability and their gravitational effects on astronomical objects in their vicinity. Curiously, the known universe may be a black hole in itself: the mass and radius of the universe are believed to be about 10^{53} kg and 10^{26} m respectively and, since $G/c^2 \approx 10^{-27}$ m/kg, $(GM/c^2 R)_{\text{universe}} \approx 1$.

Recently a gravitational frequency shift has been detected in a laboratory experiment by measuring the change in frequency of gamma rays after they had "fallen" through a height h near the earth's surface. A body of mass m that falls a height h gains mgh of energy. If a falling photon of original frequency ν is taken to have the constant mass $h\nu/c^2$ (the frequency shift is so small that the change in mass may be neglected), its final energy $h\nu'$ is given by

$$h\nu' = h\nu + mgh = h\nu + \frac{h\nu gh}{c^2}$$

2.22
$$= h\nu \left(1 + \frac{gh}{c^2} \right)$$

For $h = 20$ m,

$$\frac{\Delta \nu}{\nu} = \frac{gh}{c^2} = \frac{9.8 \text{ m/s}^2 \times 20 \text{ m}}{(3 \times 10^8 \text{ m/s})^2}$$

$$= 2.2 \times 10^{-15}$$

A shift of this magnitude is just detectable, and the results confirm Eq. 2.22.

Problems

1. The threshold wavelength for photoelectric emission in tungsten is 2,300 Å. What wavelength of light must be used in order for electrons with a maximum energy of 1.5 eV to be ejected?

2. The threshold frequency for photoelectric emission in copper is 1.1×10^{15} Hz. Find the maximum energy of the photoelectrons (in joules and electron volts) when light of frequency 1.5×10^{15} Hz is directed on a copper surface.

3. The work function of sodium is 2.3 eV. What is the maximum wavelength of light that will cause photoelectrons to be emitted from sodium? What will the maximum kinetic energy of the photoelectrons be if 2,000 Å light falls on a sodium surface?

4. Find the wavelength and frequency of a 100-MeV photon.

5. Find the energy of a 7,000-Å photon.

6. Under favorable circumstances the human eye can detect 10^{-18} J of electromagnetic energy. How many 6,000-Å photons does this represent?

7. A 1,000-W radio transmitter operates at a frequency of 880 kHz. How many photons per second does it emit?

8. How many photons per second are emitted by a 10-W yellow lamp? (Assume the light is monochromatic with a wavelength of 6,000 Å.)

9. Light from the sun arrives at the earth at the rate of about 1,400 W/m² of area perpendicular to the direction of the light. (a) Find the maximum pressure (in lb/in.²) this light can exert on the earth's surface. (b) Assume that sunlight consists exclusively of 6,000-Å photons. How many photons per m² arrive at that part of the earth directly facing the sun in each second? (c) The average radius of the earth's orbit is 1.5×10^{11} m. What is the power output of the sun in watts, and how many photons per second does it emit? (d) How many photons per m³ are there near the earth?

10. What is the wavelength of the X rays emitted when 100-keV electrons strike a target? What is their frequency?

11. An X-ray machine produces 0.1-Å X rays. What accelerating voltage does it employ?

12. The distance between adjacent atomic planes in calcite is 3×10^{-8} cm.

What is the smallest angle between these planes and an incident beam of 0.3-Å X rays at which scattered X rays can be detected?

13. A potassium chloride crystal has a density of $1.98 \times 10^3 \, kg/m^3$. The molecular weight of KCl is 74.55. Find the distance between adjacent atoms.

14. How much energy must a photon have if it is to have the momentum of a 10-MeV proton?

15. What is the frequency of an X-ray photon whose momentum is $1.1 \times 10^{-23} \, kg\text{-}m/s$?

16. Prove that it is impossible for a photon to give up all its energy and momentum to a free electron, so that the photoelectric effect can take place only when photons strike bound electrons.

17. A beam of X rays is scattered by free electrons. At 45° from the beam direction the scattered X rays have a wavelength of 0.022 Å. What is the wavelength of the X rays in the direct beam?

18. An X-ray photon whose initial frequency was $1.5 \times 10^{19} \, Hz$ emerges from a collision with an electron with a frequency of $1.2 \times 10^{19} \, Hz$. How much kinetic energy was imparted to the electron?

19. An X-ray photon of initial frequency $3 \times 10^{19} \, Hz$ collides with an electron and is scattered through 90°. Find its new frequency.

20. Find the energy of an X-ray photon which can impart a maximum energy of 50 keV to an electron.

21. A monochromatic X-ray beam whose wavelength is 0.558 Å is scattered through 46°. Find the wavelength of the scattered beam.

22. In Sec. 2.4 the X rays scattered by a crystal were assumed to undergo no change in wavelength. Show that this assumption is reasonable by calculating the Compton wavelength of a Na atom and comparing it with the typical X-ray wavelength of 1 Å.

23. As discussed in Chap. 12, certain atomic nuclei emit photons in undergoing transitions from "excited" energy states to their "ground" or normal states. These photons constitute gamma rays. When a nucleus emits a photon, it recoils in the opposite direction. (a) The $^{57}_{27}Co$ nucleus decays by K capture to $^{57}_{26}Fe$, which then emits a photon in losing 14.4 keV to reach its ground state. The mass of a $^{57}_{26}Fe$ atom is $9.5 \times 10^{-26} \, kg$. By how much is the photon energy reduced from

the full 14.4 keV available as a result of having to share energy with the recoiling atom? (*b*) In certain crystals the atoms are so tightly bound that the entire crystal recoils when a gamma-ray photon is emitted, instead of the individual atom. This phenomenon is known as the *Mössbauer effect*. By how much is the photon energy reduced in this situation if the excited $^{57}_{26}$Fe nucleus is part of a 1-g crystal? *(*c*) The essentially recoil-free emission of gamma rays in situations like that of (*b*) means that it is possible to construct a source of essentially monoenergetic and hence monochromatic photons. Such a source was used in the experiment described in the last paragraph of Sec. 2.7. What is the original frequency and the change in frequency of a 14.4-keV gamma-ray photon after it has fallen 20 m near the earth's surface?

√24. A positron collides head on with an electron and both are annihilated. Each particle had a kinetic energy of 1 MeV. Find the wavelength of the resulting photons.

WAVE PROPERTIES OF PARTICLES 3

In retrospect it may seem odd that two decades passed between the discovery in 1905 of the particle properties of waves and the speculation in 1924 that particles might exhibit wave behavior. It is one thing, however, to suggest a revolutionary hypothesis to explain otherwise mysterious data and quite another to advance an equally revolutionary hypothesis in the absence of a strong experimental mandate. The latter is just what Louis de Broglie did in 1924 when he proposed that matter possesses wave as well as particle characteristics. So different was the intellectual climate at the time from that prevailing at the turn of the century that de Broglie's notion received immediate and respectful attention, whereas the earlier quantum theory of light of Planck and Einstein created hardly any stir despite its striking empirical support. The existence of de Broglie waves was demonstrated by 1927, and the duality principle they represent provided the starting point for Schrödinger's successful development of quantum mechanics in the previous year.

3.1 DE BROGLIE WAVES

A photon of light of frequency ν has the momentum

$$p = \frac{h\nu}{c}$$

which can be expressed in terms of wavelength λ as

$$p = \frac{h}{\lambda}$$

since $\lambda\nu = c$. The wavelength of a photon is therefore specified by its momentum according to the relation

3.1
$$\lambda = \frac{h}{p}$$

Drawing upon an intuitive expectation that nature is symmetric, de Broglie asserted that Eq. 3.1 is a completely general formula that applies to material particles as well as to photons. The momentum of a particle of mass m and velocity v is

$$p = mv$$

and consequently its *de Broglie wavelength* is

3.2 $\qquad \lambda = \dfrac{h}{mv}$ **De Broglie waves**

The greater the particle's momentum, the shorter its wavelength. In Eq. 3.2 m is the relativistic mass

$$m = \frac{m_0}{\sqrt{1 - v^2/c^2}}$$

Equation 3.2 has been amply verified by experiments involving the diffraction of fast electrons by crystals, experiments analogous to those that showed X rays to be electromagnetic waves. Before we consider these experiments, it is appropriate to look into the question of what kind of wave phenomenon is involved in the matter waves of de Broglie. In a light wave the electromagnetic field varies in space and time, in a sound wave pressure varies in space and time; what is it whose variations constitute de Broglie waves?

3.2 WAVE FUNCTION

The variable quantity characterizing de Broglie waves is called the *wave function,* denoted by the symbol Ψ (the Greek letter *psi*). **The value of the wave function associated with a moving body at the particular point x, y, z in space at the time t is related to the likelihood of finding the body there at the time.** Ψ itself, however, has no direct physical significance. There is a simple reason why Ψ cannot be interpreted in terms of an experiment. The probability P that something be somewhere at a given time can have any value between two limits: 0, corresponding to the certainty of its absence, and 1, corresponding to the certainty of its presence. (A probability of 0.2, for instance, signifies a 20 percent chance of finding the body.) But the amplitude of any wave may be negative as well as positive, and a negative probability is meaningless. Hence Ψ itself cannot be an observable quantity.

This objection does not apply to $|\Psi|^2$, the square of the absolute value of the wave function. For this and other reasons $|\Psi|^2$ is known as *probability density.* **The probability of experimentally finding the body described by the wave** function Ψ at the point x, y, z at the time t is proportional to the value of $|\Psi|^2$ there at t. A large value of $|\Psi|^2$ means the strong possibility of the body's presence, while a small value of $|\Psi|^2$ means the slight possibility of its presence. As long as $|\Psi|^2$ is not actually 0 somewhere, however, there is a definite chance, however small, of detecting it there. This interpretation was first made by Max Born in 1926.

There is a big difference between the probability of an event and the event itself. Although we shall speak of the wave function Ψ that describes a particle as being spread out in space, this does not mean that the particle itself is thus spread out. When an experiment is performed to detect electrons, for instance, a whole electron is either found at a certain time and place or it is not; there is no such thing as 20 percent of an electron. However, it is entirely possible for there to be a 20 percent chance that the electron be found at that time and place, and it is this likelihood that is specified by $|\Psi|^2$.

Alternatively, if an experiment involves a great many identical bodies all described by the same wave function Ψ, the *actual density* of bodies at x, y, z at the time t is proportional to the corresponding value of $|\Psi|^2$.

While the wavelength of the de Broglie waves associated with a moving body is given by the simple formula

$$\lambda = \frac{h}{mv}$$

determining their amplitude Ψ as a function of position and time usually presents a formidable problem. We shall discuss the calculation of Ψ in Chap. 5 and then go on to apply the ideas developed there to the structure of the atom in Chap. 6. Until then we shall assume that we have whatever knowledge of Ψ is required by the situation at hand.

In the event that a wave function Ψ is complex, with both real and imaginary parts, the probability density is given by the product $\Psi^* \Psi$ of Ψ and its *complex conjugate* Ψ^*. The complex conjugate of any function is obtained by replacing $i (= \sqrt{-1})$ by $-i$ wherever it appears in the function. Every complex function Ψ can be written in the form

$$\Psi = A + iB$$

where A and B are real functions. The complex conjugate Ψ^* of Ψ is

$$\Psi^* = A - iB$$

and so

$$\Psi^* \Psi = A^2 - i^2 B^2 = A^2 + B^2$$

since $i^2 = -1$. Hence $\Psi^* \Psi$ is always a positive real quantity.

3.3 DE BROGLIE WAVE VELOCITY

With what velocity do de Broglie waves travel? Since we associate a de Broglie wave with a moving body, it is reasonable to expect that this wave travels at the same velocity v as that of the body. If we call the de Broglie wave velocity w, we may apply the usual formula

$$w = \nu\lambda$$

to determine the value of w. The wavelength λ is just the de Broglie wavelength

$$\lambda = \frac{h}{mv}$$

We shall take the frequency ν to be that specified by the quantum equation

$$E = h\nu$$

Hence

$$\nu = \frac{E}{h}$$

or, since

$$E = mc^2$$

we have

$$\nu = \frac{mc^2}{h}$$

The de Broglie wave velocity is therefore

3.3

$$w = \nu\lambda$$

$$= \frac{mc^2}{h}\frac{h}{mv}$$

$$= \frac{c^2}{v}$$

Since the particle velocity v must be less than the velocity of light c, the de Broglie wave velocity w is always greater than c! Clearly v and w are never equal for a moving body. In order to understand this unexpected result, we shall digress briefly to consider the notions of *phase velocity* and *group velocity*. (Phase velocity is sometimes called *wave velocity*.)

Let us begin by reviewing how waves are described mathematically. For clarity we shall consider a string stretched along the x axis whose vibrations are in the y direction, as in Fig. 3-1, and are simple harmonic in character. If we choose $t = 0$ when the displacement y of the string at $x = 0$ is a maximum, its displacement at any future time t at the same place is given by the formula

3.4

$$y = A \cos 2\pi\nu t$$

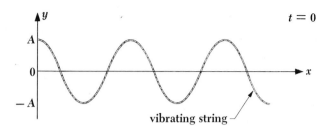

$t = 0$

vibrating string

FIGURE 3-1 Wave motion.

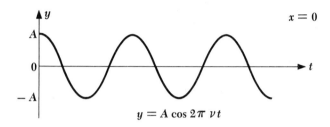

$x = 0$

$$y = A \cos 2\pi \, \nu t$$

where A is the amplitude of the vibrations (that is, their maximum displacement on either side of the x axis) and ν their frequency.

Equation 3.4 tells us what the displacement of a single point on the string is as a function of time t. A complete description of wave motion in a stretched string, however, should tell us what y is at *any* point on the string at *any* time. What we want is a formula giving y as a function of both x and t. To obtain such a formula, let us imagine that we shake the string at $x = 0$ when $t = 0$, so that a wave starts to travel down the string in the $+x$ direction (Fig. 3-2). This wave has some speed w that depends upon the properties of the string. The wave travels the distance $x = wt$ in the time t; hence the time interval between the formation of the wave at $x = 0$ and its arrival at the point x is x/w. Accordingly the displacement y of the string at x at any time t is exactly the same as the value of y at $x = 0$ *at the earlier time* $t - x/w$. By simply replacing t in Eq. 3.4 with $t - x/w$, then, we have the desired formula giving y in terms of both x and t:

3.5 $$y = A \cos 2\pi\nu \left(t - \frac{x}{w} \right)$$

As a check, we note that Eq. 3.5 reduces to Eq. 3.4 at $x = 0$.

Equation 3.5 may be rewritten

$$y = A \cos 2\pi \left(\nu t - \frac{\nu x}{w} \right)$$

Since

$$w = \nu\lambda$$

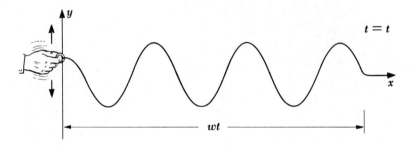

FIGURE 3-2 Wave propagation.

we have

3.6 $$y = A \cos 2\pi \left(vt - \frac{x}{\lambda} \right)$$

Equation 3.6 is often more convenient to apply then Eq. 3.5.

Perhaps the most widely used description of a wave, however, is still another form of Eq. 3.5. We define the quantities *angular frequency* ω and *wave number* k by the formulas

3.7 $\omega = 2\pi v$ **Angular frequency**

3.8 $k = \dfrac{2\pi}{\lambda}$ **Wave number**

3.9 $ = \dfrac{\omega}{w}$

The unit of ω is the rad/s and that of k is the rad/m. Angular frequency gets its name from uniform circular motion, where a particle that moves around a circle v times per second sweeps out $2\pi v$ rad/s. The wave number is equal to the number of radians corresponding to a wave train 1 m long, since there are 2π rad in one complete wave. In terms of ω and k, Eq. 3.5 becomes

3.10 $y = A \cos(\omega t - kx)$

In three dimensions k becomes a vector $\mathbf{k}$ normal to the wave fronts and x is replaced by the radius vector $\mathbf{r}$. The scalar product $\mathbf{k} \cdot \mathbf{r}$ is then used instead of kx in Eq. 3.10.

3.4 PHASE AND GROUP VELOCITIES

The amplitude of the de Broglie waves that correspond to a moving body reflects the probability that it be found at a particular place at a particular time. It is clear that de Broglie waves cannot be represented simply by a formula resembling Eq. 3.10, which describes an indefinite series of waves all with the same amplitude A. Instead, we expect the wave representation of a moving body to correspond to a *wave packet*, or *wave group*, like that shown in Fig. 3-3, whose constituent waves have amplitudes upon which the likelihood of detecting the body depends.

A familiar example of how wave groups come into being is the case of *beats*. When two sound waves of the same amplitude, but of slightly different frequencies, are produced simultaneously, the sound we hear has a frequency equal to the average of the two original frequencies and its amplitude rises and falls periodically. The amplitude fluctuations occur a number of times per second equal to the difference between the two original frequencies. If the original sounds have frequencies of, say, 440 and 442 Hz, we will hear a fluctuating sound of frequency 441 Hz with two loudness peaks, called *beats*, per second. The production of beats is illustrated in Fig. 3-4.

A way of mathematically describing a wave group, then, is in terms of a superposition of individual wave patterns, each of different wavelength, whose interference with one another results in the variation in amplitude that defines the group shape. If the speeds of the waves are the same, the speed with which the wave group travels is identical with the common wave speed. However, if the wave speed varies with wavelength, the different individual waves do not proceed together, and the wave group has a speed different from that of the waves that compose it.

It is not difficult to compute the speed u with which a wave group travels. Let us suppose that a wave group arises from the combination of two waves

FIGURE 3-3 A wave group.

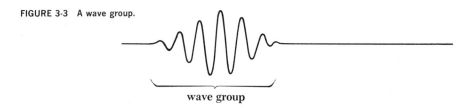

wave group

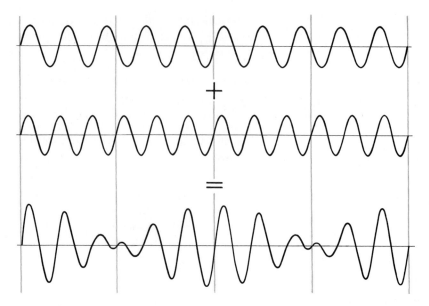

FIGURE 3-4 The production of beats.

with the same amplitude A but differing by an amount $d\omega$ in angular frequency and an amount dk in wave number. We may represent the original waves by the formulas

$$y_1 = A \cos (\omega t - kx)$$
$$y_2 = A \cos [(\omega + d\omega)t - (k + dk)x]$$

The resultant displacement y at any time t and any position x is the sum of y_1 and y_2. With the help of the identity

$$\cos \alpha + \cos \beta = 2 \cos \tfrac{1}{2}(\alpha + \beta) \cos \tfrac{1}{2}(\alpha - \beta)$$

and the relation

$$\cos (-\theta) = \cos \theta$$

we find that

$$
\begin{aligned}
y &= y_1 + y_2 \\
&= 2A \cos \tfrac{1}{2}[(2\omega + d\omega)t - (2k + dk)x] \cos \tfrac{1}{2}(d\omega\, t - dk\, x)
\end{aligned}
$$

Since $d\omega$ and dk are small compared with ω and k respectively,

$$2\omega + d\omega \approx 2\omega$$
$$2k + dk \approx 2k$$

and

3.11 $y = 2A \cos{(\omega t - kx)} \cos{\left(\dfrac{d\omega}{2} t - \dfrac{dk}{2} x\right)}$

Equation 3.11 represents a wave of angular frequency ω and wave number k that has superimposed upon it a modulation of angular frequency $\frac{1}{2}d\omega$ and of wave number $\frac{1}{2}dk$. The effect of the modulation is to produce successive wave groups, as in Fig. 3-4. The phase velocity w is

3.12 $w = \dfrac{\omega}{k}$ **Phase velocity**

while the velocity u of the wave groups is

3.13 $u = \dfrac{d\omega}{dk}$ **Group velocity**

In general, depending upon the manner in which phase velocity varies with wave number in a particular medium, the group velocity may be greater than or less than the phase velocity. If the phase velocity w is the same for all wavelengths, the group and phase velocities are the same.

The angular frequency and wave number of the de Broglie waves associated with a body of rest mass m_0 moving with the velocity v are

3.14 $\omega = 2\pi\nu$

$= \dfrac{2\pi mc^2}{h}$

$= \dfrac{2\pi m_0 c^2}{h\sqrt{1 - v^2/c^2}}$

and

3.15 $k = \dfrac{2\pi}{\lambda}$

$= \dfrac{2\pi mv}{h}$

$= \dfrac{2\pi m_0 v}{h\sqrt{1 - v^2/c^2}}$

Both ω and k are functions of the velocity v. The phase velocity w is, as we found earlier,

$w = \dfrac{\omega}{k}$

$= \dfrac{c^2}{v}$

which exceeds both the velocity of the body v and the velocity of light c, since $v < c$.

The group velocity u of the de Broglie waves associated with the body is

$$u = \frac{d\omega}{dk}$$

$$= \frac{d\omega/dv}{dk/dv}$$

Now

$$\frac{d\omega}{dv} = \frac{2\pi m_0 v}{h(1 - v^2/c^2)^{3/2}}$$

and

$$\frac{dk}{dv} = \frac{2\pi m_0}{h(1 - v^2/c^2)^{3/2}}$$

and so the group velocity is

3.16 $u = v$

The de Broglie wave group associated with a moving body travels with the same velocity as the body. The phase velocity w of the de Broglie waves evidently has no simple physical significance in itself.

3.5 THE DIFFRACTION OF PARTICLES

A wave manifestation having no analog in the behavior of Newtonian particles is diffraction. In 1927 Davisson and Germer in the United States and G. P. Thomson in England independently confirmed de Broglie's hypothesis by demonstrating that electrons exhibit diffraction when they are scattered from crystals whose atoms are spaced appropriately. We shall consider the experiment of Davisson and Germer because its interpretation is more direct.

Davisson and Germer were studying the scattering of electrons from a solid, using an apparatus like that sketched in Fig. 3-5. The energy of the electrons in the primary beam, the angle at which they are incident upon the target, and the position of the detector can all be varied. Classical physics predicts that the scattered electrons will emerge in all directions with only a moderate dependence of their intensity upon scattering angle and even less upon the energy of the primary electrons. Using a block of nickel as the target, Davisson and Germer verified these predictions.

In the midst of their work there occurred an accident that allowed air to enter their apparatus and oxidize the metal surface. To reduce the oxide to pure nickel, the target was baked in a high-temperature oven. After this treatment, the target

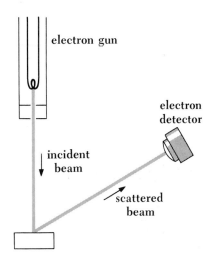

FIGURE 3-5   The Davisson-Germer experiment.

electron gun

electron
detector

incident
beam

scattered
beam

was returned to the apparatus and the measurements resumed. Now the results were very different from what had been found before the accident: instead of a continuous variation of scattered electron intensity with angle, distinct maxima and minima were observed whose positions depended upon the electron energy! Typical polar graphs of electron intensity after the accident are shown in Fig. 3-6; the method of plotting is such that the intensity at any angle is proportional to the distance of the curve at that angle from the point of scattering.

Two questions come to mind immediately: what is the reason for this new effect, and why did it not appear until after the nickel target was baked?

De Broglie's hypothesis suggested the interpretation that electron waves were being diffracted by the target, much as X rays are diffracted by planes of atoms in a crystal. This interpretation received support when it was realized that the effect of heating a block of nickel at high temperature is to cause the many

FIGURE 3-6   Results of the Davisson-Germer experiment.

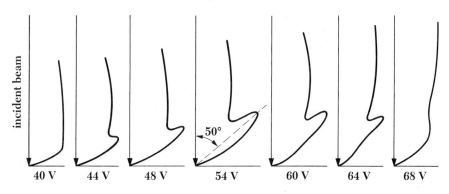

incident beam

50°

40 V   44 V   48 V   54 V   60 V   64 V   68 V

small individual crystals of which it is normally composed to form into a single large crystal, all of whose atoms are arranged in a regular lattice.

Let us see whether we can verify that de Broglie waves are responsible for the findings of Davisson and Germer. In a particular determination, a beam of 54-eV electrons was directed perpendicularly at the nickel target, and a sharp maximum in the electron distribution occurred at any angle of 50° with the original beam. The angles of incidence and scattering relative to the family of Bragg planes shown in Fig. 3-7 will both be 65°. The spacing of the planes in this family, which can be measured by X-ray diffraction, is 0.91 Å. The Bragg equation for maxima in the diffraction pattern is

$$n\lambda = 2d \sin \theta$$

Here $d = 0.91$ Å and $\theta = 65°$; assuming that $n = 1$, the de Broglie wavelength λ of the diffracted electrons is

$$\begin{aligned} n\lambda &= 2d \sin \theta \\ &= 2 \times 0.91 \text{ Å} \times 65° \\ &= 1.65 \text{ Å} \end{aligned}$$

Now we use de Broglie's formula

$$\lambda = \frac{h}{mv}$$

to calculate the expected wavelength of the electrons. The electron kinetic energy of 54 eV is small compared with its rest energy m_0c^2 of 5.1×10^5 eV, and so we can ignore relativistic considerations. Since

$$T = \tfrac{1}{2}mv^2$$

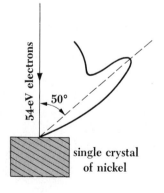

FIGURE 3-7 The diffraction of de Broglie waves by the target is responsible for the results of Davisson and Germer.

54-eV electrons

50°

single crystal of nickel

the electron momentum mv is

$$mv = \sqrt{2mT}$$
$$= \sqrt{2 \times 9.1 \times 10^{-31} \text{ kg} \times 54 \text{ eV} \times 1.6 \times 10^{-19} \text{ J/eV}}$$
$$= 40 \times 10^{-24} \text{ kg-m/s}$$

The electron wavelength is therefore

proton momentum

$mv = \sqrt{2mT}$

$mv = \sqrt{2 \times 1.6 \times 10^{-27} \times 1 \times 10^{3} \text{ eV} \times 1.6 \times 10^{-19}}$

$$\lambda = \frac{h}{mv}$$
$$= \frac{6.63 \times 10^{-34} \text{ J-s}}{4.0 \times 10^{-24} \text{ kg-m/s}}$$
$$= 1.66 \times 10^{-10} \text{ m}$$
$$= 1.66 \text{ Å}$$

in excellent agreement with the observed wavelength. The Davisson-Germer experiment thus provides direct verification of de Broglie's hypothesis of the wave nature of moving bodies.

The analysis of the Davisson-Germer experiment is actually less straightforward than indicated above, since the energy of an electron increases when it enters a crystal by an amount equal to the work function of the surface. Hence the electron speeds in the experiment were greater inside the crystal and the corresponding de Broglie wavelength shorter than the corresponding values outside. An additional complication arises from interference between waves diffracted by different families of Bragg planes, which restricts the occurrence of maxima to certain combinations of electron energy and angle of incidence rather than merely to any combination that obeys the Bragg equation.

Electrons are not the only bodies whose wave behavior can be demonstrated. The diffraction of neutrons and of whole atoms when scattered by suitable crystals has been observed, and in fact neutron diffraction, like X-ray and electron diffraction, is today a widely used tool for investigating crystal structures.

As in the case of electromagnetic waves, the wave and particle aspects of moving bodies can never be simultaneously observed, so that we cannot determine which is the "correct" description. All we can say is that in some respects a moving body exhibits wave properties and in other respects it exhibits particle properties. Which set of properties is most conspicuous depends upon how the de Broglie wavelength compares with the dimensions of the bodies involved: the 1.66 Å wavelength of a 54-eV electron is of the same order of magnitude as the lattice spacing in a nickel crystal, but the wavelength of an automobile moving at 60 mi/h is about 5×10^{-38} ft, far too small to manifest itself.

3.6 THE UNCERTAINTY PRINCIPLE

The fact that a moving body must be regarded as a de Broglie wave group rather than as a localized entity suggests that there is a fundamental limit to the accuracy with which we can measure its particle properties. Figure 3-8a shows a de Broglie wave group: the particle may be anywhere within the wave group. If the group is very narrow, as in Fig. 3-8b, the position of the particle is readily found, but the wavelength is impossible to establish. At the other extreme, a wide group, as in Fig. 3-8c, permits a satisfactory wavelength estimate, but where is the particle located?

A straightforward argument based upon the nature of wave groups permits us to relate the inherent uncertainty Δx in a measurement of particle position with the inherent uncertainty Δp in a simultaneous measurement of its momentum.

The simplest example of the formation of wave groups is that given in Sec. 3.4, where two wave trains slightly different in angular frequency ω and propagation constant k were superposed to yield the series of groups shown in Fig. 3-4. Here let us consider the wave groups that arise when the de Broglie waves

$$\Psi_1 = A \cos (\omega t - kx)$$
$$\Psi_2 = A[\cos (\omega + \Delta\omega)t - (k + \Delta k)x]$$

FIGURE 3-8 The width of a wave group is a measure of the uncertainty in the location of the particle it represents. The narrower the wave group, the greater the uncertainty in the wavelength.

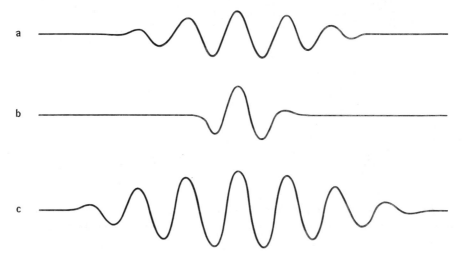

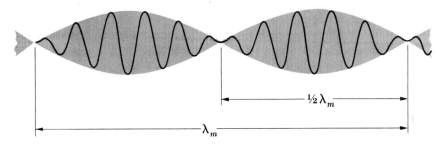

FIGURE 3-9 Wave groups that result from the interference of wave trains having the same amplitudes but different frequencies.

are combined. From a calculation identical with the one used in obtaining Eq. 3.11, we find that

3.17
$$\Psi = \Psi_1 + \Psi_2$$
$$\approx 2A \cos(\omega t - kx) \cos(\tfrac{1}{2}\Delta\omega\, t - \tfrac{1}{2}\Delta k\, x)$$

which is plotted in Fig. 3-9. The width of each group is evidently equal to half the wavelength λ_m of the modulation. It is reasonable to suppose that this width is of the same order of magnitude as the inherent uncertainty Δx in the position of the group, that is,

3.18
$$\Delta x \approx \tfrac{1}{2}\lambda_m$$

The modulation wavelength is related to its propagation constant k_m by

$$\lambda_m = \frac{2\pi}{k_m}$$

From Eq. 3.17 we see that the propagation constant of the modulation is

$$k_m = \tfrac{1}{2}\Delta k$$

with the result that

$$\lambda_m = \frac{2\pi}{\tfrac{1}{2}\Delta k}$$

and

$$\Delta x \approx \frac{2\pi}{\Delta k}$$

3.19
$$\Delta x\, \Delta k \approx 2\pi$$

A moving body corresponds to a single wave group, not a succession of them, but a single wave group can also be thought of in terms of the superposition of trains of harmonic waves. However, an infinite number of wave trains of different frequencies, wave numbers, and amplitudes is required for an isolated group of arbitrary shape.

At a certain time t, the wave group $\Psi(x)$ can be represented by the *Fourier integral*

3.20
$$\Psi(x) = \int_0^\infty g(k) \cos kx \, dk$$

where the function $g(k)$ describes how the amplitudes of the waves that contribute to $\Psi(x)$ vary with wave number k. This function is called the *Fourier transform* of $\Psi(x)$, and it specifies the wave group just as completely as $\Psi(x)$ does. Figure 3-10 contains graphs of the Fourier transforms of a pulse and of a wave group. For comparison, the Fourier transform of an infinite train of harmonic waves is also included; only a single wave number is present in this case, of course.

Strictly speaking, the wave numbers needed to represent a wave group extend from $k = 0$ to $k = \infty$, but for a group whose length Δx is finite, the waves whose amplitudes $g(k)$ are appreciable have wave numbers that lie within a finite interval Δk. As Fig. 3-10 indicates, the shorter the group, the broader the range of wave numbers needed to describe it, and vice versa. The relationship between the distance Δx and the wave-number spread Δk depends upon the shape of the wave group and upon how Δx and Δk are defined. The minimum value of the product $\Delta x \, \Delta k$ occurs when the group has the form of a gaussian function, in which case its Fourier transform happens to be a gaussian function also. If Δx and Δk are taken as the standard deviations of the respective functions $\Psi(x)$ and $g(k)$, then $\Delta x \, \Delta k = \frac{1}{2}$. In general, $\Delta x \, \Delta k$ has the order of magnitude of 1:

3.21
$$\Delta x \, \Delta k \approx 1$$

FIGURE 3-10 The wave functions and Fourier transforms for (a) a pulse, (b) a wave group, and (c) an infinite wave train. A brief disturbance requires a broader range of frequencies to describe it than a disturbance of greater duration.

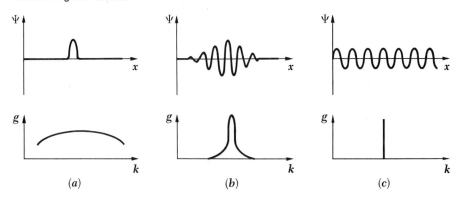

The de Broglie wavelength of a particle of momentum p is

$$\lambda = \frac{h}{p}$$

The wave number corresponding to this wavelength is

$$k = \frac{2\pi}{\lambda}$$

$$= \frac{2\pi p}{h}$$

Hence an uncertainty Δk in the wave number of the de Broglie waves associated with the particle results in an uncertainty Δp in the particle's momentum according to the formula

$$\Delta p = \frac{h\,\Delta k}{2\pi}$$

Since $\Delta x\,\Delta k \approx 1$, $\Delta k \approx 1/\Delta x$ and

3.22 $\qquad \Delta x\,\Delta p \geqslant \dfrac{h}{2\pi}$ $\qquad\qquad\qquad\qquad$ **Uncertainty principle**

ALSO CALLED INDETERMINACY & HEISENBERG PRINCIPLE

The sign $\geqslant$ is used because Δx and Δp are *irreducible minima that are consequences of the wave natures of moving bodies;* any instrumental or statistical uncertainties that arise in the actual conduct of the measurement only augment the product $\Delta x\,\Delta p$.

Equation 3.22 is one form of the *uncertainty principle* first obtained by Werner Heisenberg in 1927. It states that the product of the uncertainty Δx in the position of a body at some instant and the uncertainty Δp in its momentum at the same instant is equal to or greater than $h/2\pi$. We cannot measure simultaneously both position and momentum with perfect accuracy. If we arrange matters so that Δx is small, corresponding to the narrow wave group of Fig. 3-8b, Δp will be large. If we reduce Δp in some way, corresponding to the wide wave group of Fig. 3-8c, Δx will be large. These uncertainties are due not to inadequate apparatus but to the imprecise character in nature of the quantities involved.

The quantity $h/2\pi$ appears quite often in modern physics because, besides its connection with the uncertainty principle, $h/2\pi$ also turns out to be the basic unit of angular momentum. It is therefore customary to abbreviate $h/2\pi$ by the symbol $\hbar$:

$$\hbar = \frac{h}{2\pi} = 1.054 \times 10^{-34}\ \text{J-s}$$

In the remainder of this book we shall use $\hbar$ in place of $h/2\pi$.

The uncertainty principle can be arrived at in a variety of ways. Let us obtain it by basing our argument upon the particle nature of waves instead of upon the wave nature of particles as we did above.

Suppose that we wish to measure the position and momentum of something at a certain moment. To accomplish this, we must prod it with something else that is to carry the desired information back to us; that is, we have to touch it with our finger, illuminate it with light, or interact with it in some other way. We might be examining an electron with the help of light of wavelength λ, as in Fig. 3-11. In this process photons of light strike the electron and bounce off it. Each photon possesses the momentum h/λ, and when it collides with the electron, the electron's original momentum p is changed. The precise change cannot be predicted, but it is likely to be of the same order of magnitude as the photon momentum h/λ. Hence the *act of measurement* introduces an uncertainty of

3.23 $$\Delta p \approx \frac{h}{\lambda}$$

in the momentum of the electron. The longer the wavelength of the light we employ in "seeing" the electron, the smaller the consequent uncertainty in its momentum.

Because light has wave properties, we cannot expect to determine the electron's position with infinite accuracy under any circumstances, but we might reasonably hope to keep the irreducible uncertainty Δx in its position to

FIGURE 3-11 An electron cannot be observed without changing its momentum.

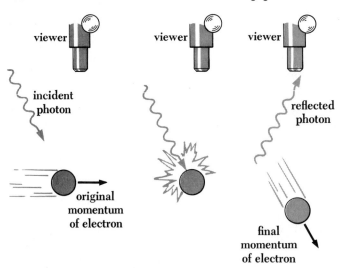

1 wavelength of the light being used. That is,

3.24 $\qquad \Delta x \approx \lambda$

The shorter the wavelength, the smaller the uncertainty in the position of the electron.

From Eqs. 3.23 and 3.24 it is clear that, if we employ light of short wavelength to improve the accuracy of the position determination, there will be a corresponding reduction in the accuracy of the momentum determination, while light of long wavelength will yield an accurate momentum value but an inaccurate position value. Substituting $\lambda = \Delta x$ into Eq. 3.23 yields

3.25 $\qquad \Delta x \, \Delta p \geqslant h$

This result is consistent with Eq. 3.22, since both Δx and Δp here were defined rather pessimistically.

Arguments like the preceding one, though superficially attractive, must as a rule be approached with caution. The above argument implies that the electron can possess a definite position and momentum at any instant, and that it is the measurement process that introduces the indeterminacy in $\Delta x \, \Delta p$. On the contrary, this indeterminacy is inherent in the nature of a moving body. The justification for the many "derivations" of this kind is, first, that they show it is impossible to imagine a way around the uncertainty principle, and second, that they present a view of the principle that can be appreciated in a more familiar context than that of wave packets.

3.7 APPLICATIONS OF THE UNCERTAINTY PRINCIPLE

Planck's constant h is so minute—only 6.63×10^{-34} J-s—that the limitations imposed by the uncertainty principle are significant only in the realm of the atom. On this microscopic scale, however, there are many phenomena that can be understood in terms of this principle; we shall consider several of them here.

One interesting question is whether electrons are present in atomic nuclei. As we shall learn later, typical nuclei are less than 10^{-14} m in radius. For an electron to be confined within such a nucleus, the uncertainty in its position may not exceed 10^{-14} m. The corresponding uncertainty in the electron's momentum is

$$\Delta p \geqslant \frac{\hbar}{\Delta x}$$

$$\geqslant \frac{1.054 \times 10^{-34} \text{ J-s}}{10^{-14} \text{ m}}$$

$$\geqslant 1.1 \times 10^{-20} \text{ kg-m/s}$$

If this is the uncertainty in the electron's momentum, the momentum itself must be at least comparable in magnitude. An electron whose momentum is 1.1×10^{-20} kg-m/s has a kinetic energy T many times greater than its rest energy m_0c^2, and we may accordingly use the extreme relativistic formula

$$T = pc$$

to find T. Substituting for p and c, we obtain

$$T = 1.1 \times 10^{-20} \text{ kg-m/s} \times 3 \times 10^8 \text{ m/s}$$
$$= 3.3 \times 10^{-12} \text{ J}$$

Since 1 eV $= 1.6 \times 10^{-19}$ J, the kinetic energy of the electron must be well over 20 MeV if it is to be a nuclear constituent. Experiments indicate that the electrons associated even with unstable atoms never have more than a fraction of this energy, and we conclude that electrons cannot be present within nuclei.

Let us now ask how much energy an electron needs to be confined to an atom. The hydrogen atom is about 5×10^{-11} m in radius, and therefore the uncertainty in the position of its electron may not exceed this figure. The corresponding momentum uncertainty is

$$\Delta p = 2.1 \times 10^{-24} \text{ kg-m/s}$$

An electron whose momentum is of this order of magnitude is nonrelativistic in behavior, and its kinetic energy is

$$T = \frac{p^2}{2m}$$
$$= \frac{(2.1 \times 10^{-24} \text{ kg-m/s})^2}{2 \times 9.1 \times 10^{-31} \text{ kg}}$$
$$= 2.4 \times 10^{-18} \text{ J}$$

or about 15 eV. This is a wholly plausible figure.

Another form of the uncertainty principle is sometimes useful. We might wish to measure the energy E emitted sometime during the time interval Δt in an atomic process. If the energy is in the form of electromagnetic waves, the limited time available restricts the accuracy with which we can determine the frequency ν of the waves. Let us assume that the uncertainty in the number of waves we count in a wave group is one wave. Since the frequency of the waves under study is equal to the number of them we count divided by the time interval, the uncertainty $\Delta \nu$ in our frequency measurement is

$$\Delta \nu = \frac{1}{\Delta t}$$

The corresponding energy uncertainty is

$$\Delta E = h \, \Delta \nu$$

and so

$$\Delta E = \frac{h}{\Delta t}$$

or

$$\Delta E \, \Delta t \geqslant h$$

A more realistic calculation changes this to

3.26 $$\Delta E \, \Delta t \geqslant \hbar$$

Equation 3.26 states that the product of the uncertainty ΔE in an energy measurement and the uncertainty Δt in the time at which the measurement was made is equal to or greater than $\hbar$.

As an example of the significance of Eq. 3.26 we can consider the radiation of light from an "excited" atom. Such an atom divests itself of its excess energy by emitting one or more photons of characteristic frequency. The average period that elapses between the excitation of an atom and the time it radiates is 10^{-8} s. Thus the photon energy is uncertain by an amount

$$\begin{aligned} \Delta E &= \frac{\hbar}{\Delta t} \\ &= \frac{1.054 \times 10^{-34} \text{ J-s}}{10^{-8} \text{ s}} \\ &= 1.1 \times 10^{-26} \text{ J} \end{aligned}$$

and the frequency of the light is uncertain by

$$\begin{aligned} \Delta \nu &= \frac{\Delta E}{h} \\ &= 1.6 \times 10^{7} \text{ Hz} \end{aligned}$$

This is the irreducible limit to the accuracy with which we can determine the frequency of the radiation emitted by an atom.

3.8   THE WAVE-PARTICLE DUALITY

Despite the abundance of experimental confirmation, many of us find it hard to appreciate how what we normally think of as a wave can also be a particle and how what we normally think of as a particle can also be a wave. The uncertainty principle provides a valuable perspective on this question which makes it possible to put such statements as those at the end of Sec. 2.2 on a more concrete basis.

Figure 3-12 shows an experimental arrangement in which light that has been diffracted by a double slit is detected on a "screen" that consists of many adjacent photoelectric cells. The photoelectric cells respond to photons, which have all the properties we associate with particles. However, when we plot the number of photons each cell counts in a certain period of time against the location of the cell, we find the characteristic pattern produced by the interference of a pair of coherent wave trains. This pattern even occurs when the light intensity is so low that, on the average, only one photon at a time is in the apparatus. The problem is, how can a photon that passes through one of the slits be affected by the presence of the other slit? In other words, how can a photon interfere with itself? This problem does not arise in the case of waves, which are spread out in space, but it would seem to have meaning in the case of photons, whose behavior suggests that they are localized in very small regions of space.

To have meaning, every question or statement in science must ultimately be reducible to an experiment. Here the relevant experiment is one that would

FIGURE 3-12 Hypothetical experiment to determine which slit each photon contributing to a double-slit interference pattern has passed through.

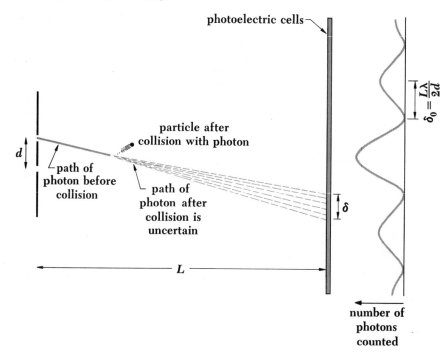

detect which of the slits a particular photon passes through on its way to the screen. Let us imagine that we introduce a cloud of small particles between the slits and the screen. A photon that passes through one of the slits strikes a particle and gives it a certain impulse which enables us to detect it (Fig. 3-12). Provided that we can establish the position of the particle with an uncertainty Δy that is less than half the space d between the slits, we can determine which slit the photon passed through. Therefore

$$\Delta y < \frac{d}{2}$$

But if we are able to limit the uncertainty in the y coordinate of the struck particle to Δy, the uncertainty Δp_y in the y component of its momentum is

3.27 $\qquad \Delta p_y \geqslant \frac{\hbar}{\Delta y} > \frac{2\hbar}{d}$

Since the collision introduces a change of Δp_y in the particle's momentum, the same change must have occurred in the photon's momentum. A change of Δp_y in the photon's momentum means a shift of

$$\delta = \frac{\Delta p_y}{p_x} L$$

in the location on the screen which the photon strikes; because $p_y \ll p$ (the width of the diffraction pattern is small compared with the distance L), $p_x \approx p$, and we can write

$$\delta = \frac{\Delta p_y}{p} L$$

The photon momentum is related to the wavelength λ of the light by Eq. 3.1,

$$p = \frac{h}{\lambda}$$

and so

$$\delta = \frac{\Delta p_y \lambda L}{h}$$

From Eq. 3.27 we have $\Delta p_y > 2\hbar/d$, which means that the shift in the photon's screen position is

3.28 $\qquad \delta = \frac{\lambda L}{\pi d}$

The distance δ_0 between a maximum (that is, a "bright line") in the interference pattern and an adjacent minimum ("dark line") is known from elementary optics to be

3.29
$$\delta_0 = \frac{\lambda L}{2d}$$

This distance is almost the same as the minimum shift involved in establishing which slit each photon passes through. What would otherwise be a pattern of alternating bright and dark lines becomes blurred owing to the interactions between the photons and the particles used to trace their paths. Thus *no interference can be observed:* the price of determining the exact path of each photon is the destruction of the interference pattern. If our interest is in the wave aspects of a phenomenon, they can be demonstrated; if our interest is in the particle aspects of the same phenomenon, they too can be demonstrated; but it is impossible to demonstrate *both* aspects in a simultaneous experiment. (Using photoelectric cells to detect an interference pattern is not a simultaneous experiment in this sense, because there is no way in which a photoelectric cell can determine through which slit a particular photon striking it has passed.)

The original question of how a photon can interfere with itself therefore turns out to be meaningless. It is important to be aware of the distinction between a legitimate question that cannot be answered because our existing knowledge is not sufficiently detailed or advanced to cope with it and a question whose very statement is in contradiction with experiment. Questions that seek to pry apart the elements of the wave-particle duality fall into the latter class in view of the uncertainty principle, whose own empirical validity has been thoroughly established.

Problems

1. Find the de Broglie wavelength of an electron whose speed is 10^8 m/s.

2. Find the de Broglie wavelength of a 1-MeV proton.

3. Nuclear dimensions are of the order of 10^{-14} m. (*a*) Find the energy in eV of an electron whose de Broglie wavelength is 10^{-15} m and which is thus capable of revealing details of nuclear structure. (*b*) Make the same calculation for a neutron.

4. Neutrons in equilibrium with matter at room temperature (300 K) have average energies of about $\frac{1}{25}$ eV. (Such neutrons are often called "thermal neutrons.") Find their de Broglie wavelength.

5. Derive a formula expressing the de Broglie wavelength (in Å) of an electron in terms of the potential difference V (in volts) through which it is accelerated.

6. Derive a formula for the de Broglie wavelength of a particle in terms of its kinetic energy T and its rest energy m_0c^2. If $T \gg m_0c^2$, how does the particle's wavelength compare with the wavelength of a photon of the same energy?

7. Assume that electromagnetic waves are a special case of de Broglie waves. Show that photons must travel with the velocity c and that the rest mass of the photon must be 0.

8. Obtain the de Broglie wavelength of a moving particle in the following way, which parallels de Broglie's original treatment. Consider a particle of rest mass m_0 as having a characteristic frequency of vibration of ν_0 specified by the relationship $h\nu_0 = m_0c^2$. The particle travels with the speed v relative to an observer. With the help of special relativity, show that the observer sees a progressive wave whose phase velocity is $w = c^2/v$ and whose wavelength is h/mv, where $m = m_0/\sqrt{1 - v^2/c^2}$.

9. The velocity of ocean waves is $\sqrt{g\lambda/2\pi}$, where g is the acceleration of gravity. Find the group velocity of these waves.

10. The velocity of ripples on a liquid surface is $\sqrt{2\pi S/\lambda\rho}$, where S is the surface tension and ρ the density of the liquid. Find the group velocity of these waves.

11. The position and momentum of a 1-keV electron are simultaneously determined. If its position is located to within 1 Å, what is the percentage of uncertainty in its momentum?

12. An electron microscope uses 40-keV electrons. Find its ultimate resolving power on the assumption that this is equal to the wavelength of the electrons.

13. Compare the uncertainties in the velocities of an electron and a proton confined in a 10-Å box.

14. Wavelengths can be determined with accuracies of one part in 10^6. What is the uncertainty in the position of a 1-Å X-ray photon when its wavelength is simultaneously measured?

15. At a certain time t a measurement establishes the position of an electron with an accuracy of $\pm 10^{-11}$ m. Find the uncertainty in the electron's momentum at t and, from this, the uncertainty in its position 1 s later. If the latter uncertainty is not $\pm 10^{-11}$ m, account for the difference in terms of the concept of a moving particle as a wave packet.

16. (a) How much time is needed to measure the kinetic energy of an electron whose speed is 10 m/s with an uncertainty of no more than 0.1 percent? How

far will the electron have traveled in this period of time? (*b*) Make the same calculations for a 1-g insect whose speed is the same. What do these sets of figures indicate?

17. The atoms in a solid possess a certain minimum *zero-point energy* even at 0 K, while no such restriction holds for the molecules in an ideal gas. Use the uncertainty principle to explain these statements.

18. Verify that the uncertainty principle can be expressed in the form $\Delta L\, \Delta\theta \geqslant \hbar$, where ΔL is the uncertainty in the angular momentum of a body and $\Delta\theta$ is the uncertainty in its angular position. [*Hint:* Consider a particle moving in a circle.]

THE ATOM 2

ATOMIC STRUCTURE 4

Far in the past people began to suspect that matter, despite its appearance of being continuous, possesses a definite structure on a microscopic level beyond the direct reach of our senses. This suspicion did not take on a more concrete form until a little over a century and a half ago; since then the existence of atoms and molecules, the ultimate particles of matter in its common forms, has been amply demonstrated, and their own ultimate particles, electrons, protons, and neutrons, have been identified and studied as well. In this chapter and in others to come our chief concern will be the structure of the atom, since it is this structure that is responsible for nearly all the properties of matter that have shaped the world around us.

Every atom consists of a small nucleus of protons and neutrons with a number of electrons some distance away. It is tempting to think of the electrons as circling the nucleus as planets do the sun, but classical electromagnetic theory denies the possibility of stable electron orbits. In an effort to resolve this paradox, Niels Bohr applied quantum ideas to atomic structure in 1913 to obtain a model which, despite its serious inadequacies and subsequent replacement by a quantum-mechanical description of greater accuracy and usefulness, nevertheless remains a convenient mental picture of the atom. While it is not the general policy of this book to go deeply into hypotheses that have had to be discarded, we shall discuss Bohr's theory of the hydrogen atom because it provides a valuable transition to the more abstract quantum theory of the atom. For this reason our account of the Bohr theory differs somewhat from the original one given by Bohr, though all the results are identical.

4.1 ATOMIC MODELS

While the scientists of the nineteenth century accepted the idea that the chemical elements consist of atoms, they knew virtually nothing about the atoms themselves. The discovery of the electron and the realization that all atoms contain

101

electrons provided the first important insight into atomic structure. Electrons contain negative electrical charges, while atoms themselves are electrically neutral: every atom must therefore contain enough positively charged matter to balance the negative charge of its electrons. Furthermore, electrons are thousands of times lighter than whole atoms; this suggests that the positively charged constituent of atoms is what provides them with nearly all their mass. When J. J. Thomson proposed in 1898 that atoms are uniform spheres of positively charged matter in which electrons are embedded, his hypothesis then seemed perfectly reasonable. Thomson's plum-pudding model of the atom—so called from its resemblance to that raisin-studded delicacy—is sketched in Fig. 4-1. Despite the importance of the problem, 13 years passed before a definite experimental test of the plum-pudding model was made. This experiment, as we shall see, compelled the abandonment of this apparently plausible model, leaving in its place a concept of atomic structure incomprehensible in the light of classical physics.

The most direct way to find out what is inside a plum pudding is to plunge a finger into it, a technique not very different from that used by Geiger and Marsden to find out what is inside an atom. In their classic experiment, performed in 1911 at the suggestion of Ernest Rutherford, they employed as probes the fast *alpha particles* spontaneously emitted by certain radioactive elements. Alpha particles are helium atoms that have lost two electrons, leaving them with a charge of $+2e$; we shall examine their origin and properties in more detail later. Geiger and Marsden placed a sample of an alpha-particle-emitting substance behind a lead screen that had a small hole in it, as in Fig. 4-2, so that a narrow beam of alpha particles was produced. This beam was directed at a thin gold

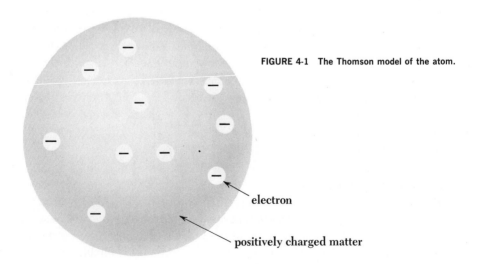

FIGURE 4-1   The Thomson model of the atom.

electron

positively charged matter

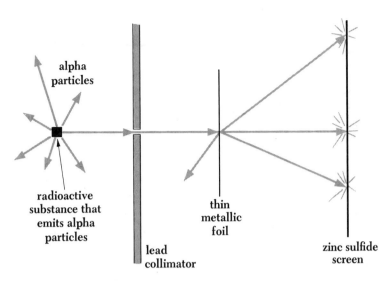

alpha
particles

radioactive
substance that
emits alpha
particles

lead
collimator

thin
metallic
foil

zinc sulfide
screen

FIGURE 4-2 The Rutherford scattering experiment.

foil. A moveable zinc sulfide screen, which gives off a visible flash of light when struck by an alpha particle, was placed on the other side of the foil. It was anticipated that most of the alpha particles would go right through the foil, while the remainder would at most suffer only slight deflections. This behavior follows from the Thomson atomic model, in which the charges within an atom are assumed to be uniformly distributed throughout its volume. If the Thomson model is correct, only weak electric forces are exerted on alpha particles passing through a thin metal foil, and their initial momenta should be enough to carry them through with only minor departures from their original paths.

What Geiger and Marsden actually found was that, while most of the alpha particles indeed emerged without deviation, some were scattered through very large angles. A few were even scattered in the backward direction. Since alpha particles are relatively heavy (over 7,000 times more massive than electrons) and those used in this experiment traveled at high speed, it was clear that strong forces had to be exerted upon them to cause such marked deflections. To explain the results, Rutherford was forced to picture an atom as being composed of a tiny *nucleus*, in which its positive charge and nearly all of its mass are concentrated, with its electrons some distance away (Fig. 4-3). Considering an atom as largely empty space, it is easy to see why most alpha particles go right through a thin foil. When an alpha particle approaches a nucleus, however, it encounters an intense electric field and is likely to be scattered through a considerable angle. The atomic electrons, being so light, do not appreciably affect the motion of incident alpha particles.

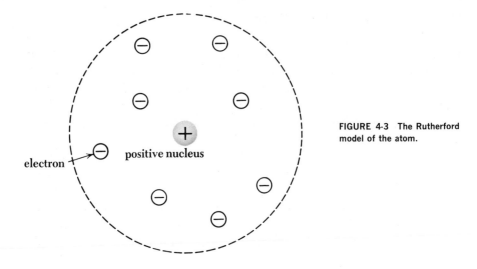

FIGURE 4-3 The Rutherford model of the atom.

Numerical estimates of electric-field intensities within the Thomson and Rutherford models emphasize the difference between them. If we assume with Thomson that the positive charge within a gold atom is spread evenly throughout its volume, and if we neglect the electrons completely, the electric-field intensity at the atom's surface (where it is a maximum) is about 10^{13} V/m. On the other hand, if we assume with Rutherford that the positive charge within a gold atom is concentrated in a small nucleus at its center, the electric-field intensity at the surface of the nucleus exceeds 10^{21} V/m—a factor of 10^8 greater. Such a strong field can deflect or even reverse the direction of an energetic alpha particle that comes near the nucleus, while the feebler field of the Thomson atom cannot.

The experiments of Geiger and Marsden and later work of a similar kind also supplied information about the nuclei of the atoms that composed the various target foils. The deflection an alpha particle experiences when it passes near a nucleus depends upon the magnitude of the nuclear charge, and so comparing the relative scattering of alpha particles by different foils provides a way of estimating the nuclear charges of the atoms involved. All of the atoms of any one element were found to have the same unique nuclear charge, and this charge increased regularly from element to element in the periodic table. The nuclear charges always turned out to be multiples of $+e$; the number of unit positive charges in the nuclei of an element is today called the *atomic number* of the element. We know now that protons, each with a charge $+e$, are responsible for the charge on a nucleus, and so the atomic number of an element is the same as the number of protons in the nuclei of its atoms.

*4.2 ALPHA-PARTICLE SCATTERING

Rutherford arrived at a formula, describing the scattering of alpha particles by thin foils on the basis of his atomic model, that agreed with the experimental results. The derivation of this formula both illustrates the application of fundamental physical laws in a novel setting and introduces certain notions, such as that of the *cross section* for an interaction, that are important in many other aspects of modern physics.

Rutherford began by assuming that the alpha particle and the nucleus it interacts with are both small enough to be considered as point masses and charges; that the electrostatic repulsive force between alpha particle and nucleus (which are both positively charged) is the only one acting; and that the nucleus is so massive compared with the alpha particle that it does not move during their interaction. Owing to the variation of the electrostatic force with $1/r^2$, where r is the instantaneous separation between alpha particle and nucleus, the alpha particle's path is a hyperbola with the nucleus at the outer focus (Fig. 4-4). The *impact parameter b* is the minimum distance to which the alpha particle would approach the nucleus if there were no force between them, and the *scattering angle* θ is the angle between the asymptotic direction of approach of the alpha particle and the asymptotic direction in which it recedes. Our first task is to find a relationship between b and θ.

FIGURE 4-4 Rutherford scattering.

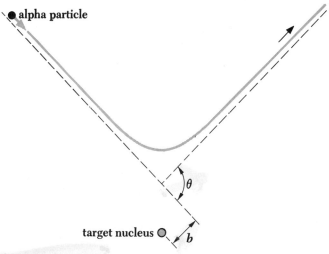

θ = scattering angle
b = impact parameter

As a result of the impulse $\int \mathbf{F}\,dt$ given it by the nucleus, the momentum of the alpha particle changes by $\Delta \mathbf{p}$ from the initial value $\mathbf{p}_1$ to the final value $\mathbf{p}_2$. That is,

4.1
$$\Delta \mathbf{p} = \mathbf{p}_2 - \mathbf{p}_1$$
$$= \int \mathbf{F}\,dt$$

Because the nucleus remains stationary during the passage of the alpha particle, by hypothesis, the alpha-particle kinetic energy remains constant; hence the *magnitude* of its momentum also remains constant, and

$$p_1 = p_2 = mv$$

Here v is the alpha-particle velocity far from the nucleus. From Fig. 4-5 we see that, according to the law of sines,

$$\frac{\Delta p}{\sin \theta} = \frac{mv}{\sin \dfrac{(\pi - \theta)}{2}}$$

Since

$$\sin \frac{1}{2}(\pi - \theta) = \cos \frac{\theta}{2}$$

and

$$\sin \theta = 2 \sin \frac{\theta}{2} \cos \frac{\theta}{2}$$

FIGURE 4-5 Geometrical relationships in Rutherford scattering.

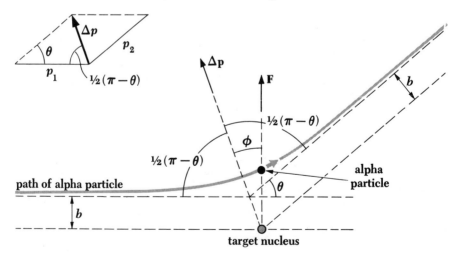

106   THE ATOM

we have for the momentum change

4.2 $\qquad \Delta p = 2mv \sin\dfrac{\theta}{2}$

Because the impulse $\int \mathbf{F}\, dt$ is in the same direction as the momentum change $\Delta \mathbf{p}$, its magnitude is

4.3 $\qquad |\int \mathbf{F}\, dt| = \int F \cos\phi\, dt$

where ϕ is the instantaneous angle between $\mathbf{F}$ and $\Delta \mathbf{p}$ along the path of the alpha particle. Inserting Eqs. 4.2 and 4.3 in Eq. 4.1,

$$2mv \sin\frac{\theta}{2} = \int_0^\infty F \cos\phi\, dt$$

To change the variable on the right-hand side from t to ϕ, we note that the limits of integration will change to $-\frac{1}{2}(\pi - \theta)$ and $+\frac{1}{2}(\pi - \theta)$, corresponding to ϕ at $t = 0$ and $t = \infty$ respectively, and so

4.4 $\qquad 2mv \sin\dfrac{\theta}{2} = \displaystyle\int_{-(\pi-\theta)/2}^{+(\pi-\theta)/2} F \cos\phi\, \dfrac{dt}{d\phi}\, d\phi$

The quantity $d\phi/dt$ is just the angular velocity ω of the alpha particle about the nucleus (this is evident from Fig. 4-5). The electrostatic force exerted by the nucleus on the alpha particle acts along the radius vector joining them, and so there is no torque on the alpha particle and its angular momentum $m\omega r^2$ is constant. Hence

$$m\omega r^2 = \text{constant}$$
$$= mr^2 \frac{d\phi}{dt}$$
$$= mvb$$

from which we see that

$$\frac{dt}{d\phi} = \frac{r^2}{vb}$$

Substituting this expression for $dt/d\phi$ in Eq. 4.4,

4.5 $\qquad 2mv^2 b \sin\dfrac{\theta}{2} = \displaystyle\int_{-(\pi-\theta)/2}^{+(\pi-\theta)/2} Fr^2 \cos\phi\, d\phi$

As we recall, F is the electrostatic force exerted by the nucleus on the alpha particle. The charge on the nucleus is Ze, corresponding to the atomic number Z, and that on the alpha particle is $2e$. Therefore

$$F = \frac{1}{4\pi\varepsilon_0} \frac{2Ze^2}{r^2}$$

and

$$\frac{4\pi\varepsilon_0 mv^2 b}{Ze^2} \sin\frac{\theta}{2} = \int_{-(\pi-\theta)/2}^{+(\pi-\theta)/2} \cos\phi \, d\phi$$

$$= 2\cos\frac{\theta}{2}$$

The scattering angle θ is related to the impact parameter b by the equation

$$\cot\frac{\theta}{2} = \frac{2\pi\varepsilon_0 mv^2}{Ze^2} b$$

It is more convenient to specify the alpha-particle energy T instead of its mass and velocity separately; with this substitution,

4.6
$$\cot\frac{\theta}{2} = \frac{4\pi\varepsilon_0 T}{Ze^2} b$$

Figure 4-6 is a schematic representation of Eq. 4.6; the rapid decrease in θ as b increases is evident. A very near miss is required for a substantial deflection.

FIGURE 4-6 The scattering angle decreases with increasing impact parameter.

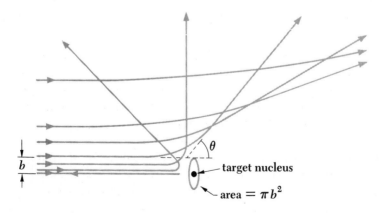

THE ATOM

Equation 4.6 cannot be directly confronted with experiment since there is no way of measuring the impact parameter corresponding to a particular observed scattering angle. An indirect strategy is required. Our first step is to note that all alpha particles approaching a target nucleus with an impact parameter from 0 to b will be scattered through an angle of θ or more, where θ is given in terms of b by Eq. 4.6. This means that an alpha particle that is initially directed anywhere within the area πb^2 around a nucleus will be scattered through θ or more (Fig. 4-6); the area πb^2 is accordingly called the *cross section* for the interaction. The general symbol for cross section is σ, and so here

4.7 $\qquad \sigma = \pi b^2$

We must keep in mind that the incident alpha particle is actually scattered before it reaches the immediate vicinity of the nucleus and hence does not necessarily pass within a distance b of it.

Now we consider a foil of thickness t that contains n atoms per unit volume. The number of target nuclei per unit area is nt, and an alpha-particle beam incident upon an area A therefore encounters ntA nuclei. The aggregate cross section for scatterings of θ or more is the number of target nuclei ntA multiplied by the cross section σ for such scattering per nucleus, or $ntA\sigma$. Hence the fraction f of incident alpha particles scattered by θ or more is the ratio between the aggregate cross section $ntA\sigma$ for such scattering and the total target area A. That is,

$$f = \frac{\text{alpha particles scattered by } \theta \text{ or more}}{\text{incident alpha particles}}$$

$$= \frac{\text{aggregate cross section}}{\text{target area}}$$

$$= \frac{ntA\sigma}{A}$$

$$= nt\pi b^2$$

Substituting for b from Eq. 4.6,

4.8 $\qquad f = \pi nt \left(\dfrac{Ze^2}{4\pi\varepsilon_0 T} \right)^2 \cot^2 \dfrac{\theta}{2}$

In the above calculation it was assumed that the foil is sufficiently thin so that the cross sections of adjacent nuclei do not overlap and that a scattered alpha particle receives its entire deflection from an encounter with a single nucleus.

Let us use Eq. 4.8 to determine what fraction of a beam of 7.7-MeV alpha particles is scattered through angles of more than 45° when incident upon a gold foil 3×10^{-7} m thick. (These values are typical of the alpha-particle energies and foil thicknesses used by Geiger and Marsden; for comparison, a human hair is about 10^{-4} m in diameter.) We begin by finding n, the number of gold atoms per unit volume in the foil, from the relationship

$$\frac{\text{Atoms}}{\text{Volume}} = \frac{(\text{atoms/kmol}) \times (\text{mass/volume})}{\text{mass/kmol}}$$

$$n = \frac{N_0 \rho}{w}$$

where N_0 is Avogadro's number, ρ the density of gold, and w its atomic weight. Since $N_0 = 6.03 \times 10^{26}$ atoms/kmol, $\rho = 1.93 \times 10^4$ kg/m^3, and $w = 197$, we have

$$n = \frac{6.03 \times 10^{26} \text{ atoms/kmol} \times 1.93 \times 10^4 \text{ kg/m}^3}{197 \text{ kg/kmol}}$$

$$= 5.91 \times 10^{28} \text{ atoms/m}^3$$

The atomic number Z of gold is 79, a kinetic energy of 7.7 MeV is equal to 1.23×10^{-12} J, and $\theta = 45°$; from these figures we find that

$$f = 7 \times 10^{-5}$$

of the incident alpha particles are scattered through 45° or more—only 0.007 percent! A foil this thin is quite transparent to alpha particles.

In an actual experiment, a detector measures alpha particles scattered between θ and $\theta + d\theta$, as in Fig. 4-7. The fraction of incident alpha particles so scattered is found by differentiating Eq. 4.8 with respect to θ, an operation that yields

4.9 $$df = -\pi n t \left(\frac{Ze^2}{4\pi\varepsilon_0 T} \right)^2 \cot\frac{\theta}{2} \csc^2\frac{\theta}{2} \, d\theta$$

(The minus sign expresses the fact that f decreases with increasing θ.) In the experiment, a fluorescent screen was placed a distance r from the foil, and the scattered alpha particles were detected by means of the scintillations they caused. Those alpha particles scattered between θ and $\theta + d\theta$ reach a zone of a sphere of radius r whose width is $rd\theta$. The zone radius itself is $r \sin\theta$, and so the area dS of the screen struck by these particles is

$$dS = (2\pi r \sin\theta)(rd\theta)$$

$$= 2\pi r^2 \sin\theta \, d\theta$$

$$= 4\pi r^2 \sin\frac{\theta}{2} \cos\frac{\theta}{2} \, d\theta$$

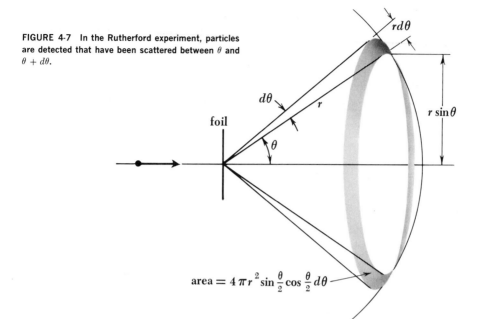

foil

$$r\,d\theta$$

$$d\theta$$

$$r$$

$$\theta$$

$$r\sin\theta$$

$$\text{area} = 4\,\pi\,r^{2}\sin\frac{\theta}{2}\cos\frac{\theta}{2}\,d\theta$$

If a total of N_i alpha particles strike the foil during the course of the experiment, the number scattered into $d\theta$ at θ is $N_i\,df$. The number $N(\theta)$ per unit area striking the screen at θ, which is the quantity actually measured, is

$$N(\theta) = \frac{N_i\,|df|}{dS}$$

$$= \frac{N_i\pi nt\left(\dfrac{Ze^2}{4\pi\varepsilon_0 T}\right)^2 \cot\dfrac{\theta}{2}\csc^2\dfrac{\theta}{2}\,d\theta}{4\pi r^2 \sin\dfrac{\theta}{2}\cos\dfrac{\theta}{2}\,d\theta}$$

4.10 $$N(\theta) = \frac{N_i ntZ^2 e^4}{(8\pi\varepsilon_0)^2 r^2 T^2 \sin^4(\theta/2)}$$ **Rutherford scattering formula**

Equation 4.10 is the *Rutherford scattering formula*.

According to Eq. 4.10, the number of alpha particles per unit area arriving at the fluorescent screen a distance r from the scattering foil should be directly proportional to the thickness t of the foil, the number of foil atoms per unit volume n, and the square of the atomic number Z of the foil atoms, and it should be inversely proportional to the square of the kinetic energy T of the alpha particles and to $\sin^4(\theta/2)$, where θ is the scattering angle. These predictions agreed with the measurements of Geiger and Marsden mentioned earlier, which led Rutherford to conclude that his assumptions, chief among them the hypothesis of the nuclear atom, were correct. Rutherford is therefore credited with the "discovery" of the nucleus. Figure 4-8 shows how $N(\theta)$ varies with θ.

4.4 NUCLEAR DIMENSIONS

When we say that the experimental data on the scattering of alpha particles by thin foils verifies our assumption that atomic nuclei are point particles, what is really meant is that their dimensions are insignificant compared with the

FIGURE 4-8 Rutherford scattering.

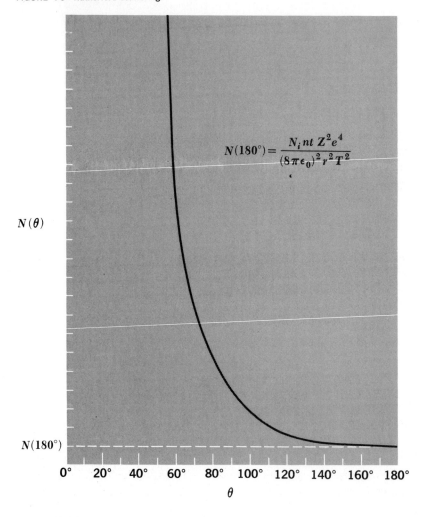

$$N(180°) = \frac{N_i \, nt \, Z^2 e^4}{(8\pi\epsilon_0)^2 \, r^2 \, T^2}$$

$N(\theta)$

$N(180°)$

0°   20°   40°   60°   80°   100°   120°   140°   160°   180°

θ

minimum distance to which the incident alpha particles approach the nuclei. Rutherford scattering therefore permits us to determine an upper limit to nuclear dimensions. Let us compute the distance of closest approach r_0 of the most energetic alpha particles employed in the early experiments. An alpha particle will have its smallest r_0 when its impact parameter is $b = 0$, corresponding to a head-on approach followed by a 180° scattering. At the instant of closest approach the initial kinetic energy T of the particle is entirely converted to electrostatic potential energy, and so at that instant

4.11
$$T = \frac{1}{4\pi\varepsilon_0} \frac{2Ze^2}{r_0}$$

since the charge of the alpha particle is $2e$ and that of the nucleus Ze. Hence

$$r_0 = \frac{2Ze^2}{4\pi\varepsilon_0 T}$$

The maximum T found in alpha particles of natural origin is 7.7 MeV, which is

$$7.7 \times 10^6 \text{ eV} \times 1.6 \times 10^{-19} \text{ J/eV} = 1.2 \times 10^{-12} \text{ J}$$

Since $1/4\pi\varepsilon_0 = 9 \times 10^9 \text{ N-m}^2/\text{C}^2$

$$r_0 = \frac{2 \times 9 \times 10^9 \text{ N-m}^2/\text{C}^2 \times (1.6 \times 10^{-19} \text{ C})^2 Z}{1.2 \times 10^{-12} \text{ J}}$$

$$= 3.8 \times 10^{-16} Z \text{ m}$$

The atomic number of gold, a typical foil material, is $Z = 79$, so that

$$r_0(\text{Au}) = 3.0 \times 10^{-14} \text{ m}$$

The radius of the gold nucleus is therefore less than 3.0×10^{-14} m, well under $\frac{1}{10,000}$ the radius of the atom as a whole.

In more recent years particles of much higher energies than 7.7 MeV have been artificially accelerated, and it has been found that the Rutherford scattering formula does indeed eventually fail to agree with experiment. We shall discuss these experiments and the information they provide on actual nuclear dimensions in Chap. 11.

4.5 ELECTRON ORBITS

The Rutherford model of the atom, so convincingly confirmed by experiment, postulates a tiny, massive, positively charged nucleus surrounded at a relatively great distance by enough electrons to render the atom, as a whole, electrically neutral. Thomson visualized the electrons in his model atom as embedded in the positively charged matter that fills it, and thus as being unable to move. The electrons in Rutherford's model atom, however, cannot be stationary, because there is nothing that can keep them in place against the electrostatic force attracting them to the nucleus. If the electrons are in motion around the nucleus, however, dynamically stable orbits (comparable with those of the planets about the sun) are possible (Fig. 4-9).

Let us examine the classical dynamics of the hydrogen atom, whose single electron makes it the simplest of all atoms. We shall assume a circular electron orbit for convenience, though it might as reasonably be assumed elliptical in shape. The centripetal force

$$F_c = \frac{mv^2}{r}$$

holding the electron in an orbit r from the nucleus is provided by the electrostatic force

$$F_e = \frac{1}{4\pi\varepsilon_0}\frac{e^2}{r^2}$$

between them, and the condition for orbit stability is

$$F_c = F_e$$

4.12
$$\frac{mv^2}{r} = \frac{1}{4\pi\varepsilon_0}\frac{e^2}{r^2}$$

FIGURE 4-9 Force balance in the hydrogen atom.

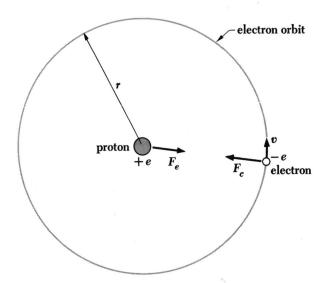

The electron velocity v is therefore related to its orbit radius r by the formula

4.13
$$v = \frac{e}{\sqrt{4\pi\varepsilon_0 mr}}$$

The total energy E of the electron in a hydrogen atom is the sum of its kinetic energy

$$T = \frac{1}{2}mv^2$$

and its potential energy

$$V = -\frac{e^2}{4\pi\varepsilon_0 r}$$

(The minus sign signifies that the force on the electron is in the $-r$ direction.) Hence

$$E = T + V$$
$$= \frac{mv^2}{2} - \frac{e^2}{4\pi\varepsilon_0 r}$$

Substituting for v from Eq. 4.12,

$$E = \frac{e^2}{8\pi\varepsilon_0 r} - \frac{e^2}{4\pi\varepsilon_0 r}$$

4.14
$$E = -\frac{e^2}{8\pi\varepsilon_0 r} \quad = \quad \text{TOTAL ENERGY OF ELECTRON}$$

The total energy of an atomic electron is negative; this is necessary if it is to be bound to the nucleus. If E were greater than zero, the electron would have too much energy to remain in a closed orbit about the nucleus.

Experiments indicate that 13.6 eV is required to separate a hydrogen atom into a proton and an electron; that is, its binding energy E is -13.6 eV. Since 13.6 eV $= 2.2 \times 10^{-18}$ J, we can find the orbital radius of the electron in a hydrogen atom from Eq. 4.14:

$$r = -\frac{e^2}{8\pi\varepsilon_0 E}$$
$$= -\frac{(1.6 \times 10^{-19} \text{ C})^2}{8\pi \times 8.85 \times 10^{-12} \text{ F/m} \times (-2.2 \times 10^{-18} \text{ J})}$$
$$= 5.3 \times 10^{-11} \text{ m}$$

An atomic radius of this order of magnitude agrees with estimates made in other ways.

The above analysis is a straightforward application of Newton's laws of motion and Coulomb's law of electric force—both pillars of classical physics—and is in accord with the experimental observation that atoms are stable. However, it is *not* in accord with electromagnetic theory—another pillar of classical physics—which predicts that accelerated electric charges radiate energy in the form of electromagnetic waves. An electron pursuing a curved path is accelerated and therefore should continuously lose energy, rapidly spiraling into the nucleus (Fig. 4-10). Whenever they have been directly tested, the predictions of electromagnetic theory have always agreed with experiment, yet atoms do not collapse. This contradiction can mean only one thing: The laws of physics that are valid in the macroscopic world do not hold true in the microscopic world of the atom.

The reason for the failure of classical physics to yield a meaningful analysis of atomic structure is that it approaches nature exclusively in terms of the abstract concepts of "pure" particles and "pure" waves. As we learned in the two preceding chapters, particles and waves have many properties in common, though the smallness of Planck's constant renders the wave-particle duality imperceptible in the macroscopic world. The validity of classical physics decreases as the scale of the phenomena under study decreases, and full allowance must be made for the particle behavior of waves and the wave behavior of particles if the atom is to be understood. In the remainder of this chapter we shall see how the Bohr

FIGURE 4-10 An atomic electron should, classically, spiral rapidly into the nucleus as it radiates energy due to its acceleration.

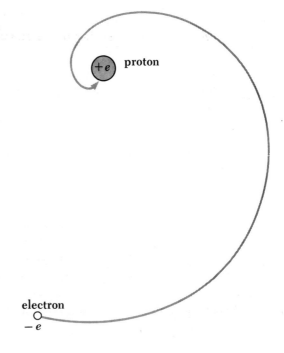

atomic model, which combines classical and modern notions, accomplishes part of the latter task. Not until we consider the atom from the point of view of quantum mechanics, which makes no compromise with intuitive notions acquired in our daily lives, will we find a really successful theory of the atom.

An interesting question arises at this point. In our derivation of the Rutherford scattering formula we made use of the same laws of physics that proved such dismal failures when applied to atomic stability. Is it not therefore possible, even likely, that the formula is not correct, and that the atom in reality does not resemble the Rutherford model of a small central nucleus surrounded by distant electrons? This question is not a trivial one, and it is, in a way, a curious coincidence that the quantum-mechanical analysis of alpha-particle scattering from thin foils results in precisely the same formula that Rutherford obtained. To verify that a classical calculation ought to be at least approximately correct, we note that the de Broglie wavelength of an alpha particle whose speed is 2×10^7 m/s is

$$\lambda = \frac{h}{mv} = \frac{6.63 \times 10^{-34} \text{ J-s}}{6.6 \times 10^{-27} \text{ kg} \times 2 \times 10^7 \text{ m/s}}$$
$$= 5 \times 10^{-15} \text{ m}$$

As we saw in Sec. 4.4, the closest an alpha particle with this wavelength ever gets to a gold nucleus is 3×10^{-14} m, which is 6 de Broglie wavelengths, and so it is reasonable to regard the alpha particle as a classical particle in the interaction. We are therefore correct in thinking of the atom in terms of Rutherford's model, though the dynamics of the atomic electrons—which is another matter entirely—requires a nonclassical approach.

4.6 ATOMIC SPECTRA

The ability of the Bohr theory of the atom to explain the origin of spectral lines is among its most spectacular accomplishments, and so it is appropriate to preface our exposition of the theory itself with a look at atomic spectra.

We have already mentioned that heated solids emit radiation in which all wavelengths are present, though with different intensities. We shall learn in Chap. 9 that the observed features of this radiation can be explained on the basis of the quantum theory of light independent of the details of the radiation process itself or of the nature of the solid. From this fact it follows that, when a solid is heated to incandescence, we are witnessing the collective behavior of a great many interacting atoms rather than the characteristic behavior of the individual atoms of a particular element.

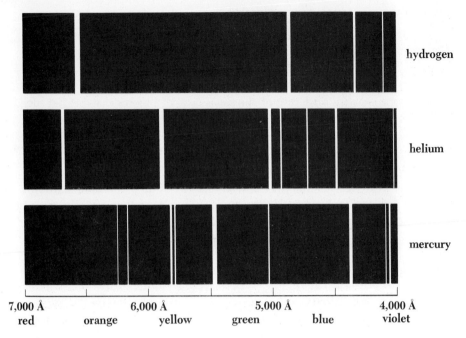

| 7,000 Å | 6,000 Å | 5,000 Å | 4,000 Å |
| red | orange | yellow | green | blue | violet |

FIGURE 4-11 Portions of the emission spectra of hydrogen, helium, and mercury.

At the other extreme, the atoms or molecules in a rarefied gas are so far apart on the average that their only mutual interactions occur during occasional collisions. Under these circumstances we would expect any emitted radiation to be characteristic of the individual atoms or molecules present, an expectation that is realized experimentally. When an atomic gas or vapor at somewhat less than atmospheric pressure is suitably "excited," usually by the passage of an electric current through it, the emitted radiation has a spectrum which contains certain discrete wavelengths only. Figure 4-11 shows the atomic spectra of several elements; they are called *emission line spectra*. Every element displays a unique line spectrum when a sample of it in the vapor phase is excited;

FIGURE 4-12 A portion of the band spectrum of PN.

$\longleftarrow \lambda$

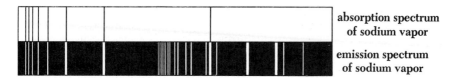

absorption spectrum
of sodium vapor

emission spectrum
of sodium vapor

FIGURE 4-13 The dark lines in the absorption spectrum of an element correspond to bright lines in its emission spectrum.

spectroscopy is therefore a useful tool for analyzing the composition of an unknown substance.

The spectrum of an excited molecular gas or vapor contains *bands* which consist of many separate lines very close together (Fig. 4-12). Bands owe their origin to rotations and vibrations of the atoms in an electronically excited molecule, and we shall consider their interpretation in a later chapter.

When white light is passed through a gas, it is found to absorb light of certain of the wavelengths present in its emission spectrum. The resulting *absorption line spectrum* consists of a bright background crossed by dark lines corresponding to the missing wavelengths (Fig. 4-13); emission spectra consist of bright lines on a dark background. The dark Fraunhofer lines in the solar spectrum occur because the luminous part of the sun, which radiates almost exactly according to theoretical predictions for any object heated to 5800 K, is surrounded by an envelope of cooler gas which absorbs light of certain wavelengths only.

In the latter part of the nineteenth century it was discovered that the wavelengths present in atomic spectra fall into definite sets called *spectral series*. The wavelengths in each series can be specified by a simple empirical formula, with remarkable similarity among the formulas for the various series that comprise the complete spectrum of an element. The first such spectral series was found by J. J. Balmer in 1885 in the course of a study of the visible part of the hydrogen spectrum. Figure 4-14 shows the *Balmer series*. The line with the longest

FIGURE 4-14 The Balmer series of hydrogen.

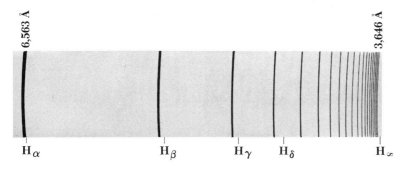

H_α H_β H_γ H_δ H_∞

wavelength, 6,563 Å, is designated H_α, the next, whose wavelength is 4,863 Å, is designated H_β, and so on. As the wavelength decreases, the lines are found closer together and weaker in intensity until the *series limit* at 3,646 Å is reached, beyond which there are no further separate lines but only a faint continuous spectrum. Balmer's formula for the wavelengths of this series is

4.15 $$\frac{1}{\lambda} = R\left(\frac{1}{2^2} - \frac{1}{n^2}\right) \qquad n = 3, 4, 5, \ldots \qquad \text{Balmer}$$

VISIBLE FOR H_2

The quantity R, known as the *Rydberg constant,* has the value

$$R = 1.097 \times 10^7 \text{ m}^{-1}$$
$$= 1.097 \times 10^{-3} \text{ Å}^{-1}$$

The H_α line corresponds to $n = 3$, the H_β line to $n = 4$, and so on. The series limit corresponds to $n = \infty$, so that it occurs at a wavelength of $4/R$, in agreement with experiment.

The Balmer series contains only those wavelengths in the visible portion of the hydrogen spectrum. The spectral lines of hydrogen in the ultraviolet and infrared regions fall into several other series. In the ultraviolet the *Lyman series* contains the wavelengths specified by the formula

ULTRAVIOLET FOR H_2

4.16 $$\frac{1}{\lambda} = R\left(\frac{1}{1^2} - \frac{1}{n^2}\right) \qquad n = 2, 3, 4, \ldots \qquad \text{Lyman}$$

In the infrared, three spectral series have been found whose component lines have the wavelengths specified by the formulas

INFRARED FOR H_2

4.17 $$\frac{1}{\lambda} = R\left(\frac{1}{3^2} - \frac{1}{n^2}\right) \qquad n = 4, 5, 6, \ldots \qquad \text{Paschen}$$

4.18 $$\frac{1}{\lambda} = R\left(\frac{1}{4^2} - \frac{1}{n^2}\right) \qquad n = 5, 6, 7, \ldots \qquad \text{Brackett}$$

4.19 $$\frac{1}{\lambda} = R\left(\frac{1}{5^2} - \frac{1}{n^2}\right) \qquad n = 6, 7, 8, \ldots \qquad \text{Pfund}$$

The above spectral series of hydrogen are plotted in terms of wavelength in Fig. 4-15; the Brackett series evidently overlaps the Paschen and Pfund series. The value of R is the same in Eqs. 4.15 to 4.19.

The existence of such remarkable regularities in the hydrogen spectrum, together with similar regularities in the spectra of more complex elements, poses a definitive test for any theory of atomic structure.

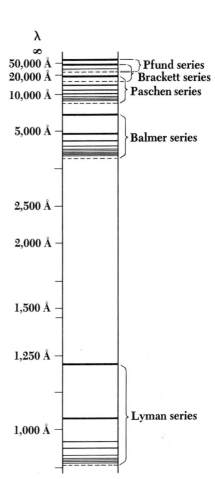

λ

50,000 Å ⎫ Pfund series
20,000 Å ⎬ Brackett series
10,000 Å ⎬ Paschen series

5,000 Å ⎬ Balmer series

2,500 Å

2,000 Å

FIGURE 4-15 The spectral series of hydrogen.

1,500 Å

1,250 Å

⎬ Lyman series

1,000 Å

4.7 THE BOHR ATOM

We saw in Sec. 4-5 that the principles of classical physics are incompatible with the observed stability of the hydrogen atom. The electron in this atom is obliged to whirl around the nucleus to keep from being pulled into it and yet must radiate electromagnetic energy continuously. Because other apparently paradoxical phenomena, like the photoelectric effect and the diffraction of electrons, find explanation in terms of quantum concepts, it is appropriate to inquire whether this might not also be true for the atom.

Let us start by examining the wave behavior of an electron in orbit around a hydrogen nucleus. The de Broglie wavelength of this electron is

$$\lambda = \frac{h}{mv}$$

where the electron speed v is that given by Eq. 4.13:

$$v = \frac{e}{\sqrt{4\pi\varepsilon_0 mr}}$$

Hence

4.20
$$\lambda = \frac{h}{e}\sqrt{\frac{4\pi\varepsilon_0 r}{m}}$$

By substituting 5.3×10^{-11} m for the radius r of the electron orbit, we find the electron wavelength to be

$$\lambda = \frac{6.63 \times 10^{-34}\text{ J-s}}{1.6 \times 10^{-19}\text{ C}}\sqrt{\frac{4\pi \times 8.85 \times 10^{-12}\text{ F/m} \times 5.3 \times 10^{-11}\text{ m}}{9.1 \times 10^{-31}\text{ kg}}}$$

$$= 33 \times 10^{-11}\text{ m}$$

This wavelength is exactly the same as the circumference of the electron orbit,

$$2\pi r = 33 \times 10^{-11}\text{ m}$$

The orbit of the electron in a hydrogen atom corresponds to one complete electron wave joined on itself (Fig. 4-16).

The fact that the electron orbit in a hydrogen atom is one electron wavelength in circumference provides the clue we need to construct a theory of the atom. If we consider the vibrations of a wire loop (Fig. 4-17), we find that their wavelengths always fit an integral number of times into the loop's circumference so that each wave joins smoothly with the next. If the wire were perfectly rigid, these vibrations would continue indefinitely. Why are these the only vibrations possible in a wire loop? If a fractional number of wavelengths is placed around the loop, as in Fig. 4-18, destructive interference will occur as the waves travel around the loop, and the vibrations will die out rapidly. By considering the behavior of electron waves in the hydrogen atom as analogous to the vibrations of a wire loop, then, we may postulate that **an electron can circle a nucleus indefinitely without radiating energy provided that its orbit contains an integral number of de Broglie wavelengths.**

This postulate is the clue to understanding the atom. It combines both the particle and wave characters of the electron into a single statement, since the electron wavelength is computed from the orbital speed required to balance the

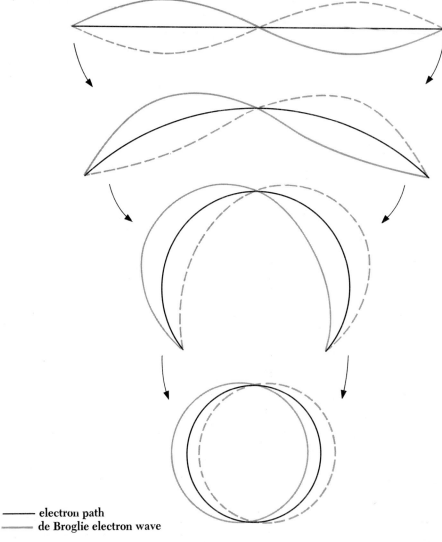

———— electron path
———— de Broglie electron wave

FIGURE 4-16 The orbit of the electron in a hydrogen atom corresponds to a complete electron de Broglie wave joined on itself.

electrostatic attraction of the nucleus. While we can never observe these antithetical characters simultaneously, they are inseparable in nature.

It is a simple matter to express the condition that an electron orbit contain an integral number of de Broglie wavelengths. The circumference of a circular orbit of radius r is $2\pi r$, and so we may write the condition for orbit stability as

4.21 $n\lambda = 2\pi r_n$ $n = 1, 2, 3, \ldots$

where r_n designates the radius of the orbit that contains n wavelengths. The integer n is called the *quantum number* of the orbit. Substituting for λ, the electron wavelength given by Eq. 4.20, yields

$$\frac{nh}{e}\sqrt{\frac{4\pi\varepsilon_0 r_n}{m}} = 2\pi r_n$$

and so the stable electron orbits are those whose radii are given by

4.22 $r_n = \dfrac{n^2 h^2 \varepsilon_0}{\pi m e^2}$ $n = 1, 2, 3, \ldots$

FIGURE 4-17   The vibrations of a wire loop.

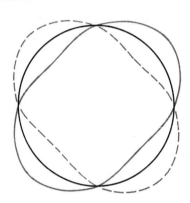

circumference = 2 wavelengths

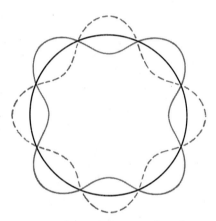

circumference = 4 wavelengths

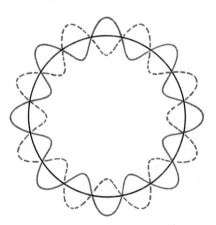

circumference = 8 wavelengths

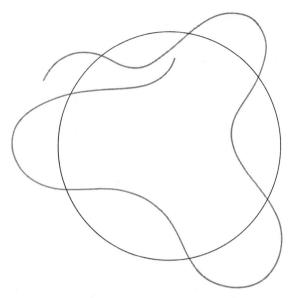

FIGURE 4-18 A fractional number of wavelengths cannot persist because destructive interference will occur.

The radius of the innermost orbit is customarily called the *Bohr radius* of the hydrogen atom and is denoted by the symbol a_0:

$$a_0 = r_1 = 5.3 \times 10^{-11} \text{ m}$$
$$= 0.53 \text{ Å}$$

The other radii are given in terms of a_0 by the formula

$$r_n = n^2 a_0$$

so that the spacing between adjacent orbits increases progressively.

4.8 ENERGY LEVELS AND SPECTRA

The various permitted orbits involve different electron energies. The electron energy E_n is given in terms of the orbit radius r_n by Eq. 4.14 as

$$E_n = -\frac{e^2}{8\pi\varepsilon_0 r_n}$$

Substituting for r_n from Eq. 4.22, we see that

4.23 $$E_n = -\frac{me^4}{8\varepsilon_0^2 h^2}\left(\frac{1}{n^2}\right) \qquad n = 1, 2, 3, \ldots$$ **Energy levels**

The energies specified by Eq. 4.23 are called the *energy levels* of the hydrogen atom and are plotted in Fig. 4-19. These levels are all negative, signifying that the electron does not have enough energy to escape from the atom. The lowest energy level E_1 is called the *ground state* of the atom, and the higher levels E_2, E_3, E_4, . . . are called *excited states*. As the quantum number n increases, the corresponding energy E_n approaches closer and closer to 0; in the limit of $n = \infty$, $E_\infty = 0$ and the electron is no longer bound to the nucleus to form an atom. (A positive energy for a nucleus-electron combination means that the electron is not bound to the nucleus and has no quantum conditions to fulfill; such a combination does not constitute an atom, of course.)

It is now necessary for us to confront directly the equations we have developed with experiment. An especially striking experimental result is that atoms exhibit line spectra in both emission and absorption; do these spectra follow from our atomic model?

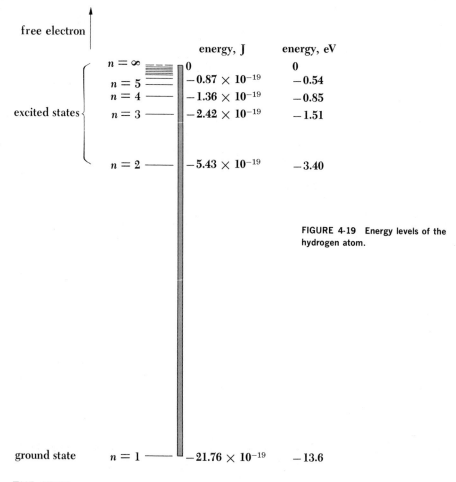

FIGURE 4-19 Energy levels of the hydrogen atom.

The presence of definite, discrete energy levels in the hydrogen atom suggests a connection with line spectra. Let us tentatively assert that, when an electron in an excited state drops to a lower state, the lost energy is emitted as a single photon of light. According to our model, electrons cannot exist in an atom except in certain specific energy levels. The jump of an electron from one level to another, with the difference in energy between the levels being given off all at once in a photon rather than in some more gradual manner, fits in well with this model. If the quantum number of the initial (higher energy) state is n_i and the quantum number of the final (lower energy) state is n_f, we are asserting that

$$\text{Initial energy} - \text{final energy} = \text{photon energy}$$

4.24
$$E_i - E_f = h\nu$$

where ν is the frequency of the emitted photon.

The initial and final states of a hydrogen atom that correspond to the quantum numbers n_i and n_f have, from Eq. 4.23, the energies

$$\text{Initial energy} = E_i = -\frac{me^4}{8\varepsilon_0^2 h^2}\left(\frac{1}{n_i^2}\right)$$

$$\text{Final energy} = E_f = -\frac{me^4}{8\varepsilon_0^2 h^2}\left(\frac{1}{n_f^2}\right)$$

Hence the energy difference between these states is

$$E_i - E_f = \frac{me^4}{8\varepsilon_0^2 h^2}\left(-\frac{1}{n_i^2}\right) - \left(-\frac{1}{n_f^2}\right)$$

$$= \frac{me^4}{8\varepsilon_0^2 h^2}\left(\frac{1}{n_f^2} - \frac{1}{n_i^2}\right)$$

The frequency ν of the photon released in this transition is

$$\nu = \frac{E_i - E_f}{h}$$

4.25
$$= \frac{me^4}{8\varepsilon_0^2 h^3}\left(\frac{1}{n_f^2} - \frac{1}{n_i^2}\right)$$

In terms of photon wavelength λ, since

$$\lambda = \frac{c}{\nu}$$

we have

$$\frac{1}{\lambda} = \frac{\nu}{c}$$

4.26
$$\frac{1}{\lambda} = \frac{me^4}{8\varepsilon_0^2 ch^3}\left(\frac{1}{n_f^2} - \frac{1}{n_i^2}\right)$$

Hydrogen spectrum

Equation 4.26 states that the radiation emitted by excited hydrogen atoms should contain certain wavelengths only. These wavelengths, furthermore, fall into definite sequences that depend upon the quantum number n_f of the final energy level of the electron. Since the initial quantum number n_i must always be greater than the final quantum number n_f in each case, in order that there be an excess of energy to be given off as a photon, the calculated formulas for the first five series are

$$n_f = 1: \quad \frac{1}{\lambda} = \frac{me^4}{8\varepsilon_0{}^2ch^3}\left(\frac{1}{1^2} - \frac{1}{n^2}\right) \qquad n = 2, 3, 4, \ldots \qquad \textbf{Lyman}$$

$$n_f = 2: \quad \frac{1}{\lambda} = \frac{me^4}{8\varepsilon_0{}^2ch^3}\left(\frac{1}{2^2} - \frac{1}{n^2}\right) \qquad n = 3, 4, 5, \ldots \qquad \textbf{Balmer}$$

$$n_f = 3: \quad \frac{1}{\lambda} = \frac{me^4}{8\varepsilon_0{}^2ch^3}\left(\frac{1}{3^2} - \frac{1}{n^2}\right) \qquad n = 4, 5, 6, \ldots \qquad \textbf{Paschen}$$

$$n_f = 4: \quad \frac{1}{\lambda} = \frac{me^4}{8\varepsilon_0{}^2ch^3}\left(\frac{1}{4^2} - \frac{1}{n^2}\right) \qquad n = 5, 6, 7, \ldots \qquad \textbf{Brackett}$$

$$n_f = 5: \quad \frac{1}{\lambda} = \frac{me^4}{8\varepsilon_0{}^2ch^3}\left(\frac{1}{5^2} - \frac{1}{n^2}\right) \qquad n = 6, 7, 8, \ldots \qquad \textbf{Pfund}$$

These sequences are identical in form with the empirical spectral series discussed earlier. The Lyman series corresponds to $n_f = 1$; the Balmer series corresponds to $n_f = 2$; the Paschen series corresponds to $n_f = 3$; the Brackett series corresponds to $n_f = 4$; and the Pfund series corresponds to $n_f = 5$.

We still cannot consider our assertion that the line spectrum of hydrogen originates in electron transitions from high to low energy states as proved, however. The final step is to compare the value of the constant term in the above equations with that of the Rydberg constant R of the empirical equations 4.15 to 4.19. The value of this constant term is

$$\frac{me^4}{8\varepsilon_0{}^2ch^3} = \frac{9.1 \times 10^{-31}\,\text{kg} \times (1.6 \times 10^{-19}\,\text{C})^4}{8 \times (8.85 \times 10^{-12}\,\text{F/m})^2 \times 3 \times 10^8\,\text{m/s} \times (6.63 \times 10^{-34}\,\text{J-s})^3}$$
$$= 1.097 \times 10^7\,\text{m}^{-1}$$

which is indeed the same as R! This theory of the hydrogen atom, which is essentially that developed by Bohr in 1913, therefore agrees both qualitatively and quantitatively with experiment. Figure 4-20 shows schematically how spectral lines are related to atomic energy levels.

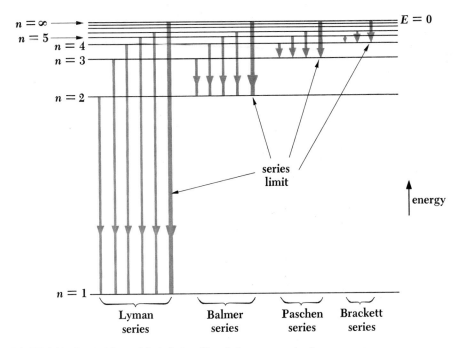

$n = \infty$

$n = 5$

$n = 4$

$n = 3$

$n = 2$

$E = 0$

series
limit

energy

$n = 1$

Lyman
series

Balmer
series

Paschen
series

Brackett
series

FIGURE 4-20 Spectral lines originate in transitions between energy levels.

4.9 NUCLEAR MOTION

In the preceding analysis, we assumed that the hydrogen nucleus (a single proton) remains stationary while the orbital electron revolves around it. What must actually happen is that both nucleus and electron revolve around their common center of mass, which is very close to the nucleus because the nuclear mass is much greater than that of the electron (Fig. 4-21). Because the nucleus and the electron are always on opposite sides of the center of mass, their linear momenta are in opposite directions, and linear momentum is conserved by the atom.

A system of this kind is equivalent to a single particle of mass m' that revolves around the position of the heavier particle. (This equivalence is demonstrated in most mechanics textbooks; see Sec. 8.8.) If m is the electron mass and M the nuclear mass, the m' is given by

4.27 $$m' = \frac{mM}{m + M}$$

The quantity m' is called the *reduced mass* of the electron because its value is less than m. To correct for the motion of the nucleus in the hydrogen atom,

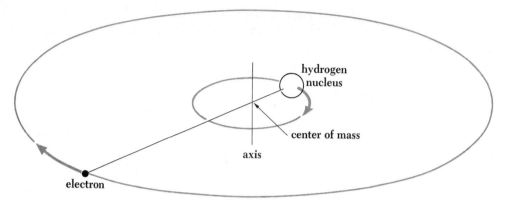

FIGURE 4-21 Both the electron and nucleus of a hydrogen atom revolve around a common center of mass.

then, all we need to do is to imagine that the electron is replaced by a particle of mass m' and charge $-e$. The energy levels of the atom therefore become

4.28
$$E_n = -\frac{m'e^4}{8\varepsilon_0^2 h^2}\left(\frac{1}{n^2}\right)$$

Owing to motion of the nucleus, all the energy levels of hydrogen are changed by the fraction

$$\frac{m'}{m} = \frac{M}{M+m}$$
$$= \frac{1{,}836}{1{,}837}$$
$$= 0.99945$$

an increase of 0.055 percent since the energies E_n, being smaller in absolute value, are therefore less negative. The use of Eq. 4.28 in place of 4.23 removes a small but definite discrepancy between the predicted wavelengths of the various spectral lines of hydrogen and the actual experimentally determined wavelengths. The value of the Rydberg constant R to eight significant figures without correcting for nuclear motion is 1.0973731×10^7 m^{-1}; the correction lowers it to 1.0967758×10^7 m^{-1}.

The notion of reduced mass played an important part in the discovery of *deuterium*, an isotope of hydrogen whose atomic mass is almost exactly double that of ordinary hydrogen owing to the presence of a neutron as well as a proton in the nucleus. Because of the greater nuclear mass, the spectral lines of deuterium are all shifted slightly to wavelengths shorter than those of ordinary hydrogen. The H_α line of deuterium, for example, has a wavelength of 6,561 Å, while that of hydrogen is 6,563 Å: a small but definite difference, sufficient for the identification of deuterium.

4.10 ATOMIC EXCITATION

There are two principal mechanisms that can excite an atom to an energy level above its ground state, thereby enabling it to radiate. One mechanism is a collision with another particle during which part of their joint kinetic energy is absorbed by the atom. An atom excited in this way will return to its ground state in an average of 10^{-8} s by emitting one or more photons. To produce an electric discharge in a rarefied gas, an electric field is established which accelerates electrons and atomic ions until their kinetic energies are sufficient to excite atoms they happen to collide with. Neon signs and mercury-vapor lamps are familiar examples of how a strong electric field applied between electrodes in a gas-filled tube leads to the emission of the characteristic spectral radiation of that gas, which happens to be reddish light in the case of neon and bluish light in the case of mercury vapor.

A different excitation mechanism is involved when an atom absorbs a photon of light whose energy is just the right amount to raise the atom to a higher energy level. For example, a photon of wavelength 1,217 Å is emitted when a hydrogen atom in the $n = 2$ state drops to the $n = 1$ state; the absorption of a photon of wavelength 1,217 Å by a hydrogen atom initially in the $n = 1$ state will therefore bring it up to the $n = 2$ state. This process explains the origin of absorption spectra. When white light, which contains all wavelengths, is passed through hydrogen gas, photons of those wavelengths that correspond to transitions between energy levels are absorbed. The resulting excited hydrogen atoms reradiate their excitation energy almost at once, but these photons come off in random directions with only a few in the same direction as the original beam of white light. The dark lines in an absorption spectrum are therefore never completely black, but only appear so by contrast with the bright background. We expect the absorption spectrum of any element to be identical with its emission spectrum, then, which agrees with observation.

Atomic spectra are not the only means of investigating the presence of discrete energy levels within atoms. A series of experiments based on the first of the excitation mechanisms of the previous section was performed by Franck and Hertz starting in 1914. These experiments provided a direct demonstration that atomic energy levels do indeed exist and, furthermore, that these levels are the same as those suggested by observations of line spectra.

Franck and Hertz bombarded the vapors of various elements with electrons of known energy, using an apparatus like that shown in Fig. 4-22. A small potential difference V_0 is maintained between the grid and collecting plate, so that only electrons having energies greater than a certain minimum contribute to the current i through the ammeter. As the accelerating potential V is increased, more and more electrons arrive at the plate and i rises (Fig. 4-22). If kinetic energy is conserved in a collision between an electron and one of the atoms in the vapor, the electron merely bounces off in a direction different from its original one. Because an atom is so much heavier than an electron, the latter loses almost no kinetic energy in the process. After a certain critical electron energy is reached, however, the plate current drops abruptly. The interpretation of this effect is that an electron colliding with one of the atoms gives up some or all of its kinetic energy in exciting the atom to an energy level above its ground state. Such a collision is called *inelastic*, in contrast to an *elastic* collision in which kinetic energy is conserved. The critical electron energy corresponds to the excitation energy of the atom.

Then, as the accelerating potential V is raised further, the plate current again increases, since the electrons now have sufficient energy left after experiencing an inelastic collision to reach the plate. Eventually another sharp drop in plate current i occurs, which is interpreted as arising from the excitation of the same higher energy level in another atom. As Fig. 4-23 indicates, a series of critical potentials for a particular atomic species is obtained in this way. Thus the highest potentials result from several inelastic collisions and are multiples of the lower ones.

To check the interpretation of critical potentials as being due to discrete atomic energy levels, Franck and Hertz observed the emission spectra of vapors during electron bombardment. In the case of mercury vapor, for example, they found that a minimum electron energy of 4.9 eV was required to excite the 2,536-Å

FIGURE 4-22   Apparatus for the Franck-Hertz experiment.

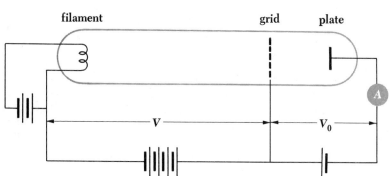

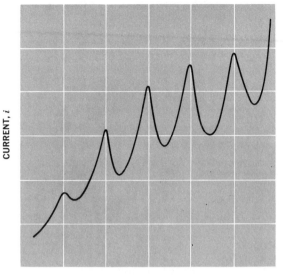

FIGURE 4-23 Results of the Franck-Hertz experiment, showing critical potentials.

CURRENT, i

ACCELERATING POTENTIAL, V

spectral line of mercury—and a photon of 2,536-Å light has an energy of just 4.9 eV. The Franck-Hertz experiments were performed shortly after Bohr announced his theory of the hydrogen atom, and they provided independent confirmation of his basic ideas.

4.11 THE CORRESPONDENCE PRINCIPLE

The principles of quantum physics, so different from those of classical physics in the microscopic world that lies beyond the reach of our senses, must nevertheless yield results identical with those of classical physics in the domain where experiment indicates the latter to be valid. We have already seen that this fundamental requirement is satisfied by the theory of relativity, the quantum theory of radiation, and the wave theory of matter; we shall now show that it is satisfied also by Bohr's theory of the atom.

According to electromagnetic theory, an electron moving in a circular orbit radiates electromagnetic waves whose frequencies are equal to its frequency of revolution and to harmonics (that is, integral multiples) of that frequency. In a hydrogen atom the electron's speed is

$$v = \frac{e}{\sqrt{4\pi\varepsilon_0 mr}}$$

according to Eq. 4.13, where r is the radius of its orbit. Hence the frequency of revolution f of the electron is

$$f = \frac{\text{electron speed}}{\text{orbit circumference}}$$

$$= \frac{v}{2\pi r}$$

$$= \frac{e}{2\pi \sqrt{4\pi\varepsilon_0 m r^3}}$$

The radius r_n of a stable orbit is given in terms of its quantum number n by Eq. 4.22 as

$$r_n = \frac{n^2 h^2 \varepsilon_0}{\pi m e^2}$$

and so the frequency of revolution is

4.29 $$f = \frac{m e^4}{8 \varepsilon_0^2 h^3}\left(\frac{2}{n^3}\right)$$

Under what circumstances should the Bohr atom behave classically? If the electron orbit is so large that we might expect to be able to measure it directly, quantum effects should be entirely inconspicuous. An orbit 1 cm across, for example, meets this specification; its quantum number is very close to $n = 10,000$, and, while hydrogen atoms so grotesquely large do not actually occur because their energies would be only infinitesimally below the ionization energy, they are not prohibited in theory. What does the Bohr theory predict that such an atom will radiate? According to Eq. 4.25, a hydrogen atom dropping from the n_ith energy level to the n_fth energy level emits a photon whose frequency is

$$\nu = \frac{m e^4}{8 \varepsilon_0^2 h^3}\left(\frac{1}{n_f^2} - \frac{1}{n_i^2}\right)$$

Let us write n for the initial quantum number n_i and $n - p$ (where $p = 1, 2, 3, \ldots$) for the final quantum number n_f. With this substitution,

$$\nu = \frac{m e^4}{8 \varepsilon_0^2 h^3}\left[\frac{1}{(n - p)^2} - \frac{1}{n^2}\right]$$

$$= \frac{m e^4}{8 \varepsilon_0^2 h^3}\left[\frac{2np - p^2}{n^2(n - p)^2}\right]$$

Now, when n_i and n_f are both very large, n is much greater than p, and

$$2np - p^2 \approx 2np$$
$$(n - p)^2 \approx n^2$$

so that

4.30
$$\nu = \frac{me^4}{8\varepsilon_0{}^2h^3}\left(\frac{2p}{n^3}\right)$$

When $p = 1$, the frequency ν of the radiation is exactly the same as the frequency of rotation f of the orbital electron given in Eq. 4.29. Harmonics of this frequency are radiated when $p = 2, 3, 4, \ldots$. Hence both quantum and classical pictures of the hydrogen atom make identical predictions in the limit of very large quantum numbers. When $n = 2$, Eq. 4.29 predicts a radiation frequency that differs from that given by Eq. 4.25 by almost 300 percent, while when $n = 10{,}000$, the discrepancy is only about 0.01 percent.

The requirement that quantum physics give the same results as classical physics in the limit of large quantum numbers was called by Bohr the *correspondence principle*. It has played an important role in the development of the quantum theory of matter.

Problems

° 1.   A 5-MeV alpha particle approaches a gold nucleus with an impact parameter of 2.6×10^{-13} m. Through what angle will it be scattered?

° 2.   What is the impact parameter of a 5-MeV alpha particle scattered by $10°$ when it approaches a gold nucleus?

° 3.   What fraction of a beam of 7.7-MeV alpha particles incident upon a gold foil 3×10^{-7} m thick is scattered by less than $1°$?

° 4.   What fraction of a beam of 7.7-MeV alpha particles incident upon a gold foil 3×10^{-7} m thick is scattered by $90°$ or more?

° 5.   Show that twice as many alpha particles are scattered by a foil through angles between 60 and $90°$ as are scattered through angles of $90°$ or more.

° 6.   A beam of 8.3-MeV alpha particles is directed at an aluminum foil. It is found that the Rutherford scattering formula ceases to be obeyed at scattering angles exceeding about $60°$. If the alpha particle is assumed to have a radius of 2×10^{-15} m, find the radius of the aluminum nucleus.

7.   Determine the distance of closest approach of 1-MeV protons incident upon gold nuclei.

8.   Find the distance of closest approach of 8-MeV protons incident upon gold nuclei.

9. The derivation of the Rutherford scattering formula was made nonrelativistically. Justify this approximation by computing the mass ratio between an 8-MeV alpha particle and an alpha particle at rest.

10. Find the frequency of rotation of the electron in the classical model of the hydrogen atom. In what region of the spectrum are electromagnetic waves of this frequency?

11. The electric-field intensity at a distance r from the center of a uniformly charged sphere of radius R and total charge Q is $Qr/4\pi\varepsilon_0 R^3$ when $r < R$. Such a sphere corresponds to the Thomson model of the atom. Show that an electron in this sphere executes simple harmonic motion about its center and derive a formula for the frequency of this motion. Evaluate the frequency of the electron oscillations for the case of the hydrogen atom and compare it with the frequencies of the spectral lines of hydrogen.

12. Find the wavelength of the spectral line corresponding to the transition in hydrogen from the $n = 6$ state to the $n = 3$ state.

13. Find the wavelength of the photon emitted when a hydrogen atom goes from the $n = 10$ state to its ground state.

14. How much energy is required to remove an electron in the $n = 2$ state from a hydrogen atom?

15. A beam of electrons bombards a sample of hydrogen. Through what potential difference must the electrons have been accelerated if the first line of the Balmer series is to be emitted?

16. Find the recoil speed of a hydrogen atom after it emits a photon in going from the $n = 4$ state to the $n = 1$ state.

17. How many revolutions does an electron in the $n = 2$ state of a hydrogen atom make before dropping to the $n = 1$ state? (The average lifetime of an excited state is about 10^{-8} s.)

18. The average lifetime of an excited atomic state is 10^{-8} s. If the wavelength of the spectral line associated with the decay of this state is 5,000 Å, find the width of the line.

19. At what temperature will the average molecular kinetic energy in gaseous hydrogen equal the binding energy of a hydrogen atom?

20. Calculate the angular momentum about the nucleus of an electron in the nth orbit of a hydrogen atom, and show from this that an alternate expression of Bohr's first postulate is that the angular momentum of such an atom must

be $n\hbar$. (In fact, the quantization of angular momentum in units of $\hbar$ was the starting point of Bohr's original work, since the hypothesis of de Broglie waves had not been proposed as yet. We shall see in Chap. 8 that this quantization rule holds only for the component of the angular momentum of a system in one particular direction, while the magnitude of the total angular momentum is quantized in a somewhat different way.)

21. A mixture of ordinary hydrogen and tritium, a hydrogen isotope whose nucleus is approximately three times more massive than ordinary hydrogen, is excited and its spectrum observed. How far apart in wavelength will the H_α lines of the two kinds of hydrogen be?

22. A μ^- meson ($m = 207\ m_e$) can be captured by a proton to form a "mesic atom." Find the radius of the first Bohr orbit of such an atom.

23. A μ^- meson is in the $n = 2$ state of a tritium atom. Find the energy radiated when the mesic atom drops to its ground state.

24. A positronium atom is a system consisting of a positron (positive electron) and an electron. (a) Compare the wavelength of the photon emitted in the $n = 3 \rightarrow n = 2$ transition in positronium with that of the H_α line. (b) Compare the ionization energy in positronium with that in hydrogen.

25. (a) Derive a formula for the energy levels of a *hydrogenic atom*, which is an ion such as He^+ or Li^{++} whose nuclear charge is $+Ze$ and which contains a single electron. (b) Sketch the energy levels of the He^+ ion and compare them with the energy levels of the H atom. (c) An electron joins a bare helium nucleus to form a He^+ ion. Find the wavelength of the photon emitted in this process if the electron is assumed to have had no kinetic energy when it combined with the nucleus.

26. Use the uncertainty principle to determine the ground-state radius r_1 of the hydrogen atom in the following way. First find a formula for the electron kinetic energy in terms of the momentum an electron must have if confined to a region of linear dimension r_1. Add this kinetic energy to the electrostatic potential energy of an electron the distance r_1 from a proton, and differentiate with respect to r_1 the resulting expression for the total electron energy E to find the value of r_1 for which E is a minimum. Compare the result with that given by Eq. 4.22 with $n = 1$.

QUANTUM MECHANICS 5

The Bohr theory of the atom, discussed in the previous chapter, is able to account for certain experimental data in a convincing manner, but it has a number of severe limitations. While the Bohr theory correctly predicts the spectral series of hydrogen, hydrogen isotopes, and hydrogenic atoms, it is incapable of being extended to treat the spectra of complex atoms having two or more electrons each; it can give no explanation of why certain spectral lines are more intense than others (that is, why certain transitions between energy levels have greater probabilities of occurrence); and it cannot account for the observation that many spectral lines actually consist of several separate lines whose wavelengths differ slightly. And, perhaps most important, it does not permit us to obtain what a really successful theory of the atom should make possible: an understanding of how individual atoms interact with one another to endow macroscopic aggregates of matter with the physical and chemical properties we observe.

These objections to the Bohr theory are not put forward in an unfriendly way, for the theory was one of those seminal achievements that transform scientific thought, but rather to emphasize that an approach to atomic phenomena of greater generality is required. Such an approach was developed in 1925–1926 by Erwin Schrödinger, Werner Heisenberg, and others under the apt name of *quantum mechanics*. By the early 1930s the application of quantum mechanics to problems involving nuclei, atoms, molecules, and matter in the solid state made it possible to understand a vast body of otherwise-puzzling data and—a vital attribute of any theory—led to predictions of remarkable accuracy.

5.1 INTRODUCTION TO QUANTUM MECHANICS

The fundamental difference between Newtonian mechanics and quantum mechanics lies in what it is that they describe. Newtonian mechanics is concerned with the motion of a particle under the influence of applied forces, and it takes for granted that such quantities as the particle's position, mass, velocity, acceleration, etc., can be measured. This assumption is, of course, completely valid in our everyday experience, and Newtonian mechanics provides the "correct" explanation for the behavior of moving bodies in the sense that the values it predicts for observable magnitudes agree with the measured values of those magnitudes.

Quantum mechanics, too, consists of relationships between observable magnitudes, but the uncertainty principle radically alters the definition of "observable magnitude" in the atomic realm. According to the uncertainty principle, the position and momentum of a particle cannot be accurately measured at the same time, while in Newtonian mechanics both are assumed to have definite, ascertainable values at every instant. The quantities whose relationships quantum mechanics explores are *probabilities*. Instead of asserting, for example, that the radius of the electron's orbit in a ground-state hydrogen atom is always exactly 5.3×10^{-11} m, quantum mechanics states that this is the *most probable* radius; if we conduct a suitable experiment, most trials will yield a different value, either larger or smaller, but the value most likely to be found will be 5.3×10^{-11} m.

At first glance quantum mechanics seems a poor substitute for Newtonian mechanics, but closer inspection reveals a striking fact: *Newtonian mechanics is nothing but an approximate version of quantum mechanics*. The certainties proclaimed by Newtonian mechanics are illusory, and their agreement with experiment is a consequence of the fact that macroscopic bodies consist of so many individual atoms that departures from average behavior are unnoticeable. Instead of two sets of physical principles, one for the macroscopic universe and one for the microscopic universe, there is only a single set, and quantum mechanics represents our best effort to date in formulating it.

5.2 THE WAVE EQUATION

As mentioned in Chap. 3, the quantity with which quantum mechanics is concerned is the *wave function* Ψ of a body. While Ψ itself has no physical interpretation, the square of its absolute magnitude $|\Psi|^2$ (or $\Psi\Psi^\circ$ if Ψ is complex) evaluated at a particular point at a particular time is proportional to the probability of experimentally finding the body there at that time. The problem of quantum mechanics is to determine Ψ for a body when its freedom of motion is limited by the action of external forces.

Even before we consider the actual calculation of Ψ, we can establish certain requirements it must always fulfill. For one thing, since $|\Psi|^2$ is proportional to the probability P of finding the body described by Ψ, the integral of $|\Psi|^2$ over all space must be finite—the body is *somewhere*, after all. If

5.1 $$\int_{-\infty}^{\infty} |\Psi|^2 \, dV$$

is 0, the particle does not exist, and the integral obviously cannot be ∞ and still mean anything; $|\Psi|^2$ cannot be negative or complex because of the way it is defined, and so the only possibility left is that its integral be a finite quantity if Ψ is to describe properly a real body.

It is usually convenient to have $|\Psi|^2$ be *equal* to the probability P of finding the particle described by Ψ, rather than merely be proportional to P. If $|\Psi|^2$ is to equal P, then it must be true that

5.2 $$\int_{-\infty}^{\infty} |\Psi|^2 \, dV = 1$$ **Normalization**

since

$$\int_{-\infty}^{\infty} P \, dV = 1$$

is the mathematical statement that the particle exists somewhere at all times. A wave function that obeys Eq. 5.2 is said to be *normalized*. Every acceptable wave function can be normalized by multiplying by an appropriate constant; we shall shortly see exactly how this is done.

Besides being normalizable, Ψ must be single-valued, since P can have only one value at a particular place and time. A further condition that Ψ must obey is that it and its partial derivatives $\partial\Psi/\partial x$, $\partial\Psi/\partial y$, $\partial\Psi/\partial z$ be continuous everywhere.

Schrödinger's equation, which is the fundamental equation of quantum mechanics in the same sense that the second law of motion is the fundamental equation of Newtonian mechanics, is a wave equation in the variable Ψ. Before we tackle Schrödinger's equation, let us review the general wave equation

5.3 $$\frac{\partial^2 y}{\partial x^2} = \frac{1}{v^2} \frac{\partial^2 y}{\partial t^2}$$ **Wave equation**

which governs a wave whose variable quantity is y that propagates in the x direction with the speed v. In the case of a wave in a stretched string, y is the displacement of the string from the x axis; in the case of a sound wave, y is the pressure difference; in the case of a light wave, y is either the electric or the magnetic field magnitude. The wave equation is derived in textbooks of mechanics for mechanical waves and in textbooks of electricity and magnetism for electromagnetic waves.

Solutions to the wave equation may be of many kinds, reflecting the variety of waves that can occur—a single traveling pulse, a train of waves of constant amplitude and wavelength, a train of superposed waves of the same amplitudes and wavelengths, a train of superposed waves of different amplitudes and wavelengths, a standing wave in a string fastened at both ends, and so on. All solutions must be of the form

5.4
$$y = F\left(t \pm \frac{x}{v}\right)$$

where F is any function that can be differentiated. The solutions $F(t - x/v)$ represent waves traveling in the $+x$ direction, and the solutions $F(t + x/v)$ represent waves traveling in the $-x$ direction.

Our interest here is in the wave equivalent of a "free particle," namely a particle that is not under the influence of any forces and therefore pursues a straight path at constant speed. This equivalent corresponds to the general solution of Eq. 5.3 for undamped (that is, constant amplitude A), monochromatic (constant angular frequency ω) harmonic waves in the $+x$ direction,

5.5
$$y = Ae^{-i\omega(t-x/v)}$$

In this formula y is a complex quantity, with both real and imaginary parts. Because

5.6
$$e^{-i\theta} = \cos\theta - i\sin\theta$$

Eq. 5.5 can be written in the form

5.7
$$y = A\cos\omega\left(t - \frac{x}{v}\right) - iA\sin\omega\left(t - \frac{x}{v}\right)$$

Only the real part of Eq. 5.7 has significance in the case of waves in a stretched string, where y represents the displacement of the string from its normal position (Fig. 5-1); in this case the imaginary part is discarded as irrelevant.

FIGURE 5-1 Waves in the xy plane traveling in the $+x$ direction along a stretched string lying on the x axis.

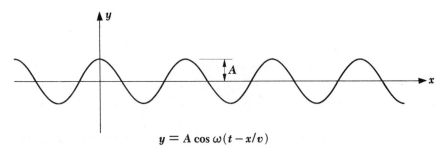

$$y = A\cos\omega(t - x/v)$$

5.3 SCHRÖDINGER'S EQUATION: TIME-DEPENDENT FORM

In quantum mechanics the wave function Ψ corresponds to the wave variable y of wave motion in general. However, Ψ, unlike y, is not itself a measurable quantity and may therefore be complex. For this reason we shall assume that Ψ is specified in the x direction by

5.8 $\qquad \Psi = Ae^{-i\omega(t-x/v)}$

When we replace ω in the above formula by $2\pi\nu$ and v by $\lambda\nu$, we obtain

5.9 $\qquad \Psi = Ae^{-2\pi i(\nu t-x/\lambda)}$

which is convenient since we already know what ν and λ are in terms of the total energy E and momentum p of the particle being described by Ψ. Since

$$E = h\nu = 2\pi\hbar\nu$$

and

$$\lambda = \frac{h}{p} = \frac{2\pi\hbar}{p}$$

we have

5.10 $\qquad \Psi = Ae^{-(i/\hbar)(Et-px)}$

Equation 5.10 is a mathematical description of the wave equivalent of an unrestricted particle of total energy E and momentum p moving in the $+x$ direction, just as Eq. 5.5 is a mathematical description of, for example, a harmonic displacement wave moving freely along a stretched string.

The expression for the wave function Ψ given by Eq. 5.10 is correct only for freely moving particles, while we are most interested in situations where the motion of a particle is subject to various restrictions. An important concern, for example, is an electron bound to an atom by the electric field of its nucleus. What we must now do is obtain the fundamental differential equation for Ψ, which we can then solve in a specific situation.

We begin by differentiating Eq. 5.10 twice with respect to x, yielding

5.11 $\qquad \dfrac{\partial^2 \Psi}{\partial x^2} = -\dfrac{p^2}{\hbar^2}\Psi$

and once with respect to t, yielding

5.12 $\qquad \dfrac{\partial \Psi}{\partial t} = -\dfrac{iE}{\hbar}\Psi$

At speeds small compared with that of light, the total energy E of a particle is the sum of its kinetic energy $p^2/2m$ and its potential energy V, where V is in general a function of position x and time t:

5.13 $$E = \frac{p^2}{2m} + V$$

Multiplying both sides of this equation by the wave function Ψ,

5.14 $$E\Psi = \frac{p^2\Psi}{2m} + V\Psi$$

From Eqs. 5.11 and 5.12 we see that

5.15 $$E\Psi = -\frac{\hbar}{i}\frac{\partial\Psi}{\partial t}$$

and

5.16 $$p^2\Psi = -\hbar^2\frac{\partial^2\Psi}{\partial x^2}$$

Substituting these expressions for $E\Psi$ and $p^2\Psi$ into Eq. 5.14, we obtain

5.17 $$i\hbar\frac{\partial\Psi}{\partial t} = -\frac{\hbar^2}{2m}\frac{\partial^2\Psi}{\partial x^2} + V\Psi$$

Time-dependent Schrödinger's equation in one dimension

Equation 5.17 is the *time-dependent form of Schrödinger's equation.* In three dimensions the time-dependent form of Schrödinger's equation is

5.18 $$i\hbar\frac{\partial\Psi}{\partial t} = -\frac{\hbar^2}{2m}\left(\frac{\partial^2\Psi}{\partial x^2} + \frac{\partial^2\Psi}{\partial y^2} + \frac{\partial^2\Psi}{\partial z^2}\right) + V\Psi$$

where the particle's potential energy V is some function of x, y, z, and t. Any restrictions that may be present on the particle's motion will affect the potential-energy function V. Once V is known, Schrödinger's equation may be solved for the wave function Ψ of the particle, from which its probability density Ψ^2 may be determined for a specified x, y, z, t.

The manner in which Schrödinger's equation was obtained starting from the wave function of a freely moving particle deserves attention. The extension of Schrödinger's equation from the special case of an unrestricted particle (potential energy $V =$ constant) to the general case of a particle subject to arbitrary forces that vary in time and space [$V = V(x,y,z,t)$] is entirely plausible, but there is no a priori way to *prove* that this extension is correct. All we can do is to postulate Schrödinger's equation, solve it for a variety of physical situations, and compare the results of the calculations with the results of experiments. If they agree, the postulate embodied in Schrödinger's equation is valid; if they disagree, the postulate must be discarded and some other approach would have to be explored. In other words, **Schrödinger's equation cannot be derived from "first principles," but represents a first principle itself.**

In practice, Schrödinger's equation has turned out to be completely accurate in predicting the results of experiments. To be sure, we must keep in mind that Eq. 5.18 can be used only for nonrelativistic problems, and a more elaborate formulation is required when particle speeds comparable with that of light are involved. Because it is in accord with experiment within its range of applicability, we are entitled to regard Schrödinger's equation as representing a successful postulate concerning certain aspects of the physical world. But for all its success, this equation remains a postulate in the same sense as the postulates of special relativity or statistical mechanics: None of these can be derived from some other principle, and each is a fundamental generalization neither more nor less valid than the empirical data it is based upon. It is worth noting in this connection that Schrödinger's equation does not represent an increase in the number of postulates required to describe the workings of the physical world, because Newton's second law of motion, regarded in classical mechanics as a postulate, can be derived from Schrödinger's equation provided that the quantities it relates are understood to be averages rather than definite values.

5.4 EXPECTATION VALUES

Once Schrödinger's equation has been solved for a particle in a given physical situation, the resulting wave function $\Psi(x,y,z,t)$ contains all the information about the particle that is permitted by the uncertainty principle. Except for those variables that happen to be quantized in certain cases, this information is in the form of probabilities and not specific numbers.

As an example, let us calculate the *expectation value* $\langle x \rangle$ of the position of a particle confined to the x axis that is described by the wave function $\Psi(x,t)$. This is the value of x we would obtain if we determined experimentally the positions of a great many particles described by the same wave function at some instant t and then averaged the results.

To make the procedure clear, we shall first answer a slightly different question: What is the average position $\bar{x}$ of a number of particles distributed along the x axis in such a way that there are N_1 particles at x_1, N_2 particles at x_2, and so on? The average position in this case is the same as the center of mass of the distribution, and so

$$\bar{x} = \frac{N_1 x_1 + N_2 x_2 + N_3 x_3 + \cdots}{N_1 + N_2 + N_3 + \cdots}$$

$$= \frac{\Sigma N_i x_i}{\Sigma N_i}$$

When we are dealing with a single particle, we must replace the number N_i of particles at x_i by the probability P_i that the particle be found in an interval dx at x_i. This probability is

$$P_i = |\Psi_i|^2 \, dx$$

where Ψ_i is the particle wave function evaluated at $x = x_i$. Making this substitution and changing the summations to integrals, we see that the expectation value of the position of the single particle is

5.19
$$\langle x \rangle = \frac{\displaystyle\int_{-\infty}^{\infty} x|\Psi|^2 \, dx}{\displaystyle\int_{-\infty}^{\infty} |\Psi|^2 \, dx}$$

If Ψ is a normalized wave function, the denominator of Eq. 5.19 is equal to the probability that the particle exists somewhere between $x = -\infty$ and $x = \infty$, and therefore has the value 1. Hence

5.20
$$\langle x \rangle = \int_{-\infty}^{\infty} x|\Psi|^2 \, dx$$

This formula states that $\langle x \rangle$ is located at the center of mass (so to speak) of $|\Psi|^2$; if $|\Psi|^2$ is plotted versus x on a graph and the area enclosed by the curve and the x axis is cut out, the balance point will be at $\langle x \rangle$.

The same procedure as that followed above can be used to obtain the expectation value $\langle G(x) \rangle$ of any quantity [for instance, potential energy $V(x)$] that is a function of the position x of a particle described by a wave function Ψ. The result is

5.21
$$\langle G(x) \rangle = \int_{-\infty}^{\infty} G(x)|\Psi|^2 \, dx \qquad\qquad \textbf{Expectation value}$$

This formula holds even if $G(x)$ varies with time, because $\langle G(x) \rangle$ in any event must be evaluated at a particular time t since Ψ is itself a function of t.

5.5   SCHRÖDINGER'S EQUATION: STEADY-STATE FORM

In a great many situations the potential energy of a particle does not depend upon time explicitly; the forces that act upon it, and hence V, vary with the position of the particle only. When this is true, Schrödinger's equation may be simplified by removing all reference to t. We note that the one-dimensional wave function Ψ of an unrestricted particle may be written

$$\Psi = Ae^{-(i/\hbar)(Et-px)}$$
$$= Ae^{-(iE/\hbar)t}e^{+(ip/\hbar)x}$$

5.22
$$= \psi e^{-(iE/\hbar)t}$$

That is, Ψ is the product of a time-dependent function $e^{-(iE/\hbar)t}$ and a position-dependent function ψ. As it happens, the time variations of *all* functions of particles acted upon by stationary forces have the same form as that of an unrestricted particle. Substituting the Ψ of Eq. 5.22 into the time-dependent form of Schrödinger's equation, we find that

$$E\psi e^{-(iE/\hbar)t} = -\frac{\hbar^2}{2m}e^{-(iE/\hbar)t}\frac{\partial^2\psi}{\partial x^2} + V\psi e^{-(iE/\hbar)t}$$

and so, dividing through by the common exponential factor,

5.23
$$\frac{\partial^2\psi}{\partial x^2} + \frac{2m}{\hbar^2}(E - V)\psi = 0$$

Steady-state Schrödinger's equation in one dimension

Equation 5.23 is the *steady-state form of Schrödinger's equation*. In three dimensions it is

$$E = i K \frac{\partial \psi}{\partial T}$$

5.24
$$\frac{\partial^2\psi}{\partial x^2} + \frac{\partial^2\psi}{\partial y^2} + \frac{\partial^2\psi}{\partial z^2} + \frac{2m}{\hbar^2}(E - V)\psi = 0$$

Steady-state Schrödinger's equation in three dimensions

In general, Schrödinger's steady-state equation can be solved only for certain values of the energy E. What is meant by this statement has nothing to do with any mathematical difficulties that may be present, but is something much more fundamental. To "solve" Schrödinger's equation for a given system means to obtain a wave function ψ that not only obeys the equation and whatever boundary conditions there are, but also fulfills the requirements for an acceptable wave function—namely, that it and its derivatives be continuous, finite, and single-valued. If there is no such wave function, the system cannot exist in a steady state. Thus energy quantization appears in wave mechanics as a natural element of the theory, and energy quantization in the physical world is revealed as a universal phenomenon characteristic of *all* stable systems.

A familiar and quite close analogy to the manner in which energy quantization occurs in solutions of Schrödinger's equation is with standing waves in a stretched string of length L that is fixed at both ends. Here, instead of a single wave propagating indefinitely in one direction, waves are traveling in both the $+x$ and $-x$ directions simultaneously subject to the condition that the displacement y always be zero at both ends of the string. An acceptable function $y(x,t)$ for the displacement must, with its derivatives, obey the same requirements of continuity, finiteness, and single-valuedness as ψ and, in addition, must be real since y represents a directly measurable quantity. The only solutions of the wave equation

$$\frac{\partial^2 y}{\partial x^2} = \frac{1}{v^2} \frac{\partial^2 y}{\partial t^2}$$

that are in accord with these various limitations are those in which the wavelengths are given by

$$\lambda_n = \frac{2L}{n+1} \qquad n = 0, 1, 2, 3, \ldots$$

as shown in Fig. 5-2. It is the *combination* of the wave equation and the restrictions placed on the nature of its solution that leads us to conclude that $y(x,t)$ can exist only for certain wavelengths λ_n.

The values of energy E_n for which Schrödinger's steady-state equation can be solved are called *eigenvalues* and the corresponding wave functions ψ_n are

FIGURE 5-2 Standing waves in a stretched string fastened at both ends.

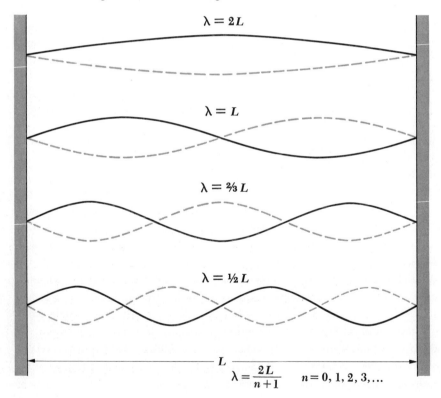

$\lambda = 2L$

$\lambda = L$

$\lambda = \frac{2}{3}L$

$\lambda = \frac{1}{2}L$

L

$$\lambda = \frac{2L}{n+1} \qquad n = 0, 1, 2, 3, \ldots$$

called *eigenfunctions*. (These terms come from the German *Eigenwert*, meaning "proper or characteristic value," and *Eigenfunktion*, or "proper or characteristic function.") The discrete energy levels of the hydrogen atom

$$E_n = -\frac{me^4}{32\pi^2\varepsilon_0{}^2\hbar^2}\left(\frac{1}{n^2}\right) \qquad n = 1, 2, 3, \ldots$$

are an example of a set of eigenvalues; we shall see in Chap. 6 why these particular values of E are the only ones that yield acceptable wave functions for the electron in the hydrogen atom.

An important example of a dynamical variable other than total energy that is found to be quantized in stable systems is angular momentum **L**. In the case of the hydrogen atom, we shall find that the eigenvalues of the magnitude of the total angular momentum are specified by

$$L_l = \sqrt{l(l + 1)}\hbar \qquad l = 0, 1, 2, \ldots, (n - 1)$$

Of course, a dynamical variable G may not be quantized. In this case measurements of G made on a number of identical systems will not yield a unique result but instead a spread of values whose average is the expectation value

$$\langle G \rangle = \int_{-\infty}^{\infty} G|\psi|^2 \, dx$$

In the hydrogen atom, the electron's position is not quantized, for instance, so that we must think of the electron as being present in the vicinity of the nucleus with a certain probability $|\psi|^2$ per unit volume but with no predictable position or even orbit in the classical sense. This probabilistic statement does not conflict with the fact that experiments performed on hydrogen atoms always show that it contains one whole electron, not 27 percent of an electron in a certain region and 73 percent elsewhere; the probability is one of *finding* the electron, and although this probability is smeared out in space, the electron itself is not.

5.6 THE PARTICLE IN A BOX: ENERGY QUANTIZATION

To solve Schrödinger's equation, even in its simpler steady-state form, usually requires sophisticated mathematical techniques. For this reason the study of quantum mechanics has traditionally been reserved for advanced students who have the required proficiency in mathematics. However, since quantum mechanics is the theoretical structure whose results are closest to experimental reality, we must explore its methods and applications if we are to achieve any understanding of modern physics. As we shall see, even a relatively limited mathematical background is sufficient for us to follow the trains of thought that have led quantum mechanics to its greatest achievements.

Our first problem using Schrödinger's equation is that of a particle bouncing back and forth between the walls of a box (Fig. 5-3). Our interest in this problem is threefold: to see how Schrödinger's equation is solved when the motion of a particle is subject to restrictions; to learn the characteristic properties of solutions of this equation, such as the limitation of particle energy to certain specific values only; and to compare the predictions of quantum mechanics with those of Newtonian mechanics.

We may specify the particle's motion by saying that it is restricted to traveling along the x axis between $x = 0$ and $x = L$ by infinitely hard walls. A particle does not lose energy when it collides with such walls, so that its total energy stays constant. From the formal point of view of quantum mechanics, the potential energy V of the particle is infinite on both sides of the box, while V is a constant—say 0 for convenience—on the inside. Since the particle cannot have an infinite amount of energy, it cannot exist outside the box, and so its wave function ψ is 0 for $x \leqslant 0$ and $x \geqslant L$. Our task is to find what ψ is within the box, namely, between $x = 0$ and $x = L$.

Within the box Schrödinger's equation becomes

5.25
$$\frac{d^2\psi}{dx^2} + \frac{2m}{\hbar^2} E\psi = 0$$

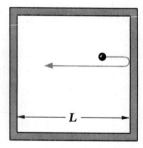

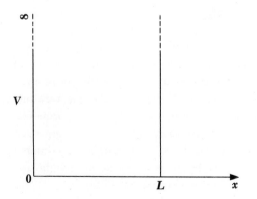

FIGURE 5-3 A particle confined to a box of width L.

since $V = 0$ there. (The total derivative $d^2\psi/dx^2$ is the same as the partial derivative $\partial^2\psi/\partial x^2$ because ψ is a function of x only in this problem.) Equation 5.25 has the two possible solutions

5.26
$$\psi = A \sin \sqrt{\frac{2mE}{\hbar^2}}\, x$$

5.27
$$\psi = B \cos \sqrt{\frac{2mE}{\hbar^2}}\, x$$

which we can verify by substitution back into Eq. 5.25; their sum is also a solution. A and B are constants to be evaluated.

These solutions are subject to the important boundary condition that $\psi = 0$ for $x = 0$ and for $x = L$. Since $\cos 0 = 1$, the second solution cannot describe the particle because it does not vanish at $x = 0$. Hence we conclude that $B = 0$. Since $\sin 0 = 0$, the first solution always yields $\psi = 0$ at $x = 0$, as required, but ψ will be 0 at $x = L$ only when

5.28
$$\sqrt{\frac{2mE}{\hbar^2}}\, L = \pi, \, 2\pi, \, 3\pi, \, \dots$$
$$= n\pi \qquad n = 1, \, 2, \, 3, \, \dots$$

This result comes about because the sines of the angles $\pi, 2\pi, 3\pi, \dots$ are all 0.

From Eq. 5.28 it is clear that the energy of the particle can have only certain values, which are the eigenvalues mentioned in the previous section. These eigenvalues, constituting the *energy levels* of the system, are

5.29
$$E_n = \frac{n^2\pi^2\hbar^2}{2mL^2} = \frac{n^2 h^2}{8mL^2} \qquad n = 1, \, 2, \, 3, \, \dots \qquad \textbf{Particle in a box}$$

The integer n corresponding to the energy level E_n is called its *quantum number*. A particle confined to a box cannot have an arbitrary energy: the fact of its confinement leads to restrictions on its wave function that permit it to have only those energies specified by Eq. 5.29.

It is significant that the particle cannot have zero energy; if it did, the wave function ψ would have to be zero everywhere in the box, and this means that the particle cannot be present there. The exclusion of $E = 0$ as a possible value for the energy of a trapped particle, like the limitation of E to a discrete set of definite values, is a quantum-mechanical result that has no counterpart in classical mechanics, where all energies, including zero, are presumed possible.

The uncertainty principle provides confirmation that $E = 0$ is not admissible. Because the particle is trapped in the box, the uncertainty in its position is

$\Delta x = L$, the width of the box. The uncertainty in its momentum must therefore be

$$\Delta p \geqslant \frac{\hbar}{L}$$

which is not compatible with $E = 0$. We note that the momentum corresponding to E_1 is, since the particle energy here is entirely kinetic,

$$p_1 = \pm \sqrt{2mE_1} = \pm \frac{\pi\hbar}{L}$$

which it in accord with the uncertainty principle.

Why are we not aware of energy quantization in our own experience? Surely a marble rolling back and forth between the sides of a level box with a smooth floor can have any speed, and therefore any energy, we choose to give it, including zero. In order to assure ourselves that Eq. 5.29 does not conflict with our direct observations while providing unique insights on a microscopic scale, we shall compute the permitted energy levels of (1) an electron in a box 1 Å wide and (2) a 10-g marble in a box 10 cm wide.

In case 1 we have $m = 9.1 \times 10^{-31}$ kg and $L = 1$ Å $= 10^{-10}$ m, so that the permitted electron energies are

$$\begin{aligned}
E_n &= \frac{n^2 \times \pi^2 \times (1.054 \times 10^{-34} \text{ J-s})^2}{2 \times 9.1 \times 10^{-31} \text{ kg} \times (10^{-10} \text{ m})^2} \\
&= 6.0 \times 10^{-18}n^2 \text{ J} \\
&= 38n^2 \text{ eV}
\end{aligned}$$

The minimum energy the electron can have is 38 eV, corresponding to $n = 1$. The sequence of energy levels continues with $E_2 = 152$ eV, $E_3 = 342$ eV, $E_4 = 608$ eV, and so on (Fig. 5-4). These energy levels are sufficiently far apart to make the quantization of electron energy in such a box conspicuous if such a box actually did exist.

In case 2 we have $m = 10$ g $= 10^{-2}$ kg and $L = 10$ cm $= 10^{-1}$ m, so that the permitted marble energies are

$$\begin{aligned}
E_n &= \frac{n^2 \times \pi^2 \times (1.054 \times 10^{-34} \text{ J-s})^2}{2 \times 10^{-2} \text{ kg} \times (10^{-1} \text{ m})^2} \\
&= 5.5 \times 10^{-64}n^2 \text{ J}
\end{aligned}$$

The minimum energy the marble can have is 5.5×10^{-64} J, corresponding to $n = 1$. A marble with this kinetic energy has a speed of only 3.3×10^{-31} m/s and is therefore experimentally indistinguishable from a stationary marble. A reasonable speed a marble might have is, say, $\frac{1}{3}$ m/s—which corresponds to the energy level of quantum number $n = 10^{30}$! The permissible energy levels are

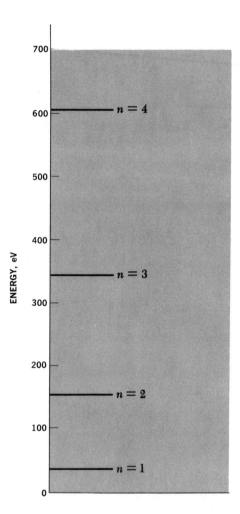

FIGURE 5-4 Energy levels of an electron confined to a box 1 Å wide.

so very close together, then, that there is no way to determine whether the marble can take on only those energies predicted by Eq. 5.29 or any energy whatever. Hence in the domain of everyday experience quantum effects are imperceptible; this accounts for the success in this domain of Newtonian mechanics.

5.7 THE PARTICLE IN A BOX: WAVE FUNCTIONS

In the previous section we found that the wave function of a particle in a box whose energy is E is

$$\psi = A \sin \sqrt{\frac{2mE}{\hbar^2}} x$$

Since the possible energies are

$$E_n = \frac{n^2\pi^2\hbar^2}{2mL^2}$$

substituting E_n for E yields

5.30
$$\psi_n = A \sin\frac{n\pi x}{L}$$

for the eigenfunctions corresponding to the energy eigenvalues E_n.

It is easy to verify that these eigenfunctions meet all the requirements we have discussed: for each quantum number n, ψ_n is a single-valued function of x, and ψ_n and $\partial\psi_n/\partial x$ are continuous. Furthermore, the integral of $|\psi_n|^2$ over all space is finite, as we can see by integrating $|\psi_n|^2\,dx$ from $x = 0$ to $x = L$ (since the particle, by hypothesis, is confined within these limits):

$$\int_{-\infty}^{\infty} |\psi_n|^2\,dx = \int_0^L |\psi_n|^2\,dx$$

$$= A^2 \int_0^L \sin^2\left(\frac{n\pi x}{L}\right) dx$$

5.31
$$= A^2\frac{L}{2}$$

To normalize ψ we must assign a value to A such that $|\psi_n|^2$ is *equal* to the probability $P\,dx$ of finding the particle between x and $x + dx$, rather than merely proportional to P. If $|\psi_n|^2$ is to equal P, then it must be true that

5.32
$$\int_{-\infty}^{\infty} |\psi_n|^2\,dx = 1$$

since

$$\int_{-\infty}^{\infty} P\,dx = 1$$

is the mathematical way of stating that the particle exists somewhere at all times. Comparing Eqs. 5.31 and 5.32, we see that the wave functions of a particle in a box are normalized if

5.33
$$A = \sqrt{\frac{2}{L}}$$

The normalized wave functions of the particle are therefore

5.34
$$\psi_n = \sqrt{\frac{2}{L}}\sin\frac{n\pi x}{L}$$

The normalized wave functions ψ_1, ψ_2, and ψ_3 together with the probability densities $|\psi_1|^2$, $|\psi_2|^2$, and $|\psi_3|^2$ are plotted in Fig. 5-5. While ψ_n may be negative as well as positive, $|\psi_n|^2$ is always positive and, since ψ_n is normalized, its value at a given x is equal to the probability P of finding the particle there. In every case $|\psi_n|^2 = 0$ at $x = 0$ and $x = L$, the boundaries of the box. At a particular point in the box the probability of the particle being present may be very different for different quantum numbers. For instance, $|\psi_1|^2$ has its maximum value of $\frac{1}{2}L$ in the middle of the box, while $|\psi_2|^2 = 0$ there: a particle in the lowest energy level of $n = 1$ is most likely to be in the middle of the box, while a particle in the next higher state of $n = 2$ is *never* there! Classical physics, of course, predicts the same probability for the particle being anywhere in the box.

The wave functions shown in Fig. 5-5 resemble the possible vibrations of a string fixed at both ends, such as those of the stretched string of Fig. 5-2. This is a consequence of the fact that waves in a stretched string and the wave representing a moving particle are described by equations of the same form, so that, when identical restrictions are placed upon each kind of wave, the formal results are identical.

FIGURE 5-5 Wave functions and probability densities of a particle confined to a box with rigid walls.

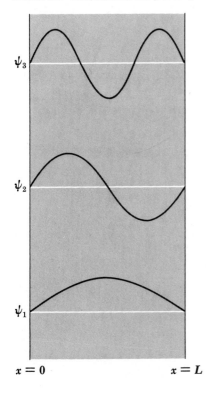

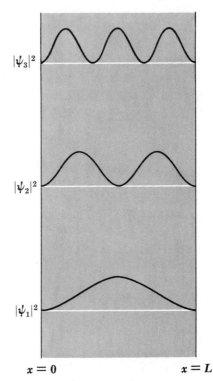

5.8 THE PARTICLE IN A NONRIGID BOX

It is interesting to solve the problem of the particle in a box when the walls of the box are no longer assumed to be infinitely rigid. In this case the potential energy V outside the box is a finite quantity; the corresponding situation in the case of a vibrating string would involve an imperfect attachment of the string at each end, so that the ends can move slightly. This problem is more difficult to treat, and we shall simply present the result here. (We shall take another look at a particle in a nonrigid box when we examine the theory of the deuteron in Chap. 11.)

The first few wave functions for a particle in such a box are shown in Fig. 5-6. The wave functions ψ_n now do *not* equal zero outside the box. Even though the particle's energy is smaller than the value of V outside the box, there is still a definite probability that it be found outside it! In other words, even though the particle does not have enough energy to break through the walls of the box according to "common sense," it may nevertheless somehow penetrate them. This peculiar situation is readily understandable in terms of the uncertainty principle. Because the uncertainty Δp in a particle's momentum is related to the uncertainty Δx in its position by the formula

$$\Delta p \, \Delta x \geqslant \hbar$$

an infinite uncertainty in particle momentum outside the box is the price of definitely establishing that the particle is never there. A particle requires an infinite amount of energy if its momentum is to have an infinite uncertainty, implying that $V = \infty$ outside the box. If V instead has a finite value outside the box, then, there is some probability—not necessarily great, but not zero

FIGURE 5-6 Wave functions and probability densities of a particle confined to a box with nonrigid walls.

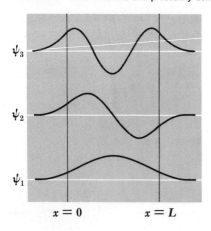

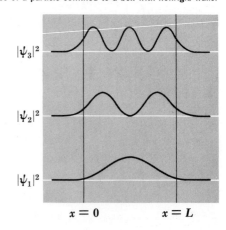

either—that the particle will "leak" out. As we shall see in Chap. 12, the quantum-mechanical prediction that particles always have some chance of escaping from confinement (since potential energies are never infinite in the real world, our original rigid-walled box has no physical counterpart) exactly fits the observed behavior of those radioactive nuclei that emit alpha particles.

When the confining box has nonrigid walls, the particle wave function ψ_n does not equal zero at the walls. The particle wavelengths that can fit into the box are therefore somewhat longer than in the case of the box with rigid walls, corresponding to lower particle momenta and hence to lower energy levels.

The condition that the potential energy V outside the box be finite has another consequence: it is now possible for a particle to have an energy E that exceeds V. Such a particle is not trapped inside the box, since it always has enough energy to penetrate its walls, and its energy is not quantized but may have any value above V. However, the particle's kinetic energy outside the box, $E - V$, is always less than its kinetic energy inside, which is just E since $V = 0$ in the box according to our original specification. Less energy means longer wavelength, and so ψ has a longer wavelength outside the box than inside.

In the optics of light waves, it is readily observed that when a light wave reaches a region where its wavelength changes (that is, a region of different index of refraction), reflection as well as transmission occurs. This is the reason we see our reflections in shop windows. The effect is common to all types of waves, and it may be shown mathematically to follow from the requirement that the wave variable (electric-field intensity E in the case of electromagnetic waves, pressure p in the case of sound waves, wave height h in the case of water waves, etc.) and its first derivative be continuous at the boundary where the wavelength change takes place.

Exactly the same considerations apply to the wave function ψ representing a moving particle. The wave function of a particle encountering a region in which it has a different potential energy, as we saw above, decreases in wavelength if V decreases and increases in wavelength if V increases. In either situation some reflection occurs at the boundaries between the regions. What does "some" reflection mean when we are discussing the motion of a single particle? Since ψ is related to the probability of finding the particle in a particular place, the partial reflection of ψ means that there is a chance that the particle will be reflected. That is, if we shoot many particles at a box with nonrigid walls, most will get through but some will be scattered.

What we have been saying, then, is that particles with enough energy to penetrate a wall nevertheless stand some chance of bouncing off instead. This prediction complements the "leaking" out of particles trapped in the box despite the fact that they have insufficient energy to penetrate its walls. Both of these predictions are unique with quantum mechanics and do not correspond to any behavior expected in classical physics. Their confirmation in numerous atomic and nuclear experiments supports the validity of the quantum-mechanical approach.

5.9 THE HARMONIC OSCILLATOR

Harmonic motion occurs when a system of some kind vibrates about an equilibrium configuration. The system may be an object supported by a spring or floating in a liquid, a diatomic molecule, an atom in a crystal lattice—there are countless examples in both the macroscopic and the microscopic realms. The condition for harmonic motion to occur is the presence of a restoring force that acts to return the system to its equilibrium configuration when it is disturbed; the inertia of the masses involved causes them to overshoot equilibrium, and the system oscillates indefinitely if no dissipative processes are also present.

In the special case of simple harmonic motion, the restoring force F on a particle of mass m is linear; that is, F is proportional to the particle's displacement x from its equilibrium position, so that

5.35 $\qquad F = -kx$

This relationship is customarily called Hooke's law. According to the second law of motion, $\mathbf{F} = m\mathbf{a}$, and so here

$$-kx = m\frac{d^2x}{dt^2}$$

5.36 $\qquad \dfrac{d^2x}{dt^2} + \dfrac{k}{m}x = 0$

There are various ways to write the solution to Eq. 5.36, a convenient one being

5.37 $\qquad x = A \cos\left(2\pi\nu t + \phi\right)$

where

5.38 $\qquad \nu = \dfrac{1}{2\pi}\sqrt{\dfrac{k}{m}}$

is the frequency of the oscillations, A is their amplitude, and ϕ, the phase constant, is a constant that depends upon the value of x at the time $t = 0$.

The importance of the simple harmonic oscillator in both classical and modern physics lies not in the strict adherence of actual restoring forces to Hooke's law, which is seldom true, but in the fact that these restoring forces reduce to Hooke's law for small displacements x. To appreciate this point we note that any force which is a function of x can be expressed in a Maclaurin's series about the equilibrium position $x = 0$ as

$$F(x) = F_{x=0} + \left(\frac{dF}{dx}\right)_{x=0} x + \frac{1}{2}\left(\frac{d^2F}{dx^2}\right)_{x=0} x^2 + \frac{1}{6}\left(\frac{d^3F}{dx^3}\right)_{x=0} x^3 + \cdots$$

Since $x = 0$ is the equilibrium position, $F_{x=0} = 0$, and since for small x the values of x^2, x^3, ... are very small compared with x, the third and higher terms of the series can be neglected. The only term of significance when x is small is therefore the second one. Hence

$$F(x) = \left(\frac{dF}{dx}\right)_{x=0} x$$

which is Hooke's law when $(dF/dx)_{x=0}$ is negative, as of course it is for any restoring force. The conclusion, then, is that *all* oscillations are simple harmonic in character when their amplitudes are sufficiently small.

The potential energy function $V(x)$ that corresponds to a Hooke's law force may be found by calculating the work needed to bring a particle from $x = 0$ to $x = x$ against such a force. The result is

5.39 $$V(x) = -\int_0^x F(x)\, dx = k\int_0^x x\, dx = \tfrac{1}{2}kx^2$$

and is plotted in Fig. 5-7. If the energy of the oscillator is E, the particle vibrates back and forth between $x = -A$ and $x = +A$, where E and A are related by $E = \tfrac{1}{2}kA^2$.

FIGURE 5-7 The potential energy of a harmonic oscillator is proportional to x^2, where x is the displacement from the equilibrium position. The amplitude A of the motion is determined by the total energy E of the oscillator, which classically can have any value.

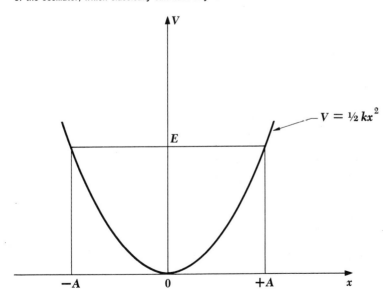

Even before we make a detailed calculation we can anticipate three quantum-mechanical modifications to this classical picture. First, there will not be a continuous spectrum of allowed energies but a discrete spectrum consisting of certain specific values only. Second, the lowest allowed energy will not be $E = 0$ but will be some definite minimum $E = E_0$. Third, there will be a certain probability that the particle can "penetrate" the potential well it is in and go beyond the limits of $-A$ and $+A$.

The actual results agree with these expectations. The energy levels of a harmonic oscillator whose classical frequency of oscillation is ν (given by Eq. 5.38) turn out to be given by the formula

5.40 $$E_n = \left(n + \frac{1}{2}\right)h\nu \qquad n = 0, 1, 2, \ldots$$ **Energy levels of harmonic oscillator**

The energy of a harmonic oscillator is thus quantized in steps of $h\nu$. The energy levels here are evenly spaced (Fig. 5-8), unlike the energy levels of a particle in a box whose spacing diverges. We note that, when $n = 0$,

5.41 $$E_0 = \frac{1}{2}h\nu$$ **Zero-point energy**

which is the lowest value the energy of the oscillator can have. This value is called the *zero-point energy* because a harmonic oscillator in equilibrium with its surroundings would approach an energy of $E = E_0$ and not $E = 0$ as the temperature approaches 0 K.

The wave functions corresponding to the first six energy levels of a harmonic oscillator are shown in Fig. 5-9. In each case the range to which a particle

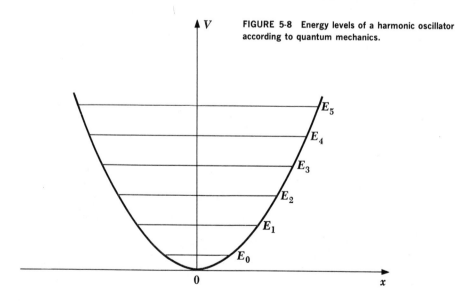

FIGURE 5-8 Energy levels of a harmonic oscillator according to quantum mechanics.

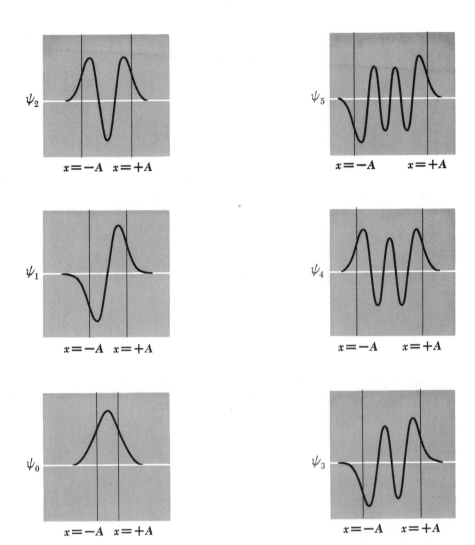

FIGURE 5-9 The first six harmonic-oscillator wave functions. The vertical lines show the limits $-A$ and $+A$ between which a classical oscillator with the same energy would vibrate.

oscillating classically with the same total energy E_n would be confined is indicated; evidently the particle is able to penetrate into classically forbidden regions—in other words, to exceed the amplitude A determined by the energy— with an exponentially decreasing probability, just as in the situation of a particle in a box with nonrigid walls.

It is interesting and instructive to compare the probability densities of a classical harmonic oscillator and a quantum-mechanical harmonic oscillator of the same energy. The upper graph of Fig. 5-10 shows this density for the classical

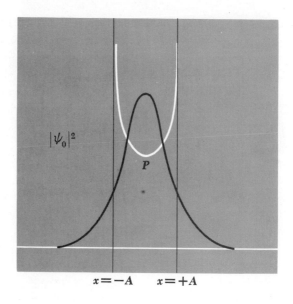

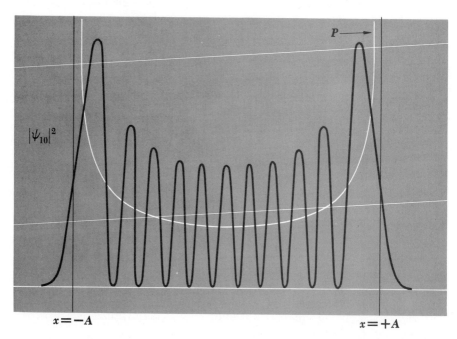

FIGURE 5-10 Probability densities for the $n = 0$ and $n = 10$ states of a quantum-mechanical harmonic oscillator. The probability densities for classical harmonic oscillators with the same energies are shown in white. In the $n = 10$ state, the wavelength is shortest at $x = 0$ and increases toward $x = \pm A$.

oscillator: The probability P of finding the particle at a given position is greatest at the end-points of its motion, where it moves slowly, and least near the equilibrium position $(x = 0)$, where it moves rapidly. Exactly the opposite behavior is manifested by a quantum-mechanical oscillator in its lowest energy state of $n = 0$. As shown, the probability density ψ_0^2 has its maximum value at $x = 0$ and drops off on either side of this position. However, this disagreement becomes less and less marked with increasing n: The lower graph of Fig. 5-10 corresponds to $n = 10$, and it is clear that ψ_{10}^2 when averaged over x has approximately the general character of the classical probability P. This is another example of the correspondence principle mentioned in Sec. 4.11: In the limit of large quantum numbers, quantum physics yields the same results as classical physics.

It might be objected that, although $|\psi_{10}|^2$ does indeed approach P when smoothed out, nevertheless $|\psi_{10}|^2$ fluctuates rapidly with x whereas P does not. However, this objection has meaning only if the fluctuations are observable, and the smaller the spacing of the peaks and hollows, the more strongly the uncertainty principle prevents their detection without altering the physical state of the oscillator. The exponential "tails" of $|\psi_{10}|^2$ beyond $x = \pm A$ also decrease in magnitude with increasing n. Thus the classical and quantum pictures begin to resemble each other more and more the larger the value of n, in agreement with the correspondence principle, although they are radically different for small n.

*5.10   THE HARMONIC OSCILLATOR: SOLUTION OF SCHRÖDINGER'S EQUATION

In this section we shall see how the preceding conclusions are obtained. Schrödinger's equation for the harmonic oscillator is, with $V = \frac{1}{2}kx^2$,

5.42
$$\frac{d^2\psi}{dx^2} + \frac{2m}{\hbar^2}\left(E - \frac{1}{2}kx^2\right)\psi = 0$$

It is convenient to simplify Eq. 5.42 by introducing the dimensionless quantities

$$y = \left(\frac{1}{\hbar}\sqrt{km}\right)^{1/2} x$$

5.43
$$= \sqrt{\frac{2\pi m\nu}{\hbar}} x$$

and

$$\alpha = \frac{2E}{\hbar}\sqrt{\frac{m}{k}}$$

5.44
$$= \frac{2E}{h\nu}$$

where ν is the classical frequency of the oscillation given by Eq. 5.38. In making these substitutions, what we have essentially done is change the units in which x and E are expressed from meters and joules, respectively, to appropriate dimensionless units. In terms of y and α Schrödinger's equation becomes

5.45
$$\frac{d^2\psi}{dy^2} + (\alpha - y^2)\psi = 0$$

We begin the solution of Eq. 5.45 by finding the asymptotic form that ψ must have as $y \to \pm\infty$. If any wave function ψ is to represent an actual particle localized in space, its value must approach zero as y approaches infinity in order that $\int_{-\infty}^{\infty} |\psi|^2 \, dy$ be a finite, nonvanishing quantity. Let us rewrite Eq. 5.45 as follows:

$$\frac{d^2\psi}{dy^2} - (y^2 - \alpha)\psi = 0$$

$$\frac{d^2\psi}{dy^2} = (y^2 - \alpha)\psi$$

$$\frac{d^2\psi/dy^2}{(y^2 - \alpha)\psi} = 1$$

As $y \to \infty$, $y^2 \gg \alpha$ and we have

5.46
$$\lim_{y\to\infty} \frac{d^2\psi/dy^2}{y^2\psi} = 1$$

A function ψ_∞ that satisfies Eq. 5.46 is

5.47
$$\psi_\infty = e^{-y^2/2}$$
since

$$\lim_{y\to\infty} \frac{d^2\psi_\infty}{dy^2} = \lim_{y\to\infty} (y^2 - 1)e^{-y^2/2} = y^2 e^{-y^2/2}$$

Equation 5.47 is the required asymptotic form of ψ.

We are now able to write

5.48
$$\psi = f(y)\psi_\infty$$
$$= f(y)e^{-y^2/2}$$

where $f(y)$ is a function of y that remains to be found. By inserting the ψ of Eq. 5.48 in Eq. 5.45 we obtain

5.49
$$\frac{d^2f}{dy^2} - 2y\frac{df}{dy} + (\alpha - 1)f = 0$$

which is the differential equation that f obeys.

The standard procedure for solving differential equations like Eq. 5.49 is to assume that $f(y)$ can be expanded in a power series in y, namely

$$f(y) = A_0 + A_1 y + A_2 y^2 + A_3 y^3 + \cdots$$

5.50
$$= \sum_{n=0}^{\infty} A_n y^n$$

and then to determine the values of the coefficients A_n. Differentiating f yields

$$\frac{df}{dy} = A_1 + 2A_2 y + 3A_3 y^2 + \cdots$$

$$= \sum_{n=1}^{\infty} nA_n y^{n-1}$$

By multiplying this equation by y we obtain

$$y\frac{df}{dy} = A_1 y + 2A_2 y^2 + 3A_3 y^3 + \cdots$$

5.51
$$= \sum_{n=0}^{\infty} nA_n y^n$$

The second derivative of f with respect to y is

$$\frac{d^2f}{dy^2} = 1 \cdot 2A_2 + 2 \cdot 3A_3 y + 3 \cdot 4A_4 y^2 + \cdots$$

$$= \sum_{n=2}^{\infty} n(n - 1)A_n y^{n-2}$$

which is equal to

5.52
$$\frac{d^2f}{dy^2} = \sum_{n=0}^{\infty} (n + 2)(n + 1)A_{n+2}y^n$$

(That the latter two series are indeed equal can be verified by working out the first terms of each.) We now substitute Eqs. 5.50 to 5.52 in Eq. 5.49 to obtain

5.53
$$\sum_{n=0}^{\infty} [(n + 2)(n + 1)A_{n+2} - (2n + 1 - \alpha)A_n]y^n = 0$$

In order for this equation to hold for all values of y, the quantity in brackets must be zero for all values of n. Hence we have the condition that

$$(n + 2)(n + 1)A_{n+2} = (2n + 1 - \alpha)A_n$$

and so

5.54
$$A_{n+2} = \frac{2n + 1 - \alpha}{(n + 2)(n + 1)} A_n$$

This *recursion formula* enables us to find the coefficients $A_2, A_3, A_4, \ldots$ in terms of A_0 and A_1. (Since Eq. 5.49 is a second-order differential equation, its solution has two arbitrary constants, which are A_0 and A_1 here.) Starting from A_0 we obtain the sequence of coefficients $A_2, A_4, A_6, \ldots$, and starting from A_1 we obtain the other sequence $A_3, A_5, A_7, \ldots$.

It is necessary for us to inquire into the behavior of

$$\psi = f(y)e^{-y^2/2}$$

as $y \to \infty$; only if $\psi \to 0$ as $y \to \infty$ can ψ be a physically acceptable wave function. Because $f(y)$ is multiplied by $e^{-y^2/2}$, ψ will meet this requirement provided that

$$\lim_{y \to \infty} f(y) < e^{y^2/2}$$

(As we shall see, it is unnecessary for us to specify just how much smaller f must be in the limit than $e^{y^2/2}$.)

A suitable way to compare the asymptotic behaviors of $f(y)$ and $e^{y^2/2}$ is to express the latter in a power series (f is already in the form of a power series) and to examine the ratio between successive coefficients of each series as $n \to \infty$. From the recursion formula of Eq. 5.54 we can tell by inspection that

$$\lim_{n \to \infty} \frac{A_{n+2}}{A_n} = \frac{2}{n}$$

Since

$$e^z = 1 + z + \frac{z^2}{2!} + \frac{z^3}{3!} + \cdots$$

we can express $e^{y^2/2}$ in a power series as

$$e^{y^2/2} = 1 + \frac{y^2}{2} + \frac{y^4}{2^2 \cdot 2!} + \frac{y^6}{2^3 \cdot 3!} + \cdots$$

$$= \sum_{n=0,2,4,\ldots}^{\infty} \frac{1}{2^{n/2}\left(\dfrac{n}{2}\right)!} y^n$$

$$= \sum_{n=0,2,4,\ldots}^{\infty} B_n y^n$$

The ratio between successive coefficients of y^n here is

$$\frac{B_{n+2}}{B_n} = \frac{2^{n/2}\left(\dfrac{n}{2}\right)!}{2^{(n+2)/2}\left(\dfrac{n+2}{2}\right)!} = \frac{2^{n/2}\left(\dfrac{n}{2}\right)!}{2 \cdot 2^{n/2}\left(\dfrac{n}{2}+1\right)\left(\dfrac{n}{2}\right)!}$$

$$= \frac{1}{2\left(\dfrac{n}{2}+1\right)} = \frac{1}{n+2}$$

In the limit of $n \to \infty$ this ratio becomes

$$\lim_{n\to\infty} \frac{B_{n+2}}{B_n} = \frac{1}{n}$$

Thus successive coefficients in the power series for f decrease *less* rapidly than those in the power series for $e^{y^2/2}$ instead of *more* rapidly, which means that $f(y)e^{-y^2/2}$ does not vanish as $y \to \infty$.

There is a simple way out of this dilemma. If the series representing f terminates at a certain value of n, so that all the coefficients A_n are zero for values of n higher than this one, ψ will go to zero as $y \to \infty$ because of the $e^{-y^2/2}$ factor. In other words, if f is a polynomial with a finite number of terms instead of an infinite series, it is acceptable. From the recursion formula

$$A_{n+2} = \frac{2n + 1 - \alpha}{(n+2)(n+1)} A_n$$

it is clear that if

5.55 $\qquad \alpha = 2n + 1$

for any value of n, then $A_{n+2} = A_{n+4} = A_{n+6} = \cdots = 0$, which is what we want.

(Equation 5.55 takes care of only one sequence of coefficients, either the sequence of even n starting with A_0 or the sequence of odd n starting with A_1. If n is even, it must be true that $A_1 = 0$ and only even powers of y appear in the polynomial, while if n is odd, it must be true that $A_0 = 0$ and only odd powers of y appear. We shall see the result later in this section, where the polynomial is tabulated for various values of n.)

The condition that $\alpha = 2n + 1$ is a necessary and sufficient condition for the wave equation 5.45 to have solutions that meet the various requirements that ψ must fulfill. From Eq. 5.44, the definition of α, we have

$$\alpha_n = \frac{2E}{h\nu} = 2n + 1$$

or

5.56 $\qquad E_n = \left(n + \frac{1}{2}\right)h\nu \qquad n = 0, 1, 2, \ldots$

This is the formula that was given as Eq. 5.40 in the preceding section.

For each choice of the parameter α_n there is a different wave function ψ_n. Each function consists of a polynomial $H_n(y)$ (called a *Hermite polynomial*) in either odd or even powers of y, the exponential factor $e^{-y^2/2}$, and a numerical coefficient which is needed for ψ_n to meet the normalization condition

$$\int_{-\infty}^{\infty} |\psi_n|^2 \, dy = 1 \qquad n = 0, 1, 2, \ldots$$

The general formula for the nth wave function is

5.57 $\qquad \psi_n = \left(\frac{2m\nu}{\hbar}\right)^{1/4} (2^n n!)^{-1/2} H_n(y) e^{-y^2/2}$

Table 5.1.

SOME HERMITE POLYNOMIALS.

n	$H_n(y)$	α_n	E_n
0	1	1	$\frac{1}{2}h\nu$
1	$2y$	3	$\frac{3}{2}h\nu$
2	$4y^2 - 2$	5	$\frac{5}{2}h\nu$
3	$8y^3 - 12y$	7	$\frac{7}{2}h\nu$
4	$16y^4 - 48y^2 + 12$	9	$\frac{9}{2}h\nu$
5	$32y^5 - 160y^3 + 120y$	11	$\frac{11}{2}h\nu$

The first six Hermite polynomials $H_n(y)$ are listed in Table 5.1, and the corresponding wave functions ψ_n are those that were plotted in Figs. 5-9 and 5-10 of the preceding section.

Problems

1. Verify that all solutions of the wave equation

$$\frac{\partial^2 y}{\partial x^2} = \frac{1}{v^2}\frac{\partial^2 y}{\partial t^2}$$

must be of the form $y = F(t \pm x/v)$ as asserted in Sec. 5.2.

2. If $\Psi_1(x,t)$ and $\Psi_2(x,t)$ are both solutions of Schrödinger's equation for a given potential $V(x)$, show that the linear combination

$$\Psi = a_1\Psi_1 + a_2\Psi_2$$

in which a_1 and a_2 are arbitrary constants is also a solution. (This result is in accord with the empirical observation of the interference of de Broglie waves, for instance in the Davisson-Germer experiment discussed in Chap. 3.)

3. Find the lowest energy of a neutron confined to a box 10^{-14} m across. (The size of a nucleus is of this order of magnitude.) 4.95×10^{10}

4. According to the correspondence principle, quantum theory should give the same results as classical physics in the limit of large quantum numbers. Show that, as $n \to \infty$, the probability of finding a particle trapped in a box between x and $x + dx$ is independent of x, which is the classical expectation.

5. Find the zero-point in electron volts of a pendulum whose period is 1 s.

6. An important property of the eigenfunctions of a system is that they are *orthogonal* to one another, which means that

$$\int_{-\infty}^{\infty} \psi_n\psi_m\, dV = 0 \qquad n \neq m$$

Verify this relationship for the eigenfunctions of a particle in a one-dimensional box with the help of the relationship $\sin\theta = (e^{i\theta} - e^{-i\theta})/2i$.

° 7. Show that the expectation values $\langle T \rangle$ and $\langle V \rangle$ of the kinetic and potential energies of a harmonic oscillator are given by $\langle T \rangle = \langle V \rangle = E_0/2$ when it is in the $n = 0$ state. (This is true for all states of a harmonic oscillator, in fact.) How does this result compare with the classical values of T and V?

*8.   Use the fact that $\alpha \geqslant 0$ (since $E \geqslant 0$) to show that the coefficients A_n of Eq. 5.50 are all zero for negative values of n.

*9.   Show that the first three harmonic-oscillator wave functions are normalized solutions of Schrödinger's equation.

10.   According to elementary classical physics, the total energy of a harmonic oscillator of mass m, frequency v, and amplitude A is $2\pi^2A^2v^2m$. Use the uncertainty principle to verify that the lowest possible energy of the oscillator is $hv/2$ by assuming that $\Delta x = A$.

11.   Which of the wave functions shown in Fig. 5-11 might conceivably have physical significance?

FIGURE 5-11

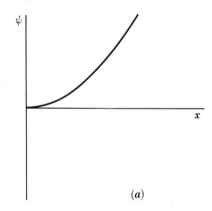

(a)

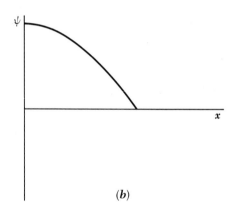

(b)

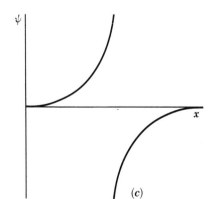

(c)

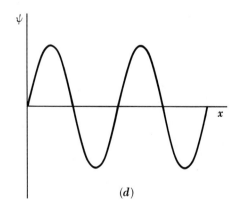

(d)

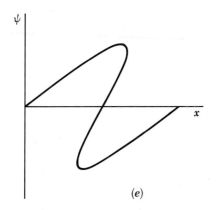 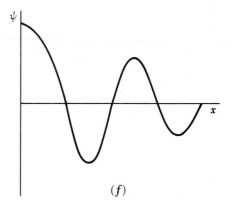

(e) (f)

FIGURE 5-11 (Continued)

QUANTUM THEORY OF THE HYDROGEN ATOM **6**

The quantum-mechanical theory of the atom, which was developed shortly after the formulation of quantum mechanics itself, represents an epochal contribution to our knowledge of the physical universe. Besides revolutionizing our approach to atomic phenomena, this theory has made it possible for us to understand such related matters as how atoms interact with one another to form stable molecules, the origin of the periodic table of the elements, and why solids are endowed with their characteristic electrical, magnetic, and mechanical properties, all topics we shall explore in later chapters. For the moment we shall concentrate on the quantum theory of the hydrogen atom and how its formal mathematical results may be interpreted in terms of familiar concepts.

6.1 SCHRÖDINGER'S EQUATION FOR THE HYDROGEN ATOM

A hydrogen atom consists of a proton, a particle of electric charge $+e$, and an electron, a particle of charge $-e$ which is 1,836 times lighter than the proton. For the sake of convenience we shall consider the proton to be stationary, with the electron moving about in its vicinity but prevented from escaping by the proton's electric field. (As in the Bohr theory, the correction for proton motion is simply a matter of replacing the electron mass m by the reduced mass m'.) Schrödinger's equation for the electron in three dimensions, which is what we must use for the hydrogen atom, is

6.1 $$\frac{\partial^2 \psi}{\partial x^2} + \frac{\partial^2 \psi}{\partial y^2} + \frac{\partial^2 \psi}{\partial z^2} + \frac{2m}{\hbar^2}(E - V)\psi = 0$$

The potential energy V here is the electrostatic potential energy

6.2 $$V = -\frac{e^2}{4\pi\varepsilon_0 r}$$

of a charge $-e$ when it is the distance r from another charge $+e$.

Since V is a function of r rather than of x, y, z, we cannot substitute Eq. 6.2 directly into Eq. 6.1. There are two alternatives: we can express V in terms of the cartesian coordinates x, y, z by replacing r by $\sqrt{x^2 + y^2 + z^2}$, or we can express Schrödinger's equation in terms of the spherical polar coordinates r, θ, ϕ defined in Fig. 6-1. As it happens, owing to the symmetry of the physical situation, doing the latter makes the problem considerably easier to solve.

The spherical polar coordinates r, θ, ϕ of the point P shown in Fig. 6-1 have the following interpretation:

r = length of radius vector from origin O to point P

$= \sqrt{x^2 + y^2 + z^2}$

θ = angle between radius vector and $+z$ axis

$=$ zenith angle

$= \cos^{-1} \dfrac{z}{\sqrt{x^2 + y^2 + z^2}}$

Spherical polar coordinates

ϕ = angle between the projection of the radius vector in the xy plane and the $+x$ axis, measured in the direction shown

$=$ azimuth angle

$= \tan^{-1} \dfrac{y}{x}$

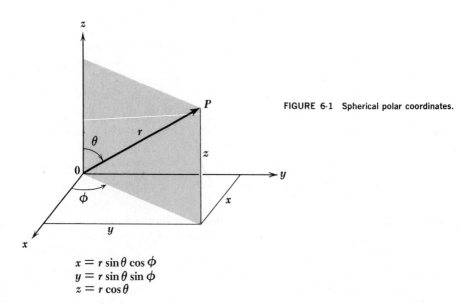

FIGURE 6-1 Spherical polar coordinates.

$x = r \sin\theta \cos\phi$
$y = r \sin\theta \sin\phi$
$z = r \cos\theta$

On the surface of a sphere whose center is at O, lines of constant zenith angle θ are like parallels of latitude on a globe (but we note that the value of θ of a point is *not* the same as its latitude; $\theta = 90°$ at the equator, for instance, but the latitude of the equator is $0°$), and lines of constant azimuth angle ϕ are like meridians of longitude (here the definitions coincide if the axis of the globe is taken as the $+z$ axis and the $+x$ axis is at $\phi = 0°$).

In spherical polar coordinates Schrödinger's equation becomes

6.3
$$\frac{1}{r^2}\frac{\partial}{\partial r}\left(r^2\frac{\partial\psi}{\partial r}\right) + \frac{1}{r^2\sin\theta}\frac{\partial}{\partial\theta}\left(\sin\theta\frac{\partial\psi}{\partial\theta}\right)$$

$$+ \frac{1}{r^2\sin^2\theta}\frac{\partial^2\psi}{\partial\phi^2} + \frac{2m}{\hbar^2}(E - V)\psi = 0$$

Substituting Eq. 6.2 for the potential energy V and multiplying the entire equation by $r^2\sin^2\theta$, we obtain

6.4
$$\sin^2\theta\frac{\partial}{\partial r}\left(r^2\frac{\partial\psi}{\partial r}\right) + \sin\theta\frac{\partial}{\partial\theta}\left(\sin\theta\frac{\partial\psi}{\partial\theta}\right) + \frac{\partial^2\psi}{\partial\phi^2}$$

$$+ \frac{2mr^2\sin^2\theta}{\hbar^2}\left(\frac{e^2}{4\pi\varepsilon_0 r} + E\right)\psi = 0 \qquad \textbf{Hydrogen atom}$$

Equation 6.4 is the partial differential equation for the wave function ψ of the electron in a hydrogen atom. Together with the various conditions ψ must obey, as discussed in Chap. 5 (for instance, that ψ have just one value at each point r, θ, ϕ), this equation completely specifies the behavior of the electron. In order to see just what this behavior is, we must solve Eq. 6.4 for ψ.

When Eq. 6.4 is solved, it turns out that three quantum numbers are required to describe the electron in a hydrogen atom, in place of the single quantum number of the Bohr theory. (In the next chapter we shall find that a fourth quantum number is needed to describe the spin of the electron.) In the Bohr model, the electron's motion is basically one-dimensional, since the only quantity that varies as it moves is its position in a definite orbit. One quantum number is enough to specify the state of such an electron, just as one quantum number is enough to specify the state of a particle in a one-dimensional box.

A particle in a three-dimensional box needs three quantum numbers for its description, since now there are three sets of boundary conditions that the particle's wave function ψ must obey: ψ must be 0 at the walls of the box in the x, y, and z directions independently. In a hydrogen atom the electron's motion is restricted by the inverse-square electric field of the nucleus instead of by the walls of a box, but nevertheless the electron is free to move in three dimensions, and it is accordingly not surprising that three quantum numbers govern its wave function also.

The three quantum numbers revealed by the solution of Eq. 6.4, together with their possible values, are as follows:

Principal quantum number $= n = 1, 2, 3, \ldots$
Orbital quantum number $\quad = l = 0, 1, 2, \ldots, n - 1$
Magnetic quantum number $= m_l = 0, \pm 1, \pm 2, \ldots, \pm l$

The principal quantum number n governs the total energy of the electron, and corresponds to the quantum number n of the Bohr theory. The orbital quantum number l governs the magnitude of the electron's angular momentum about the nucleus, and the magnetic quantum number m_l governs the direction of the angular momentum.

*6.2 SEPARATION OF VARIABLES

The virtue of writing Schrödinger's equation in spherical polar coordinates for the problem of the hydrogen atom is that in this form it may be readily separated into three independent equations, each involving only a single coordinate. The procedure is to look for solutions in which the wave function $\psi(r, \theta, \phi)$ has the form of a product of three different functions: $R(r)$, which depends upon r alone; $\Theta(\theta)$, which depends upon θ alone; and $\Phi(\phi)$, which depends upon ϕ alone. That is, we assume that

6.5 $\qquad \psi(r, \theta, \phi) = R(r)\Theta(\theta)\Phi(\phi)$ **Hydrogen atom wave function**

The function $R(r)$ describes how the wave function ψ of the electron varies along a radius vector from the nucleus, with θ and ϕ constant. The function $\Theta(\theta)$ describes how ψ varies with zenith angle θ along a meridian on a sphere centered at the nucleus, with r and ϕ constant. The function $\Phi(\phi)$ describes how ψ varies with azimuth angle ϕ along a parallel on a sphere centered at the nucleus, with r and θ constant.

From Eq. 6.5, which we may write more simply as

$$\psi = R\Theta\Phi$$

we see that

$$\frac{\partial \psi}{\partial r} = \Theta\Phi\frac{\partial R}{\partial r}$$

$$\frac{\partial \psi}{\partial \theta} = R\Phi\frac{\partial \Theta}{\partial \theta}$$

$$\frac{\partial^2 \psi}{\partial \phi^2} = R\Theta\frac{\partial^2 \Phi}{\partial \phi^2}$$

Hence, when we substitute $R\Theta\Phi$ for ψ in Schrödinger's equation for the hydrogen atom and divide the entire equation by $R\Theta\Phi$, we find that

6.6
$$\frac{\sin^2\theta}{R}\frac{\partial}{\partial r}\left(r^2\frac{\partial R}{\partial r}\right) + \frac{\sin\theta}{\Theta}\frac{\partial}{\partial\theta}\left(\sin\theta\frac{\partial\Theta}{\partial\theta}\right)$$
$$+ \frac{1}{\Phi}\frac{\partial^2\Phi}{\partial\phi^2} + \frac{2mr^2\sin^2\theta}{\hbar^2}\left(\frac{e^2}{4\pi\varepsilon_0 r} + E\right) = 0$$

The third term of Eq. 6.6 is a function of azimuth angle ϕ only, while the other terms are functions of r and θ only. Let us rearrange Eq. 6.6 to read

6.7
$$\frac{\sin^2\theta}{R}\frac{\partial}{\partial r}\left(r^2\frac{\partial R}{\partial r}\right) + \frac{\sin\theta}{\Theta}\frac{\partial}{\partial\theta}\left(\sin\theta\frac{\partial\Theta}{\partial\theta}\right)$$
$$+ \frac{2mr^2\sin^2\theta}{\hbar^2}\left(\frac{e^2}{4\pi\varepsilon_0 r} + E\right) = -\frac{1}{\Phi}\frac{\partial^2\Phi}{\partial\phi^2}$$

This equation can be correct only if both sides of it are equal to the same constant, since they are functions of *different* variables. As we shall see, it is convenient to call this constant m_l^2. The differential equation for the function Φ is therefore

6.8
$$-\frac{1}{\Phi}\frac{d^2\Phi}{d\phi^2} = m_l^2$$

When we substitute m_l^2 for the right-hand side of Eq. 6.7, divide the entire equation by $\sin^2\theta$, and rearrange the various terms, we find that

6.9
$$\frac{1}{R}\frac{\partial}{\partial r}\left(r^2\frac{\partial R}{\partial r}\right) + \frac{2mr^2}{\hbar^2}\left(\frac{e^2}{4\pi\varepsilon_0 r} + E\right)$$
$$= \frac{m_l^2}{\sin^2\theta} - \frac{1}{\Theta\sin\theta}\frac{\partial}{\partial\theta}\left(\sin\theta\frac{\partial\Theta}{\partial\theta}\right)$$

Again we have an equation in which different variables appear on each side, requiring that both sides be equal to the same constant. This constant we shall call $l(l+1)$, once more for reasons that will be apparent later. The equations for the functions Θ and R are therefore

6.10
$$\frac{m_l^2}{\sin^2\theta} - \frac{1}{\Theta\sin\theta}\frac{d}{d\theta}\left(\sin\theta\frac{d\Theta}{d\theta}\right) = l(l+1)$$

6.11
$$\frac{1}{R}\frac{d}{dr}\left(r^2\frac{dR}{dr}\right) + \frac{2mr^2}{\hbar^2}\left(\frac{e^2}{4\pi\varepsilon_0 r} + E\right) = l(l+1)$$

Equations 6.8, 6.10, and 6.11 are usually written

6.12
$$\frac{d^2\Phi}{d\phi^2} + m_l^2\Phi = 0$$

6.13
$$\frac{1}{\sin\theta}\frac{d}{d\theta}\left(\sin\theta\frac{d\Theta}{d\theta}\right) + \left[l(l+1) - \frac{m_l^2}{\sin^2\theta}\right]\Theta = 0$$

6.14
$$\frac{1}{r^2}\frac{d}{dr}\left(r^2\frac{dR}{dr}\right) + \left[\frac{2m}{\hbar^2}\left(\frac{e^2}{4\pi\varepsilon_0 r} + E\right) - \frac{l(l+1)}{r^2}\right]R = 0$$

Each of these is an ordinary differential equation for a single function of a single variable. We have therefore accomplished our task of simplifying Schrödinger's equation for the hydrogen atom, which began as a partial differential equation for a function ψ of three variables.

*6.3 QUANTUM NUMBERS

The first of the above equations, Eq. 6.12, is readily solved, with the result

6.15
$$\Phi(\phi) = Ae^{im_l\phi}$$

where A is the constant of integration. We have already stated that one of the conditions a wave function—and hence Φ, which is a component of the complete wave function ψ—must obey is that it have a single value at a given point in space. From Fig. 6-2 it is evident that ϕ and $\phi + 2\pi$ both identify the same

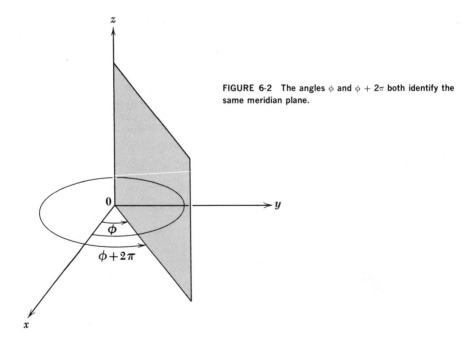

FIGURE 6-2 The angles ϕ and $\phi + 2\pi$ both identify the same meridian plane.

meridian plane. Hence it must be true that $\Phi(\phi) = \Phi(\phi + 2\pi)$, or

$$Ae^{im_l\phi} = Ae^{im_l(\phi+2\pi)}$$

which can only happen when m_l is 0 or a positive or negative integer (±1, ±2, ±3, . . .). The constant m_l is known as the *magnetic quantum number* of the hydrogen atom.

The differential equation 6.13 for $\Theta(\theta)$ has a rather complicated solution in terms of polynomials called the *associated Legendre functions.* For our present purpose, the important thing about these functions is that they exist only when the constant l is an integer equal to or greater than $|m_l|$, the absolute value of m_l. This requirement can be expressed as a condition on m_l in the form

$$m_l = 0, \pm1, \pm2, \ldots, \pm l$$

The constant l is known as the *orbital quantum number.*

The solution of the final equation, Eq. 6.14, for the radial part $R(r)$ of the hydrogen-atom wave function ψ is also complicated, being in terms of polynomials called the *associated Laguerre functions.* Equation 6.14 can be solved only when E is positive or has one of the negative values E_n (signifying that the electron is bound to the atom) specified by

6.16
$$E_n = -\frac{me^4}{32\pi^2\varepsilon_0^2\hbar^2}\left(\frac{1}{n^2}\right)$$

where n is an integer. We recognize that this is precisely the same formula for the energy levels of the hydrogen atom that Bohr obtained.

Another condition that must be obeyed in order to solve Eq. 6.14 is that n, known as the *principal quantum number,* must be equal to or greater than $l + 1$. This requirement may be expressed as a condition on l in the form

$$l = 0, 1, 2, \ldots, (n - 1)$$

Hence we may tabulate the three quantum numbers n, l, and m together with their permissible values as follows:

$$n = 1, 2, 3, \ldots \qquad \textbf{Principal quantum number}$$
6.17
$$l = 0, 1, 2, \ldots, (n - 1) \qquad \textbf{Orbital quantum number}$$
$$m_l = 0, \pm1, \pm2, \ldots, \pm l \qquad \textbf{Magnetic quantum number}$$

It is worth noting again how inevitably quantum numbers appear in quantum-mechanical theories of particles trapped in a particular region of space.

To exhibit the dependence of R, Θ, and Φ upon the quantum numbers n, l, m, we may write for the electron wave function

6.18 $\psi = R_{nl}\Theta_{lm_l}\Phi_{m_l}$

The wave functions R, Θ, and Φ together with ψ are given in Table 6.1 for $n = 1$, 2, and 3.

6.4 PRINCIPAL QUANTUM NUMBER

It is interesting to consider the interpretation of the hydrogen-atom quantum numbers in terms of the classical model of the atom. This model, as we saw in Chap. 4, corresponds exactly to planetary motion in the solar system except that the inverse-square force holding the electron to the nucleus is electrical rather than gravitational. Two quantities are *conserved*—that is, maintain a constant value at all times—in planetary motion, as Newton was able to show from Kepler's three empirical laws. These are the scalar *total energy* and the vector *angular momentum* of each planet.

Classically the total energy can have any value whatever, but it must, of course, be negative if the planet is to be trapped permanently in the solar system. In the quantum-mechanical theory of the hydrogen atom the electron energy is also a constant, but while it may have any positive value whatever, the *only* negative values it can have are specified by the formula

(6.16) $E_n = -\dfrac{me^4}{32\pi^2\varepsilon_0^2\hbar^2}\left(\dfrac{1}{n^2}\right)$

The theory of planetary motion can also be worked out from Schrödinger's equation, and it yields an energy restriction identical in form. However, the total quantum number n for any of the planets turns out to be so immense that the separation of permitted energy levels is far too small to be observable. For this reason classical physics provides an adequate description of planetary motion but fails within the atom.

The quantization of electron energy in the hydrogen atom is therefore described by the principal quantum number n.

6.5 ORBITAL QUANTUM NUMBER

The interpretation of the orbital quantum number l is a bit less obvious. Let us examine the differential equation for the radial part $R(r)$ of the wave function ψ:

(6.14) $\dfrac{1}{r^2}\dfrac{d}{dr}\left(r^2\dfrac{dR}{dr}\right) + \left[\dfrac{2m}{\hbar^2}\left(\dfrac{e^2}{4\pi\varepsilon_0 r} + E\right) - \dfrac{l(l+1)}{r^2}\right]R = 0$

Table 6.1.

NORMALIZED WAVE FUNCTIONS OF THE HYDROGEN ATOM FOR $n = 1$, 2, AND 3. The quantity $a_0 = 4\pi\epsilon_0\hbar^2/me^2 = 0.53$ Å is equal to the radius of the innermost Bohr orbit.

n	l	m_l	$\Phi(\phi)$	$\Theta(\theta)$	$R(r)$	$\psi(r,\theta,\phi)$
1	0	0	$\dfrac{1}{\sqrt{2\pi}}$	$\dfrac{1}{\sqrt{2}}$	$\dfrac{2}{a_0^{3/2}}\,e^{-r/a_0}$	$\dfrac{1}{\sqrt{\pi}\,a_0^{3/2}}\,e^{-r/a_0}$
2	0	0	$\dfrac{1}{\sqrt{2\pi}}$	$\dfrac{1}{\sqrt{2}}$	$\dfrac{1}{2\sqrt{2}\,a_0^{3/2}}\left(2-\dfrac{r}{a_0}\right)e^{-r/2a_0}$	$\dfrac{1}{4\sqrt{2\pi}\,a_0^{3/2}}\left(2-\dfrac{r}{a_0}\right)e^{-r/2a_0}$
2	1	0	$\dfrac{1}{\sqrt{2\pi}}$	$\dfrac{\sqrt{6}}{2}\cos\theta$	$\dfrac{1}{2\sqrt{6}\,a_0^{3/2}}\,\dfrac{r}{a_0}\,e^{-r/2a_0}$	$\dfrac{1}{4\sqrt{2\pi}\,a_0^{3/2}}\,\dfrac{r}{a_0}\,e^{-r/2a_0}\cos\theta$
2	1	± 1	$\dfrac{1}{\sqrt{2\pi}}\,e^{\pm i\phi}$	$\dfrac{\sqrt{3}}{2}\sin\theta$	$\dfrac{1}{2\sqrt{6}\,a_0^{3/2}}\,\dfrac{r}{a_0}\,e^{-r/2a_0}$	$\dfrac{1}{8\sqrt{\pi}\,a_0^{3/2}}\,\dfrac{r}{a_0}\,e^{-r/2a_0}\sin\theta\,e^{\pm i\phi}$
3	0	0	$\dfrac{1}{\sqrt{2\pi}}$	$\dfrac{1}{\sqrt{2}}$	$\dfrac{2}{81\sqrt{3}\,a_0^{3/2}}\left(27-18\,\dfrac{r}{a_0}+2\,\dfrac{r^2}{a_0^2}\right)e^{-r/3a_0}$	$\dfrac{1}{81\sqrt{3\pi}\,a_0^{3/2}}\left(27-18\,\dfrac{r}{a_0}+2\,\dfrac{r^2}{a_0^2}\right)e^{-r/3a_0}$
3	1	0	$\dfrac{1}{\sqrt{2\pi}}$	$\dfrac{\sqrt{6}}{2}\cos\theta$	$\dfrac{4}{81\sqrt{6}\,a_0^{3/2}}\left(6-\dfrac{r}{a_0}\right)\dfrac{r}{a_0}\,e^{-r/3a_0}$	$\dfrac{\sqrt{2}}{81\sqrt{\pi}\,a_0^{3/2}}\left(6-\dfrac{r}{a_0}\right)\dfrac{r}{a_0}\,e^{-r/3a_0}\cos\theta$
3	1	± 1	$\dfrac{1}{\sqrt{2\pi}}\,e^{\pm i\phi}$	$\dfrac{\sqrt{3}}{2}\sin\theta$	$\dfrac{4}{81\sqrt{6}\,a_0^{3/2}}\left(6-\dfrac{r}{a_0}\right)\dfrac{r}{a_0}\,e^{-r/3a_0}$	$\dfrac{1}{81\sqrt{\pi}\,a_0^{3/2}}\left(6-\dfrac{r}{a_0}\right)\dfrac{r}{a_0}\,e^{-r/3a_0}\sin\theta\,e^{\pm i\phi}$
3	2	0	$\dfrac{1}{\sqrt{2\pi}}$	$\dfrac{\sqrt{10}}{4}(3\cos^2\theta - 1)$	$\dfrac{4}{81\sqrt{30}\,a_0^{3/2}}\,\dfrac{r^2}{a_0^2}\,e^{-r/3a_0}$	$\dfrac{1}{81\sqrt{6\pi}\,a_0^{3/2}}\,\dfrac{r^2}{a_0^2}\,e^{-r/3a_0}(3\cos^2\theta - 1)$
3	2	± 1	$\dfrac{1}{\sqrt{2\pi}}\,e^{\pm i\phi}$	$\dfrac{\sqrt{15}}{2}\sin\theta\cos\theta$	$\dfrac{4}{81\sqrt{30}\,a_0^{3/2}}\,\dfrac{r^2}{a_0^2}\,e^{-r/3a_0}$	$\dfrac{1}{81\sqrt{\pi}\,a_0^{3/2}}\,\dfrac{r^2}{a_0^2}\,e^{-r/3a_0}\sin\theta\cos\theta\,e^{\pm i\phi}$
3	2	± 2	$\dfrac{1}{\sqrt{2\pi}}\,e^{\pm 2i\phi}$	$\dfrac{\sqrt{15}}{4}\sin^2\theta$	$\dfrac{4}{81\sqrt{30}\,a_0^{3/2}}\,\dfrac{r^2}{a_0^2}\,e^{-r/3a_0}$	$\dfrac{1}{162\sqrt{\pi}\,a_0^{3/2}}\,\dfrac{r^2}{a_0^2}\,e^{-r/3a_0}\sin^2\theta\,e^{\pm 2i\phi}$

This equation is solely concerned with the radial aspect of the electron's motion, that is, its motion toward or away from the nucleus; yet we notice the presence of E, the total electron energy, in it. The total energy E includes the electron's kinetic energy of orbital motion, which should have nothing to do with its radial motion.

This contradiction may be removed by the following argument. The kinetic energy T of the electron has two parts, T_{radial} due to its motion toward or away from the nucleus, and $T_{orbital}$ due to its motion around the nucleus. The potential energy V of the electron is the electrostatic energy

$$V = -\frac{e^2}{4\pi\varepsilon_0 r}$$

Hence the total energy of the electron is

$$E = T_{radial} + T_{orbital} + V$$
$$= T_{radial} + T_{orbital} - \frac{e^2}{4\pi\varepsilon_0 r}$$

Inserting this expression for E in Eq. 6.14 we obtain, after a slight rearrangement,

6.19
$$\frac{1}{r^2}\frac{d}{dr}\left(r^2\frac{dR}{dr}\right) + \frac{2m}{\hbar^2}\left[T_{radial} + T_{orbital} - \frac{\hbar^2 l(l+1)}{2mr^2}\right]R = 0$$

If the last two terms in the square brackets of this equation cancel each other out, we shall have what we want: a differential equation for $R(r)$ that involves functions of the radius vector r exclusively. We therefore require that

6.20
$$T_{orbital} = \frac{\hbar^2 l(l+1)}{2mr^2}$$

The orbital kinetic energy of the electron is

$$T_{orbital} = \tfrac{1}{2}mv^2{}_{orbital}$$

Since the angular momentum L of the electron is

$$L = mv_{orbital}r$$

we may write for the orbital kinetic energy

$$T_{orbital} = \frac{L^2}{2mr^2}$$

Hence, from Eq. 6.20,

$$\frac{L^2}{2mr^2} = \frac{\hbar^2 l(l + 1)}{2mr^2}$$

or

6.21 $$L = \sqrt{l(l + 1)}\,\hbar \qquad\qquad \text{**Electron angular momentum**}$$

Our interpretation of this result is that, since the orbital quantum number l is restricted to the values

$$l = 0, 1, 2, \ldots, (n - 1)$$

the electron can have only those particular angular momenta L specified by Eq. 6.21. Like total energy E, *angular momentum is both conserved and quantized.* The quantity

$$\hbar = h/2\pi = 1.054 \times 10^{-34} \text{ J-s}$$

is thus the natural unit of angular momentum.

In macroscopic planetary motion, once again, the quantum number describing angular momentum is so large that the separation into discrete angular-momentum states cannot be experimentally observed. For example, an electron (or, for that matter, any other body) whose orbital quantum number is 2 has the angular momentum

$$L = \sqrt{2(2 + 1)}\,\hbar$$
$$= \sqrt{6}\,\hbar$$
$$= 2.6 \times 10^{-34} \text{ J-s}$$

By contrast the orbital angular momentum of the earth is 2.7×10^{40} J-s!

It is customary to specify angular-momentum states by a letter, with s corresponding to $l = 0$, p to $l = 1$, and so on according to the following scheme:

$l = 0$	1	2	3	4	5	6 ...	
s	p	d	f	g	h	i ...	**Angular-momentum states**

This peculiar code originated in the empirical classification of spectra into series called sharp, principal, diffuse, and fundamental which occurred before the theory of the atom was developed. Thus an s state is one with no angular momentum, a p state has the angular momentum $\sqrt{2}\,\hbar$, etc.

The combination of the total quantum number with the letter that represents orbital angular momentum provides a convenient and widely used notation for atomic states. In this notation a state in which $n = 2$, $l = 0$ is a $2s$ state, for example, and one in which $n = 4$, $l = 2$ is a $4d$ state. Table 6.2 gives the designations of atomic states in hydrogen through $n = 6$, $l = 5$.

Table 6.2.

THE SYMBOLIC DESIGNATION OF ATOMIC STATES IN HYDROGEN.

	s $l=0$	p $l=1$	d $l=2$	f $l=3$	g $l=4$	h $l=5$
$n=1$	$1s$					
$n=2$	$2s$	$2p$				
$n=3$	$3s$	$3p$	$3d$			
$n=4$	$4s$	$4p$	$4d$	$4f$		
$n=5$	$5s$	$5p$	$5d$	$5f$	$5g$	
$n=6$	$6s$	$6p$	$6d$	$6f$	$6g$	$6h$

6.6 MAGNETIC QUANTUM NUMBER

The orbital quantum number l determines the *magnitude* of the electron's angular momentum. Angular momentum, however, like linear momentum, is a vector quantity, and so to describe it completely requires that its *direction* be specified as well as its magnitude. (The vector **L**, we recall, is perpendicular to the plane in which the rotational motion takes place, and its sense is given by the right-hand rule: when the fingers of the right hand point in the direction of the motion, the thumb is in the direction of **L**. This rule is illustrated in Fig. 6-3.)

What possible significance can a direction in space have for a hydrogen atom? The answer becomes clear when we reflect that an electron revolving about a nucleus is a minute current loop and has a magnetic field like that of a magnetic dipole. Hence an atomic electron that possesses angular momentum interacts with an external magnetic field **B**. The magnetic quantum number m_l specifies

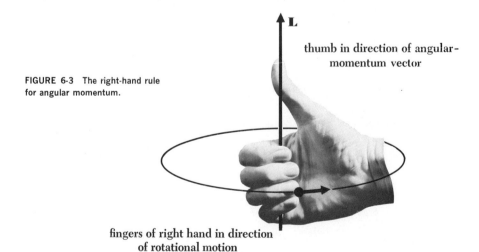

FIGURE 6-3 The right-hand rule for angular momentum.

L

thumb in direction of angular-momentum vector

fingers of right hand in direction
of rotational motion

the direction of **L** by determining the component of **L** in the field direction. This phenomenon is often referred to as *space quantization*.

If we let the magnetic-field direction be parallel to the z axis, the component of **L** in this direction is

6.22 $$L_z = m_l\hbar$$ **Space quantization**

The possible values of m_l for a given value of l range from $+l$ through 0 to $-l$, so that the number of possible orientations of the angular-momentum vector **L** in a magnetic field is $2l + 1$. When $l = 0$, L_z can have only the single value of 0; when $l = 1$, L_z may be $\hbar$, 0, or $-\hbar$; when $l = 2$, L_z may be $2\hbar$, $\hbar$, 0, $-\hbar$, or $-2\hbar$; and so on. We note that **L** can never be aligned exactly parallel or antiparallel to **B**, since L_z is always smaller than the magnitude $\sqrt{l(l + 1)}\,\hbar$ of the total angular momentum.

The space quantization of the orbital angular momentum of the hydrogen atom is shown in Fig. 6-4. We must regard an atom characterized by a certain value

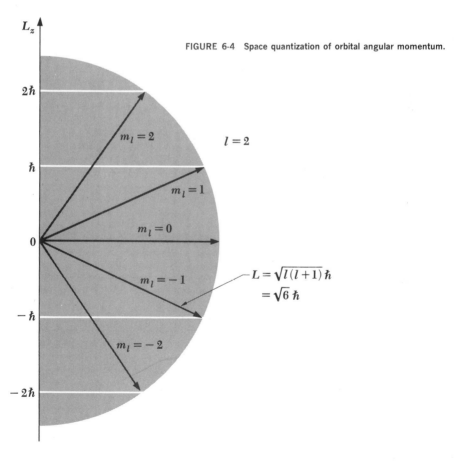

FIGURE 6-4   Space quantization of orbital angular momentum.

of m_l as standing ready to assume a certain orientation of its angular momentum **L** relative to an external magnetic field in the event it finds itself in such a field.

In the absence of an external magnetic field, the direction of the z axis is entirely arbitrary. Hence it must be true that the component of **L** in *any* direction we choose is $m_l \hbar$; the significance of an external magnetic field is that it provides an experimentally meaningful reference direction. A magnetic field is not the only such reference direction possible. For example, the line between the two H atoms in the hydrogen molecule H_2 is just as experimentally meaningful as the direction of a magnetic field, and along this line the components of the angular momenta of the H atoms are determined by their m_l values.

Why is only one component of **L** quantized? The answer is closely related to the fact that **L** can never point in any specific direction z but instead traces out a cone in space such that its projection L_z is $m_l \hbar$. The reason for the latter phenomenon is the uncertainty principle: if **L** were fixed in space, so that L_x and L_y as well as L_z had definite values, the electron would be confined to a definite plane. For instance, if **L** were in the z direction, the electron would have to be in the xy plane at all times (Fig. 6-5a). This can occur only if the electron's momentum component p_z in the z direction is infinitely uncertain, which of course is impossible if it is to be part of a hydrogen atom. However, since in reality only *one* component L_z of **L** together with its magnitude L have definite values and $|L| > |L_z|$, the electron is not limited to a single plane (Fig.

FIGURE 6-5 The uncertainty principle prohibits the angular-momentum vector L from having a definite direction in space.

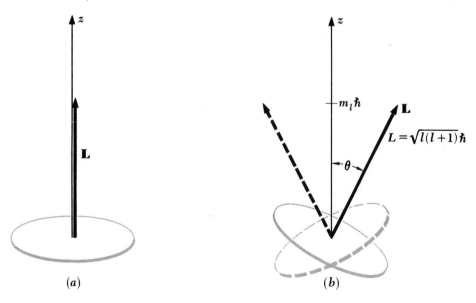

(a) (b)

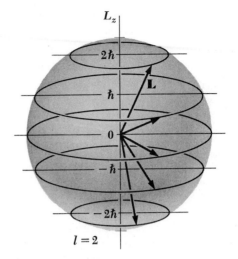

FIGURE 6-6 The angular-momentum vector L precesses constantly about the z axis.

6-5b), and there is a built-in uncertainty, as it were, in the electron's z coordinate. The direction of **L** is constantly changing, as in Fig. 6-6, and so the average values of L_x and L_y are 0, although L_z always has the specific value $m_l\hbar$.

6.7 THE NORMAL ZEEMAN EFFECT

In an external magnetic field **B**, a magnetic dipole has an amount of potential energy V_m that depends upon both the magnitude μ of its magnetic moment and the orientation of this moment with respect to the field (Fig. 6-7).

The torque τ on a magnetic dipole in a magnetic field of flux density **B** is

$$\tau = \mu B \sin \theta$$

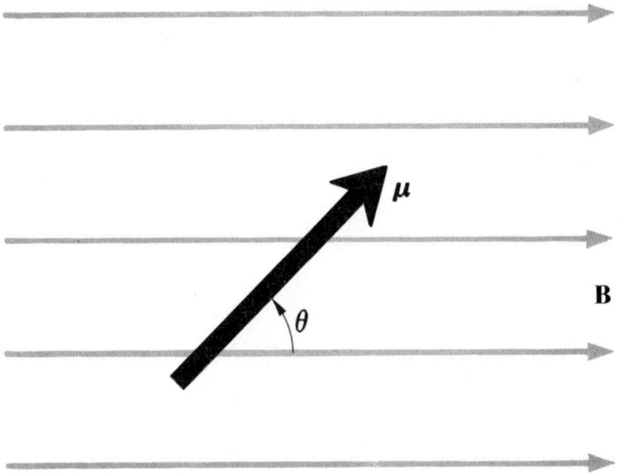

FIGURE 6-7 A magnetic dipole of moment μ at the angle θ relative to a magnetic field B.

where θ is the angle between μ and **B**. The torque is a maximum when the dipole is perpendicular to the field, and zero when it is parallel or antiparallel to it. To calculate the potential energy V_m, we must first establish a reference configuration at which V_m is zero by definition. (Since only *changes* in potential energy are ever experimentally observed, the choice of a reference configuration is arbitrary.) It is convenient to set $V_m = 0$ when $\theta = 90°$, that is, when μ is perpendicular to **B**. The potential energy at any other orientation of μ is equal to the external work that must be done to rotate the dipole from $\theta_0 = 90°$ to the angle θ characterizing that orientation. Hence

$$V_m = \int_{90°}^{\theta} \tau \, d\theta$$

$$= \mu B \int_{90°}^{\theta} \sin \theta \, d\theta$$

6.23 $$= -\mu B \cos \theta$$

When μ points in the same direction as **B**, then, $V_m = -\mu B$, its minimum value. This is a natural consequence of the fact that a magnetic dipole tends to align itself with an external magnetic field.

Since the magnetic moment of the orbital electron in a hydrogen atom depends upon its angular momentum **L**, both the magnitude of **L** and its orientation with respect to the field determine the extent of the magnetic contribution to the total energy of the atom when it is in a magnetic field. The magnetic moment of a current loop is

$$\mu = iA$$

where i is the current and A the area it encloses. An electron which makes ν rev/s in a circular orbit of radius r is equivalent to a current of $-e\nu$ (since the electronic charge is $-e$), and its magnetic moment is therefore

$$\mu = -e\nu\pi r^2$$

The linear speed v of the electron is $2\pi\nu r$, and so its angular momentum is

$$L = mvr$$
$$= 2\pi m\nu r^2$$

Comparing the formulas for magnetic moment μ and angular momentum L shows that

6.24 $$\mu = -\left(\frac{e}{2m}\right)\mathbf{L}$$ **Electron magnetic moment**

for an orbital electron. The quantity $(-e/2m)$, which involves the charge and mass of the electron only, is called its *gyromagnetic ratio*. The minus sign means that μ is in the opposite direction to **L**. While the above expression for the magnetic moment of an orbital electron has been obtained by a classical calculation, quantum mechanics yields the same result. The magnetic potential energy of an atom in a magnetic field is therefore

6.25 $$V_m = \left(\frac{e}{2m}\right) LB \cos\theta$$

which is a function of both B and θ.

From Fig. 6-5 we see that the angle θ between **L** and the z direction can have only the values specified by the relation

$$\cos\theta = \frac{m_l}{\sqrt{l(l+1)}}$$

while the permitted values of L are specified by

$$L = \sqrt{l(l+1)}\,\hbar$$

To find the magnetic energy that an atom of magnetic quantum number m_l has when it is in a magnetic field **B,** we insert the above expressions for $\cos\theta$ and L in Eq. 6.25, which yields

6.26 $$V_m = m_l\left(\frac{e\hbar}{2m}\right) B \qquad\qquad \textbf{Magnetic energy}$$

The quantity $e\hbar/2m$ is called the *Bohr magneton;* its value is 9.27×10^{-24} J/tesla (T).

In a magnetic field, then, the energy of a particular atomic state depends upon the value of m_l as well as upon that of n. A state of total quantum number n breaks up into several substates when the atom is in a magnetic field, and their energies are slightly more or slightly less than the energy of the state in the absence of the field. This phenomenon leads to a "splitting" of individual spectral lines into separate lines when atoms radiate in a magnetic field, with the spacing of the lines dependent upon the magnitude of the field. The splitting of spectral lines by a magnetic field is called the *Zeeman effect* after the Dutch physicist Zeeman, who first observed it in 1896. The Zeeman effect is a vivid confirmation of space quantization; it is further discussed in Chap. 7.

6.8 ELECTRON PROBABILITY DENSITY

In Bohr's model of the hydrogen atom the electron is visualized as revolving around the nucleus in a circular path. This model is pictured in a spherical polar coordinate system in Fig. 6-8. We see it implies that, if a suitable experiment

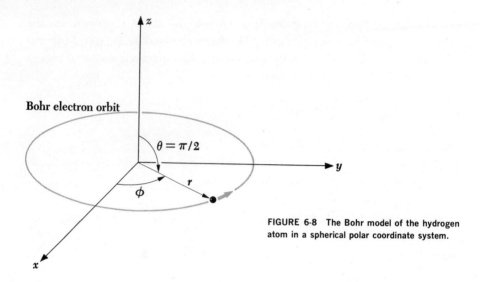

FIGURE 6-8   The Bohr model of the hydrogen atom in a spherical polar coordinate system.

were performed, the electron would always be found a distance of $r = n^2 a_0$ (where n is the quantum number of the orbit and $a_0 = 0.53$ Å is the radius of the innermost orbit) from the nucleus and in the equatorial plane $\theta = 90°$, while its azimuth angle ϕ changes with time.

The quantum theory of the hydrogen atom modifies the straightforward prediction of the Bohr model in two ways. First, no definite values for r, θ, or ϕ can be given, but only the relative probabilities for finding the electron at various locations. This imprecision is, of course, a consequence of the wave nature of the electron. Second, we cannot even think of the electron as moving around the nucleus in any conventional sense since the probability density $|\psi|^2$ is independent of time and may vary considerably from place to place.

The electron wave function ψ in a hydrogen atom is given by

$$\psi = R\Theta\Phi$$

where

$$R = R_{nl}(r)$$

describes how ψ varies with r when the orbital and total quantum numbers have the values n and l;

$$\Theta = \Theta_{lm_l}(\theta)$$

describes how ψ varies with θ when the magnetic and orbital quantum numbers have the values l and m_l; and

$$\Phi = \Phi_{m_l}(\phi)$$

190   THE ATOM

describes how ψ varies with ϕ when the magnetic quantum number is m_l. The probability density $|\psi|^2$ may therefore be written

6.27 $\qquad |\psi|^2 = |R|^2|\Theta|^2|\Phi|^2$

where it is understood that the square of any function is to be replaced by the product of it and its complex conjugate if the function is a complex quantity.

The azimuthal probability density $|\Phi|^2$, which is a measure of the likelihood of finding the electron at a particular azimuth angle ϕ, is a constant that does not depend upon ϕ at all. Hence the electron's probability density is symmetrical about the z axis regardless of the quantum state it is in, so that the electron has the same chance of being found at one angle ϕ as at another.

The radial part R of the wave function, in contrast to Φ, not only varies with r but does so in a different way for each combination of quantum numbers n and l. Figure 6-9 contains graphs of R versus r for $1s$, $2s$, $2p$, $3s$, $3p$, and $3d$ staues of the hydrogen atom. Evidently R is a maximum at $r = 0$—that is, at the nucleus itself—for all s states, while it is zero at $r = 0$ for states that possess angular momentum.

The *probability density* of the electron at the point r, θ, ϕ is proportional to $|\psi|^2$, but the *actual probability* of finding it in the infinitesimal volume element dV there is $|\psi|^2 dV$. Now in spherical polar coordinates

$$dV = r^2 \sin\theta \, dr \, d\theta \, d\phi$$

so that, since Θ and Φ are normalized functions, the actual numerical probability $P(r)dr$ of finding the electron in a hydrogen atom somewhere between r and $r + dr$ from the nucleus is

$$P(r)dr = r^2|R|^2 dr \int_0^\pi |\Theta|^2 \sin\theta \, d\theta \int_0^{2\pi} |\Phi|^2 d\phi$$

6.28 $\qquad\qquad = r^2\,|R|^2\,dr$

Equation 6.28 is plotted in Fig. 6-10 for the same states whose radial functions R were shown in Fig. 6-9; the curves are quite different as a rule. We note immediately that P is not a maximum at the nucleus for s states, as R itself is, but has its maximum at a finite distance from it. Interestingly enough, the most probable value of r for a $1s$ electron is exactly a_0, the orbital radius of a ground-state electron in the Bohr model. However, the *average* value of r for a $1s$ electron is $1.5a_0$, which seems puzzling at first sight because the energy levels are the same in both the quantum-mechanical and Bohr atomic models. This apparent discrepancy is removed when we recall that the electron energy depends upon $1/r$ rather than upon r directly, and, as it happens, the average value of $1/r$ for a $1s$ electron is exactly $1/a_0$.

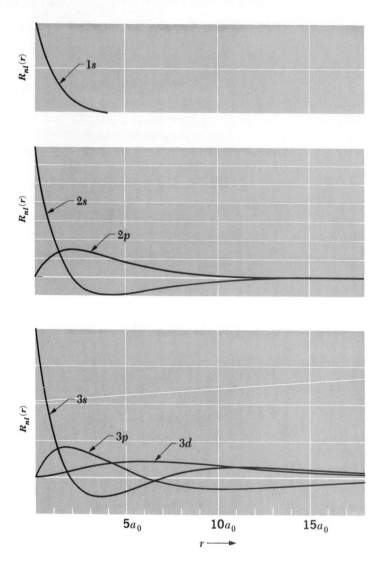

FIGURE 6-9 The variation with distance from the nucleus of the radial part of the electron wave function in hydrogen for various quantum states. The quantity $a_0 = 4\pi\varepsilon_0\hbar^2/me^2 = 0.53$ Å is the radius of the first Bohr orbit.

The function Θ varies with zenith angle θ for all quantum numbers l and m_l except $l = m_l = 0$, which are s states. The probability density $|\Theta|^2$ for an s state is a constant ($\frac{1}{2}$, in fact), which means that, since $|\Phi|^2$ is also a constant, the electron probability density $|\psi|^2$ has the same value at a given r in all directions. Electrons in other states, however, do have angular preferences, sometimes quite complicated ones. This can be seen in Fig. 6-11, in which electron probability densities as functions of r and θ are shown for several atomic states. (The quantity plotted is $|\psi|^2$, not $|\psi|^2\, dV$.) Since $|\psi|^2$ is independent of ϕ, a three-dimensional picture of $|\psi|^2$ is obtained by rotating a particular representation about a vertical axis. When this is done, the probability densities for s states are evidently spherically symmetric, while others are not. The pronounced lobe patterns characteristic of many of the states turn out to be significant in chemistry since these patterns help determine the manner in which adjacent atoms in a molecule interact; we shall refer to this notion once more in Chap. 8.

A study of Fig. 6-11 also reveals quantum-mechanical states with a remarkable resemblance to those of the Bohr model. The electron probability-density distribution for a $2p$ state with $m_l = \pm 1$, for instance, is like a doughnut in the equatorial plane centered at the nucleus, and calculation shows the most probable distance of the electron from the nucleus to be $4r_0$—precisely the radius of the Bohr orbit for the same total quantum number. Similar correspondences exist for $3d$ states with $m_l = \pm 2$, $4f$ states with $m_l = \pm 3$, and so on: in every case the highest angular momentum possible for that energy level, and in every case the angular momentum vector as near the z axis as possible so that the probability density be as close as possible to the equatorial plane. Thus the Bohr model predicts the most probable location of the electron in *one* of the several possible states in each energy level.

FIGURE 6-10 The probability of finding the electron in a hydrogen atom at a distance between r and $r + dr$ from the nucleus for the quantum states of Fig. 6-9.

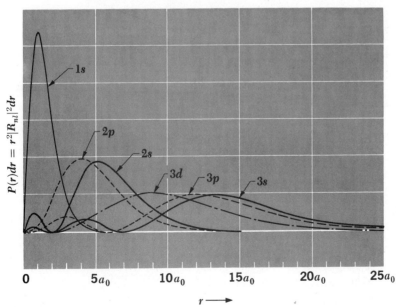

FIGURE 6-11 Photographic representation of the electron probability density distribution $|\psi|^2$ for several energy states. These may be regarded as sectional views of the distributions in a plane containing the polar axis, which is vertical and in the plane of the paper. The scale varies from figure to figure.

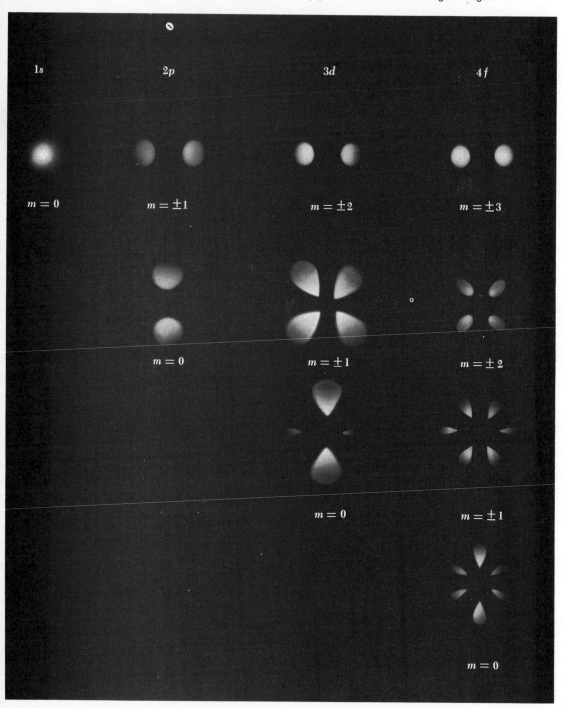

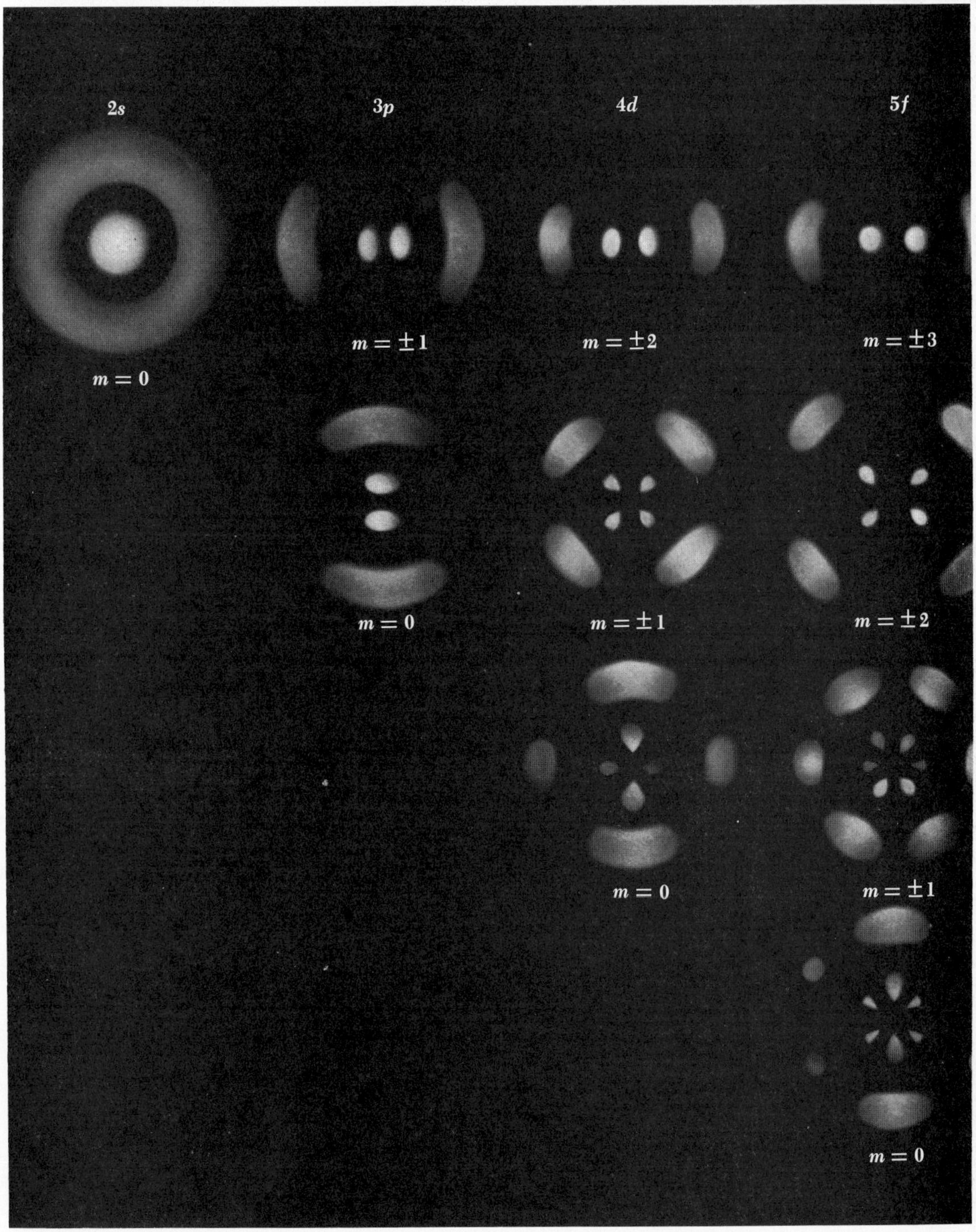

2s

3p

4d

5f

$m = 0$

$m = \pm 1$

$m = \pm 2$

$m = \pm 3$

$m = 0$

$m = \pm 1$

$m = \pm 2$

$m = 0$

$m = \pm 1$

$m = 0$

6.9 RADIATIVE TRANSITIONS

In formulating his theory of the hydrogen atom, Bohr was obliged to postulate that the frequency ν of the radiation emitted by an atom dropping from an energy level E_m to a lower level E_n is

$$\nu = \frac{E_m - E_n}{h}$$

It is not difficult to show that this relationship arises naturally from the quantum theory of the atom we have been discussing. Our starting point is the straightforward assertion that an atomic electron whose average position relative to the nucleus is constant in time does not radiate, while if its average position relative to the nucleus oscillates, electromagnetic waves are emitted whose frequency is the same as that of the oscillation. For simplicity we shall consider the component of electron motion in the x direction only.

The time-dependent wave function ψ_n of an electron in a state of quantum number n and energy E_n is the product of a time-independent wave function ψ_n and a time-varying function whose frequency is

$$\nu_n = \frac{E_n}{h}$$

Hence

6.29 $$\Psi_n = \psi_n e^{-(iE_n/\hbar)t}$$

and

6.30 $$\Psi_n^* = \psi_n^* e^{+(iE_n/\hbar)t}$$

The average position of such an electron is the expectation value (Sec. 5.4) of x, namely,

6.31 $$\langle x \rangle = \int_{-\infty}^{\infty} x\Psi_n^*\Psi_n \, dx$$

Inserting the wave functions of Eqs. 6.29 and 6.30,

$$\langle x \rangle = \int_{-\infty}^{\infty} x\psi_n^*\psi_n e^{[(iE_n/\hbar)-(iE_n/\hbar)]t} \, dx$$

6.32 $$= \int_{-\infty}^{\infty} x\psi_n^*\psi_n \, dx$$

which is constant in time since ψ_n and ψ_n^* are, by definition, functions of position only. The electron does not oscillate, and no radiation occurs. Thus quantum mechanics predicts that an atom in a specific quantum state does not radiate; this agrees with observation, though not with classical physics.

We are now in a position to consider an electron changing from one energy state to another. Let us formulate a definite problem: an atom is in its ground state when, at $t = 0$, an excitation process of some kind (a beam of radiation, say, or collisions with other particles) begins to act upon it. Subsequently we find that the atom emits radiation corresponding to a transition from an excited state of energy E_m to the ground state, and we conclude that at some time during the intervening period the atom existed in the state m. What is the frequency of the radiation?

The wave function Ψ of an electron capable of existing in states n or m may be written

6.33 $$\Psi = a\Psi_n + b\Psi_m$$

where $a*a$ is the probability that the electron is in state n and $b*b$ the probability that it is in state m. Of course, it must always be true that $a*a + b*b = 1$. At $t = 0$, $a = 1$ and $b = 0$ by hypothesis; when the electron is in the excited state, $a = 0$ and $b = 1$; and ultimately $a = 1$ and $b = 0$ once more. While the electron is in either state, there is no radiation, but when it is in the midst of the transition from m to n (that is, when both a and b have nonvanishing values), electromagnetic waves are produced. Substituting the composite wave function of Eq. 6.33 into Eq. 6.31, we obtain for the average electron position

$$\langle x \rangle = \int_{-\infty}^{\infty} x(a*\Psi_n^* + b*\Psi_m^*)(a\Psi_n + b\Psi_m)\, dx$$

6.34 $$= \int_{-\infty}^{\infty} x(a^2\Psi_n^*\Psi_n + b*a\Psi_m^*\Psi_n + a*b\Psi_n^*\Psi_m + b^2\Psi_m^*\Psi_m)\, dx$$

(Here, as before, we let $a*a = a^2$ and $b*b = b^2$.) The first and last integrals are constants, according to Eq. 6.32, and so the second and third integrals are the only ones capable of contributing to a time variation in $\langle x \rangle$.

With the help of Eqs. 6.29 to 6.31 we may expand Eq. 6.34 to obtain

6.35 $$\langle x \rangle = a^2 \int_{-\infty}^{\infty} x\psi_n^*\psi_n\, dx + b*a \int_{-\infty}^{\infty} x\psi_m^* e^{+(iE_m/\hbar)t}\psi_n e^{-(iE_n/\hbar)t}\, dx$$

$$+ a*b \int_{-\infty}^{\infty} x\psi_n^* e^{+(iE_n/\hbar)t}\psi_m e^{-(iE_m/\hbar)t}\, dx + b^2 \int_{-\infty}^{\infty} x\psi_m^*\psi_m\, dx$$

In the case of a finite bound system of two states, which is what we have here,

$$\psi_n^*\psi_m = \psi_m^*\psi_n \quad \text{and} \quad a*b = b*a$$

and so we can combine the time-varying terms of Eq. 6.35 into the single term

6.36 $$a*b \int_{-\infty}^{\infty} x\psi_n^*\psi_m [e^{(i/\hbar)(E_m - E_n)t} + e^{-(i/\hbar)(E_m - E_n)t}]\, dx$$

Now

$$e^{i\theta} + e^{-i\theta} = 2\cos\theta$$

Hence Eq. 6.36 simplifies to

$$2a*b \cos\left(\frac{E_m - E_n}{\hbar}\right) t \int_{-\infty}^{\infty} x\psi_n^*\psi_m \, dx$$

which contains the time-varying factor

$$\cos\left(\frac{E_m - E_n}{\hbar}\right) t = \cos 2\pi \left(\frac{E_m - E_n}{h}\right) t$$

$$= \cos 2\pi\nu t$$

The electron's position therefore oscillates sinusoidally at the frequency

6.37
$$\nu = \frac{E_m - E_n}{h}$$

and the full expression for $\langle x \rangle$, the average position of the electron, is

6.38
$$\langle x \rangle = a^2 \int_{-\infty}^{\infty} x\psi_n^*\psi_n \, dx + b^2 \int_{-\infty}^{\infty} x\psi_m^*\psi_m \, dx$$

$$+ 2a*b \cos 2\pi\nu t \int_{-\infty}^{\infty} x\psi_n^*\psi_m \, dx$$

When the electron is in state n or state m, the probabilities b^2 or a^2 respectively are zero, and the expectation value of the electron's position is constant. When the electron is undergoing a transition between these states, its position oscillates with the frequency ν. This frequency is identical with that postulated by Bohr and verified by experiment, and, as we have seen, Eq. 6.37 can be derived using quantum mechanics without making any special assumptions.

It is interesting to note that the frequency of the radiation is the same frequency as the beats we might imagine to be produced if the electron simultaneously existed in both the n and m states, whose characteristic frequencies are respectively E_n/h and E_m/h.

6.10 SELECTION RULES

It was not necessary for us to know the values of the probabilities a and b as functions of time, nor the electron wave functions ψ_n and ψ_m, in order to determine ν. We must know these quantities, however, if we wish to compute the likelihood for a given transition to occur. The general condition necessary for an atom in an excited state to radiate is that the integral

$$\int_{-\infty}^{\infty} x \psi_n \psi_m^* \, dx$$

not be zero, since the intensity of the radiation is proportional to it. Transitions for which this integral is finite are called *allowed transitions*, while those for which it is zero are called *forbidden transitions*.

In the case of the hydrogen atom, three quantum numbers are needed to specify the initial and final states involved in a radiative transition. If the total, orbital, and magnetic quantum numbers of the initial state are n', l', m_l' respectively and those of the final state are n, l, m_l, and the coordinate u represents either the x, y, or z coordinate, the condition for an allowed transition is

6.39
$$\int_{-\infty}^{\infty} u \psi_{n,l,m_l} \psi_{n',l',m_l'}^* \, du \neq 0$$

When u is taken as x, for example, the radiation referred to is that which would be produced by an ordinary dipole antenna lying along the x axis. Since the wave functions ψ_{n,l,m_l} for the hydrogen atom are known, Eq. 6.39 can be evaluated for $u = x$, $u = y$, and $u = z$ for all pairs of states differing in one or more quantum numbers. When this is done, it is found that the only transitions that can occur are those in which the orbital quantum number l changes by $+1$ or -1 and the magnetic quantum number m_l does not change or changes by $+1$ or -1; in other words, the condition for an allowed transition is that

6.40 $\Delta l = \pm 1$

Selection rules

6.41 $\Delta m_l = 0, \pm 1$

The change in total quantum number n is not restricted. Equations 6.40 and 6.41 are known as the *selection rules* for allowed transitions.

In order to get an intuitive idea of the physical basis for these selection rules, let us refer again to Fig. 6-11. There we see that, for instance, a transition from a 2p state to a 1s state involves a change from one probability-density distribution to another such that the oscillating charge during the transitions behaves like an electric dipole antenna. On the other hand, a transition from a 2s state to a 1s state involves a change from a spherically symmetric probability-density distribution to another spherically symmetric distribution, which means that the oscillations that take place are like those of a charged sphere that alternately expands and contracts. Oscillations of this kind do not lead to the radiation of electromagnetic waves.

The selection rule requiring that l change by ± 1 if an atom is to radiate means that an emitted photon carries off angular momentum equal to the difference between the angular momenta of the atom's initial and final states. The classical analog of a photon with angular momentum is a circularly polarized electromagnetic wave, so that this notion is not unique with quantum theory.

The preceding analysis of radiative transitions in an atom is based on a mixture of classical and quantum concepts. As we have seen, the expectation value of the position of an atomic electron oscillates at the frequency ν of Eq. 6.37 while passing from an initial eigenstate to another one of lower energy. Classically such an oscillating charge gives rise to electromagnetic waves of the same frequency ν, and indeed the observed radiation has this frequency. However, classical concepts are not always reliable guides to atomic processes, and a deeper treatment is required. Such a treatment, called *quantum electrodynamics*, modifies the preceding picture by showing that a single photon of energy $h\nu$ is emitted during the transition from state m to state n instead of electric-dipole radiation in all directions except that of the electron's motion, which would be the classical prediction.

In addition, quantum electrodynamics provides an explanation for the mechanism that causes the "spontaneous" transition of an atom from one energy state to a lower one. All electric and magnetic fields turn out to fluctuate constantly about the $\mathbf{E}$ and $\mathbf{B}$ that would be expected on purely classical grounds. Such fluctuations occur even when electromagnetic waves are absent and when, classically, $\mathbf{E} = \mathbf{B} = 0$. It is these fluctuations (often called "vacuum fluctuations" and analogous in a sense to the zero-point vibrations of a harmonic oscillator) that induce the "spontaneous" emission of photons by atoms in excited states.

Problems

1. Verify that Eqs. 6.1 and 6.3 are equivalent.

2. Show that

$$\Theta_{20}(\theta) = \frac{\sqrt{10}}{4}(3 \cos^2 \theta - 1)$$

is a solution of Eq. 6.13 and that it is normalized.

3. Show that

$$R_{10}(r) = \frac{2}{a_0^{3/2}} e^{-r/a_0}$$

is a solution of Eq. 6.14 and that it is normalized.

4. In Sec. 6.8 it is stated that the probability P of finding the electron in a

hydrogen atom between r and $r + dr$ from the nucleus is $P \, dr = r^2 |R_{nl}|^2 \, dr$. Verify that $P \, dr$ has its maximum value for a 1s electron at $r = a_0$, the Bohr radius.

√5. According to Fig. 6-10, $P \, dr$ has *two* maxima for a 2s electron. Find the values of r at which these maxima occur.

6. The wave function for a hydrogen atom in a 2p state varies with direction as well as with distance from the nucleus. In the case of a 2p electron for which $m_l = 0$, where does P have its maximum in the z direction? In the xy plane?

√7. The probability of finding an atomic electron whose radial wave function is $R(r)$ outside a sphere of radius r_0 centered on the nucleus is

$$\int_{r_0}^{\infty} |R(r)|^2 r^2 \, dr$$

The wave function $R_{10}(r)$ of Prob. 3 corresponds to the ground state of a hydrogen atom, and a_0 there is the radius of the Bohr orbit corresponding to that state. (a) Calculate the probability of finding a ground-state electron in a hydrogen atom at a distance greater than a_0 from the nucleus. (b) When the electron in a ground-state hydrogen atom is $2a_0$ from the nucleus, all its energy is potential energy. According to classical physics, the electron therefore cannot ever exceed the distance $2a_0$ from the nucleus. Find the probability that $r > 2a_0$ for the electron in a ground-state hydrogen atom.

8. *Unsöld's theorem* states that, for any value of the orbital quantum number l, the probability densities summed over all possible states from $m_l = -l$ to $m_l = +l$ yield a constant independent of angles θ or ϕ; that is,

$$\sum_{m_l=-l}^{+l} |\Theta|^2 |\Phi|^2 = \text{constant}$$

This theorem means that every closed subshell atom or ion (Sec. 7.5) has a spherically symmetric distribution of electric charge. Verify Unsöld's theorem for $l = 0$, $l = 1$, and $l = 2$ with the help of Table 6.1.

9. Find the percentage difference between L and the maximum value of L_z for an atomic electron in p, d, and f states.

°10. The selection rule for transitions between states in a harmonic oscillator is $\Delta n = \pm 1$. (a) Justify this rule on classical grounds. (b) Verify from the relevant wave functions that the transitions $n = 1 \rightarrow n = 0$ and $n = 1 \rightarrow n = 2$ are possible for a harmonic oscillator while $n = 1 \rightarrow n = 3$ is prohibited.

11. With the help of the wave functions listed in Table 6.1 verify that $\Delta l = \pm 1$ for $n = 2 \rightarrow n = 1$ transitions in the hydrogen atom.

MANY-ELECTRON ATOMS 7

Despite the accuracy with which the quantum theory accounts for certain of the properties of the hydrogen atom, and despite the elegance and essential simplicity of this theory, it cannot approach a complete description of this atom or of other atoms without the further hypothesis of electron spin and the exclusion principle associated with it. In this chapter we shall be introduced to the role of electron spin in atomic phenomena and to the reason why the exclusion principle is the key to understanding the structures of complex atomic systems.

7.1 ELECTRON SPIN

Let us begin by citing two of the most conspicuous shortcomings of the theory developed in the preceding chapter. The first is the experimental fact that many spectral lines actually consist of two separate lines that are very close together. An example of this *fine structure* is the first line of the Balmer series of hydrogen, which arises from transitions between the $n = 3$ and $n = 2$ levels in hydrogen atoms. Here the theoretical prediction is for a single line of wavelength 6,563 Å, while in reality there are two lines 1.4 Å apart—a small effect, but a conspicuous failure for the theory.

The second major discrepancy between the simple quantum theory of the atom and the experimental data occurs in the Zeeman effect. In Sec. 6.7 we saw that a hydrogen atom of magnetic quantum number m_l has the magnetic energy

7.1
$$V_m = m_l \frac{e\hbar}{2m} B$$

when it is located in a magnetic field of flux density B. Now m_l can have the $2l + 1$ values of $+l$ through 0 to $-l$, so a state of given orbital quantum number l is split into $2l + 1$ substates differing in energy by $(e\hbar/2m)B$ when the atom is in a magnetic field. However, because changes in m_l are restricted to $\Delta m_l = 0$, ± 1, a given spectral line that arises from a transition between two states of different l is split into only three components, as shown in Fig. 7-1. The *normal Zeeman effect*, then, consists of the splitting of a spectral line of frequency ν_0 into three components whose frequencies are

$$\nu_1 = \nu_0 - \frac{e\hbar}{2m}\frac{B}{h} = \nu_0 - \frac{e}{4\pi m}B$$

7.2 $\qquad\qquad \nu_2 = \nu_0$ **Normal Zeeman effect**

$$\nu_3 = \nu_0 + \frac{e\hbar}{2m}\frac{B}{h} = \nu_0 + \frac{e}{4\pi m}B$$

While the normal Zeeman effect is indeed observed in the spectra of a few elements under certain circumstances, more often it is not: four, six, or even more components may appear, and even when three components are present their spacing may not agree with Eq. 7.2. Several anomalous Zeeman patterns are shown in Fig. 7-2 together with the predictions of Eq. 7.2.

FIGURE 7-1 The normal Zeeman effect.

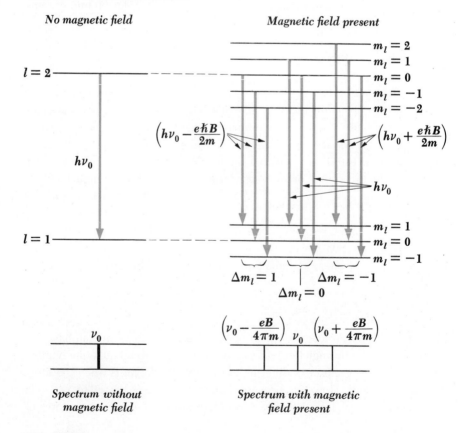

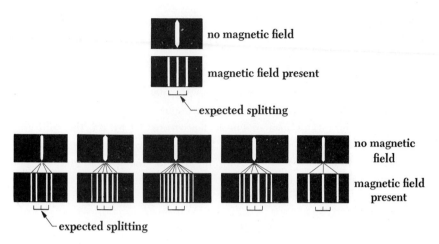

FIGURE 7-2 The normal and anomalous Zeeman effects in various spectral lines.

In an effort to account for both fine structure in spectral lines and the anomalous Zeeman effect, S. A. Goudsmit and G. E. Uhlenbeck proposed in 1925 that **the electron possesses an intrinsic angular momentum independent of any orbital angular momentum it might have and, associated with this angular momentum, a certain magnetic moment.** What Goudsmit and Uhlenbeck had in mind was a classical picture of an electron as a charged sphere spinning on its axis. The rotation involves angular momentum, and because the electron is negatively charged, it has a magnetic moment μ_s opposite in direction to its angular momentum vector $\mathbf{L}_s$. The notion of electron spin proved to be successful in explaining not only fine structure and the anomalous Zeeman effect but a wide variety of other atomic effects as well.

Of course, the idea that electrons are spinning charged spheres is hardly in accord with quantum mechanics, but in 1928 Dirac was able to show on the basis of a relativistic quantum-theoretical treatment that particles having the charge and mass of the electron must have just the intrinsic angular momentum and magnetic moment attributed to them by Goudsmit and Uhlenbeck.

The quantum number s is used to describe the spin angular momentum of the electron. The only value s can have is $s = \frac{1}{2}$; this restriction follows from Dirac's theory and, as we shall see below, may also be obtained empirically from spectral data. The magnitude S of the angular momentum due to electron spin is given in terms of the spin quantum number s by the formula

$$S = \sqrt{s(s + 1)}\,\hbar$$

7.3
$$= \frac{\sqrt{3}}{2}\,\hbar$$

which is the same formula as that giving the magnitude L of the orbital angular momentum in terms of the orbital quantum number l:

$$L = \sqrt{l(l + 1)}\,\hbar$$

The space quantization of electron spin is described by the spin magnetic quantum number m_s. Just as the orbital angular-momentum vector can have the $2l + 1$ orientations in a magnetic field from $+l$ to $-l$, the spin angular-momentum vector can have the $2s + 1 = 2$ orientations specified by $m_s = +\frac{1}{2}$ and $m_s = -\frac{1}{2}$ (Fig. 7-3). The component S_z of the spin angular momentum of an electron along a magnetic field in the z direction is determined by the

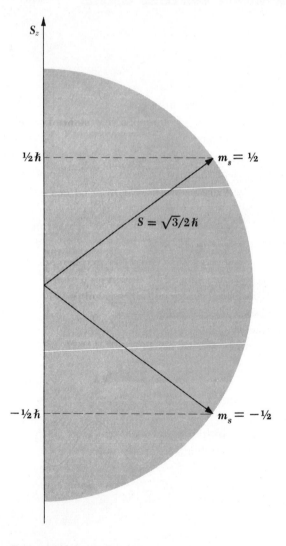

FIGURE 7-3 The two possible orientations of the spin angular-momentum vector.

spin magnetic quantum number, so that

7.4
$$S_z = m_s \hbar$$
$$= \pm \tfrac{1}{2}\hbar$$

The gyromagnetic ratio characteristic of electron spin is almost exactly twice that characteristic of electron orbital motion. Thus, taking this ratio as equal to 2, the spin magnetic moment μ_s of an electron is related to its spin angular momentum $\mathbf{S}$ by

7.5
$$\mu_s = -\frac{e}{m}\mathbf{S}$$

The possible components of μ_s along any axis, say the z axis, are therefore limited to

7.6
$$\mu_{sz} = \pm\frac{e\hbar}{2m}$$

We recognize the quantity $(e\hbar/2m)$ as the Bohr magneton.

Space quantization was first explicitly demonstrated by O. Stern and W. Gerlach in 1921. They directed a beam of neutral silver atoms from an oven through a set of collimating slits into an inhomogeneous magnetic field, as shown in Fig. 7-4. A photographic plate recorded the configuration of the beam after its passage through the field. In its normal state, the entire magnetic moment of a silver atom is due to the spin of one of its electrons. In a uniform magnetic field, such a dipole would merely experience a torque tending to align it with the field. In an inhomogeneous field, however, each "pole" of the dipole is subject to a force of different magnitude, and therefore there is a resultant force on the dipole that varies with its orientation relative to the field. Classically, all orientations should be present in a beam of atoms, which would result merely in a broad trace on the photographic plate instead of the thin line formed in the absence of any magnetic field. Stern and Gerlach found, however, that the initial beam split into two distinct parts, corresponding to the two opposite spin orientations in the magnetic field that are permitted by space quantization.

7.2 SPIN-ORBIT COUPLING

The fine-structure doubling of spectral lines may be explained on the basis of a magnetic interaction between the spin and orbital angular momenta of atomic electrons. This spin-orbit coupling can be understood in terms of a straightforward classical model. An electron revolving about a proton finds itself in a magnetic field because, in its own frame of reference, the proton is circling about *it*.

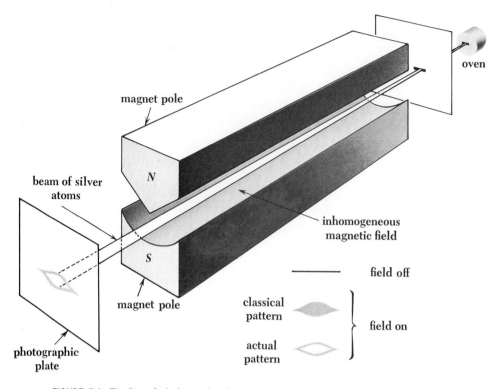

FIGURE 7-4 The Stern-Gerlach experiment.

This magnetic field then acts upon the electron's own spin magnetic moment to produce a kind of internal Zeeman effect. The energy V_m of a magnetic dipole of moment μ in a magnetic field of flux density $\mathbf{B}$ is, in general,

7.7 $$V_m = -\mu B \cos \theta$$

where θ is the angle between μ and $\mathbf{B}$. The quantity $\mu \cos \theta$ is the component of μ parallel to $\mathbf{B}$, which in the case of the spin magnetic moment of the electron is μ_{sz}. Hence, letting

$$\mu \cos \theta = \mu_{sz} = \pm \frac{e\hbar}{2m}$$

we find that

7.8 $$V_m = \pm \frac{e\hbar}{2m} B$$

Depending upon the orientation of its spin vector, the energy of an electron in a given atomic quantum state will be higher or lower by $(e\hbar/2m)B$ than its energy in the absence of spin-orbit coupling. The result is the splitting of every quantum state (except s states) into two separate substates and, consequently, the splitting of every spectral line into two component lines.

The assignment of $s = \frac{1}{2}$ is the only one that conforms to the observed fine-structure doubling. The fact that what should be single states are in fact twin states imposes the condition that the $2s + 1$ possible orientations of the spin angular-momentum vector $\mathbf{S}$ must total 2. Hence

$$2s + 1 = 2$$
$$s = \frac{1}{2}$$

To check whether the observed fine structure in spectral lines corresponds to the energy shifts predicted by Eq. 7.8, we must compute the magnitude B of the magnetic field experienced by an atomic electron. An estimate is easy to obtain. A circular wire loop of radius r that carries the current i has a magnetic field of flux density

$$B = \frac{\mu_0 i}{2r}$$

at its center. An orbital electron, say in a hydrogen atom, "sees" itself circled f times each second by a proton of charge $+e$, for a resulting flux density of

$$B = \frac{\mu_0 f e}{2r}$$

In the ground-state Bohr atom $f = 6.8 \times 10^{15}$ and $r = 5.3 \times 10^{-11}$ m, so that

$$B \approx 13 \text{ T}$$

which is a very strong magnetic field. The value of the Bohr magneton is

$$\frac{e\hbar}{2m} = 9.27 \times 10^{-24} \text{ J/T}$$

Hence the magnetic energy V_m of such an electron is, from Eq. 7.8,

$$V_m = \frac{e\hbar}{2m} B$$
$$\approx 9.27 \times 10^{-24} \text{ J/T} \times 13 \text{ T}$$
$$\approx 1.2 \times 10^{-22} \text{ J}$$

The wavelength shift corresponding to such a change in energy is about 2 Å for a spectral line of unperturbed wavelength 6,563 Å, somewhat more than the observed splitting of the line originating in the $n = 3 \rightarrow n = 2$ transition. However, the flux density of the magnetic field at orbits of higher order is less than for ground-state orbits, which accounts for the discrepancy.

7.3 THE EXCLUSION PRINCIPLE

In the normal configuration of a hydrogen atom, the electron is in its lowest quantum state. What are the normal configurations of more complex atoms? Are all 92 electrons of a uranium atom in the same quantum state, to be envisioned perhaps as circling the nucleus crowded together in a single Bohr orbit? Many lines of evidence make this hypothesis unlikely. One example is the great difference in chemical behavior exhibited by certain elements whose atomic structures differ by just one electron: for instance, the elements having atomic numbers 9, 10, and 11 are respectively the halogen gas fluorine, the inert gas neon, and the alkali metal sodium. Since the electronic structure of an atom controls its interactions with other atoms, it is hard to understand why the chemical properties of the elements should change so abruptly with a small change in atomic number if all the electrons in an atom exist together in the same quantum state.

In 1925 Wolfgang Pauli discovered the fundamental principle that governs the electronic configurations of atoms having more than one electron. His *exclusion principle* states that **no two electrons in an atom can exist in the same quantum state.** Each electron in an atom must have a different set of quantum numbers n, l, m_l, m_s.

Pauli was led to the exclusion principle by a study of atomic spectra. It is possible to determine the various states of an atom from its spectrum, and the quantum numbers of these states can be inferred. In the spectra of every element but hydrogen a number of lines are *missing* that correspond to transitions to and from states having certain combinations of quantum numbers. Thus no transitions are observed in helium to or from the ground-state configuration in which the spins of both electrons are in the same direction to give a total spin of 1, although transitions *are* observed to and from the other ground-state configuration, in which the spins are in opposite directions to give a total spin of 0. In the absent state the quantum numbers of *both* electrons would be $n = 1$, $l = 0$, $m_l = 0$, $m_s = \frac{1}{2}$, while in the state known to exist one of the electrons has $m_s = \frac{1}{2}$ and the other $m_s = -\frac{1}{2}$. Pauli showed that every unobserved atomic state involves two or more electrons with identical quantum numbers, and the exclusion principle is a statement of this empirical finding.

Before we explore the role of the exclusion principle in determining atomic structures, let us look into its quantum-mechanical implications. We saw in the previous chapter that the complete wave function ψ of the electron in a hydrogen atom can be expressed as the product of three separate wave functions, each describing that part of ψ which is a function of one of the three coordinates r, θ, ϕ. It is possible to show in an analogous way that the complete wave function

$\psi(1, 2, 3, \ldots, n)$ of a system of n particles can be approximately expressed as the product of the wave functions $\psi(1), \psi(2), \psi(3), \ldots, \psi(n)$ of the individual particles. That is,

7.9 $\qquad \psi(1, 2, 3, \ldots, n) = \psi(1)\,\psi(2)\,\psi(3)\ldots\psi(n)$

We shall use this result to investigate the kinds of wave functions that can be used to describe a system of two identical particles.

Let us suppose that one of the particles is in quantum state a and the other in state b. Because the particles are identical, it should make no difference in the probability density $|\psi|^2$ of the system if the particles are exchanged, with the one in state a replacing the one in state b and vice versa. Symbolically, we require that

7.10 $\qquad |\psi|^2(1,2) = |\psi|^2(2,1)$

Hence the wave function $\psi(2,1)$, representing the exchanged particles, can be given by either

7.11 $\qquad \psi(2,1) = \psi(1,2)$ $\qquad\qquad\qquad\qquad\qquad$ **Symmetric**

or

7.12 $\qquad \psi(2,1) = -\psi(1,2)$ $\qquad\qquad\qquad\qquad$ **Antisymmetric**

and still fulfill Eq. 7.10. The wave function of the system is not itself a measurable quantity, and so it can be altered in sign by the exchange of the particles. Wave functions that are unaffected by an exchange of particles are said to be *symmetric*, while those that reverse sign upon such an exchange are said to be *antisymmetric*.

If particle 1 is in state a and particle 2 is in state b, the wave function of the system is, according to Eq. 7.9,

7.13a $\qquad \psi_I = \psi_a(1)\,\psi_b(2)$

while if particle 2 is in state a and particle 1 is in state b, the wave function is

7.13b $\qquad \psi_{II} = \psi_a(2)\,\psi_b(1)$

Because the two particles are in fact indistinguishable, we have no way of knowing at any moment whether ψ_I or ψ_{II} describes the system. The likelihood that ψ_I is correct at any moment is the same as the likelihood that ψ_{II} is correct. Equivalently, we can say that the system spends half the time in the configuration whose wave function is ψ_I and the other half in the configuration whose wave function is ψ_{II}. Therefore a linear combination of ψ_I and ψ_{II} is the proper description of the system. There are two such combinations possible—

the symmetric one

7.14
$$\psi_S = \frac{1}{\sqrt{2}}[\psi_a(1)\,\psi_b(2) + \psi_a(2)\,\psi_b(1)]$$

and the antisymmetric one

7.15
$$\psi_A = \frac{1}{\sqrt{2}}[\psi_a(1)\,\psi_b(2) - \psi_a(2)\,\psi_b(1)]$$

The factor $1/\sqrt{2}$ is required to normalize ψ_S and ψ_A. Exchanging particles 1 and 2 leaves ψ_S unaffected, while it reverses the sign of ψ_A. Both ψ_S and ψ_A obey Eq. 7.10.

There are a number of important distinctions between the behavior of particles in systems whose wave functions are symmetric and that of particles in systems whose wave functions are antisymmetric. The most obvious is that, in the former case, both particles 1 and 2 can simultaneously exist in the same state, with $a = b$, while in the latter case, if we set $a = b$, we find that $\psi_A = 0$: the two particles *cannot* be in the same quantum state. Comparing this conclusion with Pauli's empirical exclusion principle, which states that no two electrons in an atom can be in the same quantum state, we conclude that systems of electrons are described by wave functions that reverse sign upon the exchange of any pair of them.

The results of various experiments show that *all* particles which have a spin of $\frac{1}{2}$ have wave functions that are antisymmetric to an exchange of any pair of them. Such particles, which include protons and neutrons as well as electrons, obey the exclusion principle when they are in the same system; that is, when they move in a common force field, each member of the system must be in a different quantum state. Particles of spin $\frac{1}{2}$ are often referred to as *Fermi particles* or *fermions* because, as we shall learn in Chap. 9, the behavior of aggregates of them is governed by a statistical distribution law discovered by Fermi and Dirac.

Particles whose spins are 0 or an integer have wave functions that are symmetric to an exchange of any pair of them. These particles do not obey the exclusion principle. Particles of 0 or integral spin are often referred to as *Bose particles* or *bosons* because the statistical distribution law that describes aggregates of them was discovered by Bose and Einstein. Photons, alpha particles, and helium atoms are Bose particles.

There are other important consequences of the symmetry or antisymmetry of particle wave functions besides that expressed in the exclusion principle. It is these consequences that make it useful to classify particles according to the nature of their wave functions rather than simply according to whether or not they obey the exclusion principle.

7.4 ELECTRON CONFIGURATIONS

Two basic rules determine the electron structures of many-electron atoms:

1. A system of particles is stable when its total energy is a minimum.
2. Only one electron can exist in any particular quantum state in an atom.

Before we apply these rules to actual atoms, let us examine the variation of electron energy with quantum state.

While the various electrons in a complex atom certainly interact directly with one another, much about atomic structure can be understood by simply considering each electron as though it exists in a constant mean force field. For a given electron this field is approximately the electric field of the nuclear charge Ze decreased by the partial shielding of those other electrons that are closer to the nucleus. The electrons that have the same principal quantum number n are, on the average, roughly the same distance from the nucleus. These electrons therefore interact with virtually the same electric field and have similar energies. It is conventional to speak of such electrons as occupying the same atomic *shell*. Shells are denoted by capital letters according to the following scheme:

$$n = 1 \quad 2 \quad 3 \quad 4 \quad 5 \ldots \qquad \textbf{Atomic shells}$$
$$K \quad L \quad M \quad N \quad O \ldots$$

The energy of an electron in a particular shell also depends to a certain extent upon its orbital quantum number l, though this dependence is not so great as that upon n. In a complex atom the degree to which the full nuclear charge is shielded from a given electron by intervening shells of other electrons varies with its probability-density distribution. An electron of small l is more likely to be found near the nucleus (where it is poorly shielded by the other electrons) than one of higher l (see Fig. 6-11), which results in a lower total energy (that is, higher binding energy) for it. The electrons in each shell accordingly increase in energy with increasing l. This effect is illustrated in Fig. 7-5, which is a plot of the binding energies of various atomic electrons as a function of atomic number.

Electrons that share a certain value of l in a shell are said to occupy the same *subshell*. All the electrons in a subshell have almost identical energies, since the dependence of electron energy upon m_l and m_s is comparatively minor.

The occupancy of the various subshells in an atom is usually expressed with the help of the notation introduced in the previous chapter for the various quantum states of the hydrogen atom. As indicated in Table 6.2, each subshell is identified by its total quantum number n followed by the letter corresponding to its orbital quantum number l. A superscript after the letter indicates the

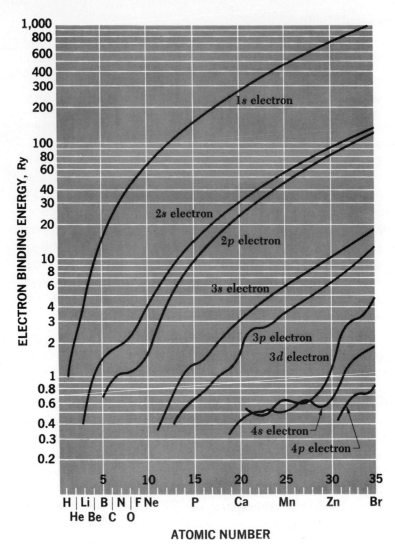

FIGURE 7-5 The binding energies of atomic electrons in Ry. (1 Ry = 1 Rydberg = 13.6 eV = ground-state energy of H atom.)

number of electrons in that subshell. For example, the electron configuration of sodium is written

$$1s^2 2s^2 2p^6 3s^1$$

which means that the $1s$ ($n = 1$, $l = 0$) and $2s$ ($n = 2$, $l = 0$) subshells contain two electrons each, the $2p$ ($n = 2$, $l = 1$) subshell contains six electrons, and the $3s$ ($n = 3$, $l = 0$) subshell contains one electron.

7.5 THE PERIODIC TABLE

When the elements are listed in order of atomic number, elements with similar chemical and physical properties recur at regular intervals. This empirical observation, known as the *periodic law*, was first formulated by Dmitri Mendeleev about a century ago. A tabular arrangement of the elements exhibiting this recurrence of properties is called a *periodic table*. Table 7.1 is perhaps the simplest form of periodic table; though more elaborate periodic tables have been devised to exhibit the periodic law in finer detail, Table 7.1 is adequate for our purposes.

Elements with similar properties form the *groups* shown as vertical columns in Table 7.1. Thus group I consists of hydrogen plus the alkali metals, all of which are extremely active chemically and all of which have valences of +1. Group VII consists of the halogens, volatile, active nonmetals that have valences of −1 and form diatomic molecules in the gaseous state. Group VIII consists of the inert gases, elements so inactive that not only do they almost never form compounds with other elements, but their atoms do not join together into molecules like the atoms of other gases.

The horizontal rows in Table 7.1 are called *periods*. Across each period is a more or less steady transition from an active metal through less active metals and weakly active nonmetals to highly active nonmetals and finally to an inert gas. Within each column there are also regular changes in properties, but they are far less conspicuous than those in each period. For example, increasing atomic number in the alkali metals is accompanied by greater chemical activity, while the reverse is true in the halogens.

A series of *transition elements* appears in each period after the third between the group II and group III elements. The transition elements are metals with a considerable chemical resemblance to one another but no pronounced resemblance to the elements in the major groups. Fifteen of the transition elements in period 6 are virtually indistinguishable in their properties, and are known as the *lanthanide* elements (or *rare earths*). A similar group of closely related metals, the *actinide* elements, is found in period 7.

The notion of electron shells and subshells fits perfectly into the pattern of the periodic table, which is just a mirror of the atomic structures of the elements. Let us see how this pattern arises.

The exclusion principle places definite limits on the number of electrons that can occupy a given subshell. A subshell is characterized by a certain total quantum number n and orbital quantum number l, where

$$l = 0, 1, 2, \ldots, (n - 1)$$

There are $2l + 1$ different values of the magnetic quantum number m_l for any

Table 7.1.

THE PERIODIC TABLE OF THE ELEMENTS. The number above the symbol of each element is its atomic number, and that below is its atomic mass in u. The elements whose atomic masses are given in parentheses do not occur in nature, but have been prepared artificially in nuclear reactions. The atomic mass in such a case is the mass number of the most long-lived radioactive isotope of the element.

Period	Group I	Group II											Group III	Group IV	Group V	Group VI	Group VII	Group VIII
1	1 H 1.00																	2 He 4.00
2	3 Li 6.94	4 Be 9.01											5 B 10.81	6 C 12.01	7 N 14.01	8 O 16.00	9 F 19.00	10 Ne 20.18
3	11 Na 22.99	12 Mg 24.31											13 Al 26.98	14 Si 28.09	15 P 30.98	16 S 32.07	17 Cl 35.46	18 Ar 39.94
4	19 K 39.10	20 Ca 40.08	21 Sc 44.96	22 Ti 47.90	23 V 50.94	24 Cr 52.00	25 Mn 54.94	26 Fe 55.85	27 Co 58.93	28 Ni 58.71	29 Cu 63.54	30 Zn 65.37	31 Ga 69.72	32 Ge 72.59	33 As 74.92	34 Se 78.96	35 Br 79.91	36 Kr 83.8
5	37 Rb 85.47	38 Sr 87.66	39 Y 88.91	40 Zr 91.22	41 Nb 92.91	42 Mo 95.94	43 Tc (99)	44 Ru 101.1	45 Rh 102.91	46 Pd 106.4	47 Ag 107.87	48 Cd 112.40	49 In 114.82	50 Sn 118.69	51 Sb 121.75	52 Te 127.60	53 I 126.90	54 Xe 131.30
6	55 Cs 132.91	56 Ba 137.34	57–71 °	72 Hf 178.49	73 Ta 180.95	74 W 183.85	75 Re 186.2	76 Os 190.2	77 Ir 192.2	78 Pt 195.09	79 Au 197.0	80 Hg 200.59	81 Tl 204.37	82 Pb 207.19	83 Bi 208.98	84 Po (210)	85 At (210)	86 Rn 222
7	87 Fr (223)	88 Ra 226.05	89–103 °°															

° Rare earths														
57 La 138.91	58 Ce 140.12	59 Pr 140.91	60 Nd 144.24	61 Pm (145)	62 Sm 150.35	63 Eu 152.0	64 Gd 157.25	65 Tb 158.92	66 Dy 162.50	67 Ho 164.92	68 Er 167.26	69 Tm 168.93	70 Yb 173.04	71 Lu 174.97

°° Actinides														
89 Ac 227	90 Th 232.04	91 Pa 231	92 U 238.03	93 Np (237)	94 Pu (242)	95 Am (243)	96 Cm (247)	97 Bk (249)	98 Cf (251)	99 Es (254)	100 Fm (253)	101 Md (256)	102 No (254)	103 Lw (257)

l, since

$$m_l = 0, \pm1, \pm2, \ldots, \pm l$$

and two possible values of the spin magnetic quantum number m_s ($+\frac{1}{2}$ and $-\frac{1}{2}$) for any m_l. Hence each subshell can contain a maximum of $2(2l + 1)$ electrons and each shell a maximum of

$$\sum_{l=0}^{l=n-1} 2(2l + 1) = 2[1 + 3 + 5 + \cdots + 2(n - 1) + 1]$$

$$= 2[1 + 3 + 5 + \cdots + 2n - 1]$$

The quantity in the brackets contains n terms whose average value is $\frac{1}{2}[1 + (2n - 1)]$, so that the maximum number of electrons in the nth shell is

$$2 \times \frac{n}{2}[1 + (2n - 1)] = 2n^2$$

An atomic shell or subshell that contains its full quota of electrons is said to be *closed*. A closed s subshell ($l = 0$) holds two electrons, a closed p subshell ($l = 1$) six electrons, a closed d subshell ($l = 2$) ten electrons, and so on.

The total orbital and spin angular momenta of the electrons in a closed subshell are zero, and their effective charge distributions are perfectly symmetrical (see Prob. 8 of Chap. 6). The electrons in a closed shell are all very tightly bound, since the positive nuclear charge is large relative to the negative charge of the inner shielding electrons (Fig. 7-6). Since an atom containing only closed shells has no dipole moment, it does not attract other electrons, and its electrons cannot be readily detached. Such atoms we expect to be passive chemically, like the inert gases—and the inert gases all turn out to have closed-shell electron configurations or their equivalents.

Those atoms with but a single electron in their outermost shells tend to lose this electron, which is relatively far from the nucleus and is shielded by the inner electrons from all but an effective nuclear charge of $+e$. Hydrogen and the alkali metals are in this category and accordingly have valences of $+1$. Atoms whose outer shells lack a single electron of being closed tend to acquire such an electron through the attraction of the imperfectly shielded strong nuclear charge, which accounts for the chemical behavior of the halogens. In this manner the similarities of the members of the various groups of the periodic table may be accounted for.

Table 7.2 shows the electron configurations of the elements. The origin of the transition elements evidently lies in the tighter binding of s electrons than d or f electrons in complex atoms, discussed in the previous section. The first element to exhibit this effect is potassium, whose outermost electron is in a $4s$

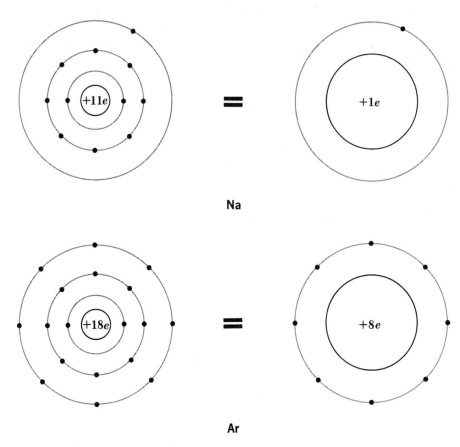

Na

Ar

FIGURE 7-6 Electron shielding in sodium and argon. Each outer electron in an Ar atom is acted upon by an effective nuclear charge 8 times greater than that acting upon the outer electron in a Na atom, even though the outer electrons in both cases are in the $M(n = 3)$ shell.

instead of a $3d$ substate. The difference in binding energy between $3d$ and $4s$ electrons is not very great, as can be seen in the configurations of chromium and copper. In both of these elements an additional $3d$ electron is present at the expense of a vacancy in the $4s$ subshell. In this connection another glance at Fig. 7.5 will be instructive.

The order in which electron subshells begin to be filled in atoms is

$$1s, 2s, 2p, 3s, 3p, 4s, 3d, 4p, 5s, 4d, 5p, 6s, 4f, 5d, 6p, 7s, 6d$$

as we can see from Table 7.2 and Fig. 7-7. The remarkable similarities in chemical behavior among the lanthanides and actinides are easy to understand on the basis of this sequence. All of the lanthanides have the same $5s^2 5p^6 6s^2$ configurations but have incomplete $4f$ subshells. The addition of $4f$ electrons has virtually no effect on the chemical properties of the lanthanide elements,

Table 7.2.

ELECTRON CONFIGURATIONS OF THE ELEMENTS.

	K	L		M			N				O				P			Q
	$1s$	$2s$	$2p$	$3s$	$3p$	$3d$	$4s$	$4p$	$4d$	$4f$	$5s$	$5p$	$5d$	$5f$	$6s$	$6p$	$6d$	$7s$
1 H	1																	
2 He	2																	
3 Li	2	1																
4 Be	2	2																
5 B	2	2	1															
6 C	2	2	2															
7 N	2	2	3															
8 O	2	2	4															
9 F	2	2	5															
10 Ne	2	2	6															
11 Na	2	2	6	1														
12 Mg	2	2	6	2														
13 Al	2	2	6	2	1													
14 Si	2	2	6	2	2													
15 P	2	2	6	2	3													
16 S	2	2	6	2	4													
17 Cl	2	2	6	2	5													
18 A	2	2	6	2	6													
19 K	2	2	6	2	6		1											
20 Ca	2	2	6	2	6		2											
21 Sc	2	2	6	2	6	1	2											
22 Ti	2	2	6	2	6	2	2											
23 V	2	2	6	2	6	3	2											
24 Cr	2	2	6	2	6	5	1											
25 Mn	2	2	6	2	6	5	2											
26 Fe	2	2	6	2	6	6	2											
27 Co	2	2	6	2	6	7	2											
28 Ni	2	2	6	2	6	8	2											
29 Cu	2	2	6	2	6	10	1											
30 Zn	2	2	6	2	6	10	2											
31 Ga	2	2	6	2	6	10	2	1										
32 Ge	2	2	6	2	6	10	2	2										
33 As	2	2	6	2	6	10	2	3										
34 Se	2	2	6	2	6	10	2	4										
35 Br	2	2	6	2	6	10	2	5										
36 Kr	2	2	6	2	6	10	2	6										
37 Rb	2	2	6	2	6	10	2	6			1							
38 Sr	2	2	6	2	6	10	2	6			2							
39 Y	2	2	6	2	6	10	2	6	1		2							
40 Zr	2	2	6	2	6	10	2	6	2		2							
41 Nb	2	2	6	2	6	10	2	6	4		1							
42 Mo	2	2	6	2	6	10	2	6	5		1							
43 Tc	2	2	6	2	6	10	2	6	5		2							
44 Ru	2	2	6	2	6	10	2	6	7		1							
45 Rh	2	2	6	2	6	10	2	6	8		1							
46 Pd	2	2	6	2	6	10	2	6	10									
47 Ag	2	2	6	2	6	10	2	6	10		1							
48 Cd	2	2	6	2	6	10	2	6	10		2							
49 In	2	2	6	2	6	10	2	6	10		2	1						
50 Sn	2	2	6	2	6	10	2	6	10		2	2						
51 Sb	2	2	6	2	6	10	2	6	10		2	3						

Table 7.2   (Continued)

	K	L		M			N				O				P			Q
	1s	2s	2p	3s	3p	3d	4s	4p	4d	4f	5s	5p	5d	5f	6s	6p	6d	7s
52 Te	2	2	6	2	6	10	2	6	10		2	4						
53 I	2	2	6	2	6	10	2	6	10		2	5						
54 Xe	2	2	6	2	6	10	2	6	10		2	6						
55 Cs	2	2	6	2	6	10	2	6	10		2	6			1			
56 Ba	2	2	6	2	6	10	2	6	10		2	6			2			
57 La	2	2	6	2	6	10	2	6	10		2	6	1		2			
58 Ce	2	2	6	2	6	10	2	6	10	2	2	6			2			
59 Pr	2	2	6	2	6	10	2	6	10	3	2	6			2			
60 Nd	2	2	6	2	6	10	2	6	10	4	2	6			2			
61 Pm	2	2	6	2	6	10	2	6	10	5	2	6			2			
62 Sm	2	2	6	2	6	10	2	6	10	6	2	6			2			
63 Eu	2	2	6	2	6	10	2	6	10	7	2	6			2			
64 Gd	2	2	6	2	6	10	2	6	10	7	2	6	1		2			
65 Tb	2	2	6	2	6	10	2	6	10	9	2	6			2			
66 Dy	2	2	6	2	6	10	2	6	10	10	2	6			2			
67 Ho	2	2	6	2	6	10	2	6	10	11	2	6			2			
68 Er	2	2	6	2	6	10	2	6	10	12	2	6			2			
69 Tm	2	2	6	2	6	10	2	6	10	13	2	6			2			
70 Yb	2	2	6	2	6	10	2	6	10	14	2	6			2			
71 Lu	2	2	6	2	6	10	2	6	10	14	2	6	1		2			
72 Hf	2	2	6	2	6	10	2	6	10	14	2	6	2		2			
73 Ta	2	2	6	2	6	10	2	6	10	14	2	6	3		2			
74 W	2	2	6	2	6	10	2	6	10	14	2	6	4		2			
75 Re	2	2	6	2	6	10	2	6	10	14	2	6	5		2			
76 Os	2	2	6	2	6	10	2	6	10	14	2	6	6		2			
77 Ir	2	2	6	2	6	10	2	6	10	14	2	6	7		2			
78 Pt	2	2	6	2	6	10	2	6	10	14	2	6	9		1			
79 Au	2	2	6	2	6	10	2	6	10	14	2	6	10		1			
80 Hg	2	2	6	2	6	10	2	6	10	14	2	6	10		2			
81 Tl	2	2	6	2	6	10	2	6	10	14	2	6	10		2	1		
82 Pb	2	2	6	2	6	10	2	6	10	14	2	6	10		2	2		
83 Bi	2	2	6	2	6	10	2	6	10	14	2	6	10		2	3		
84 Po	2	2	6	2	6	10	2	6	10	14	2	6	10		2	4		
85 At	2	2	6	2	6	10	2	6	10	14	2	6	10		2	5		
86 Rn	2	2	6	2	6	10	2	6	10	14	2	6	10		2	6		
87 Fr	2	2	6	2	6	10	2	6	10	14	2	6	10		2	6		1
88 Ra	2	2	6	2	6	10	2	6	10	14	2	6	10		2	6		2
89 Ac	2	2	6	2	6	10	2	6	10	14	2	6	10		2	6	1	2
90 Th	2	2	6	2	6	10	2	6	10	14	2	6	10		2	6	2	2
91 Pa	2	2	6	2	6	10	2	6	10	14	2	6	10	2	2	6	1	2
92 U	2	2	6	2	6	10	2	6	10	14	2	6	10	3	2	6	1	2
93 Np	2	2	6	2	6	10	2	6	10	14	2	6	10	4	2	6	1	2
94 Pu	2	2	6	2	6	10	2	6	10	14	2	6	10	5	2	6	1	2
95 Am	2	2	6	2	6	10	2	6	10	14	2	6	10	6	2	6	1	2
96 Cm	2	2	6	2	6	10	2	6	10	14	2	6	10	7	2	6	1	2
97 Bk	2	2	6	2	6	10	2	6	10	14	2	6	10	8	2	6	1	2
98 Cf	2	2	6	2	6	10	2	6	10	14	2	6	10	10	2	6		2
99 E	2	2	6	2	6	10	2	6	10	14	2	6	10	11	2	6		2
100 Fm	2	2	6	2	6	10	2	6	10	14	2	6	10	12	2	6		2
101 Md	2	2	6	2	6	10	2	6	10	14	2	6	10	13	2	6		2
102 No	2	2	6	2	6	10	2	6	10	14	2	6	10	14	2	6		2
103 Lw	2	2	6	2	6	10	2	6	10	14	2	6	10	14	2	6	1	2

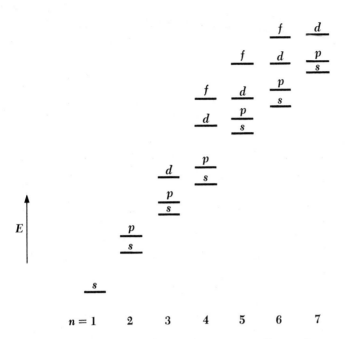

FIGURE 7-7 The sequence of quantum states in an atom. Not to scale.

which are determined by the outer electrons. Similarly, all of the actinides have $6s^2 6p^6 7s^2$ configurations, and differ only in the numbers of their $5f$ and $6d$ electrons.

These irregularities in the binding energies of atomic electrons are also responsible for the lack of completely full outer shells in the heavier inert gases. Helium ($Z = 2$) and neon ($Z = 10$) contain closed K and L shells respectively, but argon ($Z = 18$) has only 8 electrons in its M shell, corresponding to closed $3s$ and $3p$ subshells. The reason the $3d$ subshell is not filled next is simply that $4s$ electrons have higher binding energies than $3d$ electrons, as we have said, and so the $4s$ subshell is filled first in potassium and calcium. As the $3d$ subshell is filled in successively heavier transition elements, there are still one or two outer $4s$ electrons that make possible chemical activity. Not until krypton ($Z = 36$) is another inert gas reached, and here a similarly incomplete outer shell occurs with only the $4s$ and $4p$ subshells filled. Following krypton is rubidium ($Z = 37$), which skips both the $4d$ and $4f$ subshells to have a $5s$ electron. The next inert gas is xenon ($Z = 54$), which has filled $4d$, $5s$, and $5p$ subshells, but now even the inner $4f$ subshell is empty as well as the $5d$ and $5f$ subshells. The same pattern recurs with the remainder of the inert gases.

While we have sketched the origins of only a few of the chemical and physical properties of the elements in terms of their electron configurations, many more can be quantitatively understood by similar reasoning.

7.6 HUND'S RULE

In general, the electrons in an atom remain unpaired—that is, have parallel spins—whenever possible. This principle is called *Hund's rule*. The ferromagnetism of iron, cobalt, and nickel is a consequence of Hund's rule: their 3d subshells are only partially occupied, and the electrons in these subshells do not pair off to permit their spin magnetic moments to cancel out. In iron, for instance, five of the six 3d electrons have parallel spins, so that each iron atom has a large resultant magnetic moment. We shall examine other consequences of Hund's rule in Chap. 8 in connection with molecular bonding.

The origin of Hund's rule lies in the mutual repulsion of atomic electrons. Because of this repulsion, the farther apart the electrons in an atom are, the lower the energy of the atom. Electrons in the same subshell with the same spin must have different m_l values and accordingly are described by wave functions whose spatial distributions are different. Electrons with parallel spins are therefore more separated in space than if they paired off, and this arrangement, having less energy, is the more stable one.

*7.7 TOTAL ANGULAR MOMENTUM

Each electron in an atom has a certain orbital angular momentum **L** and a certain spin angular momentum **S**, both of which contribute to the total angular momentum **J** of the atom. Like all angular momenta, **J** is quantized, with a magnitude given by

7.16 $$J = \sqrt{J(J + 1)}\,\hbar$$ **Total atomic angular momentum**

and a component J_z in the z direction given by

7.17 $$J_z = M_J\hbar$$ **z component of total atomic angular momentum**

where J and M_J are the quantum numbers governing J and J_z. Our task in the remainder of this chapter is to look into the properties of **J** and their effect on atomic phenomena. We shall do this in terms of the semiclassical *vector model* of the atom, which provides a more intuitively accessible framework for understanding angular-momentum considerations than does a purely quantum-mechanical approach.

Let us first consider an atom whose total angular momentum is provided by a single electron. Atoms of the elements in group I of the periodic table—hydrogen, lithium, sodium, and so on—are of this kind since they have single electrons outside closed inner shells (except for hydrogen, which has no inner electrons) and the exclusion principle assures that the total angular momentum and magnetic moment of a closed shell are zero. Also in this category are the ions He^+, Be^+, Mg^+, B^{++}, Al^{++}, and so on.

The magnitude L of the orbital angular momentum $\mathbf{L}$ of an atomic electron is determined by its orbital quantum number l according to the formula

7.18 $\qquad L = \sqrt{l(l + 1)}\,\hbar$

while the component L_z of $\mathbf{L}$ along the z axis is determined by the magnetic quantum number m_l according to the formula

7.19 $\qquad L_z = m_l \hbar$

Similarly the magnitude S of the spin angular momentum $\mathbf{S}$ is determined by the spin quantum number s (which has the sole value $+\frac{1}{2}$) according to the formula

7.20 $\qquad S = \sqrt{s(s + 1)}\,\hbar$

while the component S_z of $\mathbf{S}$ along the z axis is determined by the magnetic spin quantum number m_s according to the formula

7.21 $\qquad S_z = m_s \hbar$

Because $\mathbf{L}$ and $\mathbf{S}$ are vectors, they must be added vectorially to yield the total angular momentum $\mathbf{J}$:

7.22 $\qquad \mathbf{J} = \mathbf{L} + \mathbf{S}$

It is customary to use the symbols j and m_j for the quantum numbers that describe J and J_z for a single electron, so that

7.23 $\qquad J = \sqrt{j(j + 1)}\,\hbar$

7.24 $\qquad J_z = m_j \hbar$

To obtain the relationships among the various angular-momentum quantum numbers, it is simplest to start with the z components of the vectors $\mathbf{J}$, $\mathbf{L}$, and $\mathbf{S}$. Since J_z, L_z, and S_z are scalar quantities,

$$J_z = L_z \pm S_z$$
$$m_j \hbar = m_l \hbar \pm m_s \hbar$$

and

7.25 $\qquad m_j = m_l \pm m_s$

The possible values of m_l range from $+l$ through 0 to $-l$, and those of m_s are $\pm s$. The quantum number l is always an integer or 0 while $s = \frac{1}{2}$, and as a result m_j must be half-integral. The possible values of m_j also range from $+j$ through 0 to $-j$ in integral steps, and so, for any value of l,

7.26 $j = l \pm s$

Like m_j, j is always half-integral.

Because of the simultaneous quantization of **J**, **L**, and **S** they can have only certain specific relative orientations. This is a general conclusion; in the case of a one-electron atom, there are only two relative orientations possible. One of these corresponds to $j = l + s$, so that $J > L$, and the other to $j = l - s$, so that $J < L$. Figure 7-8 shows the two ways in which **L** and **S** can combine to form **J** when $l = 1$. Evidently the orbital and spin angular-momentum vectors can never be exactly parallel or antiparallel to each other or to the total angular-momentum vector.

The angular momenta **L** and **S** interact magnetically, as we saw in Sec. 7.2, and as a result exert torques on each other. If there is no external magnetic field, the total angular momentum **J** is conserved in magnitude and direction, and the effect of the internal torques can only be the precession of **L** and **S** around the direction of their resultant **J** (Fig. 7-9). However, if there is an external magnetic field **B** present, then **J** precesses about the direction of **B** while **L** and **S** continue precessing about **J**, as in Fig. 7-10. The precession of **J** about **B** is what gives rise to the anomalous Zeeman effect, since different orientations of **J** involve slightly different energies in the presence of **B**.

Atomic nuclei also have intrinsic angular momenta and magnetic moments, as we shall see in Chap. 11, and these contribute to the total atomic angular momenta and magnetic moments. These contributions are small because nuclear

FIGURE 7-8   The two ways in which L and S can be added to form J when $l = 1$, $s = \frac{1}{2}$.

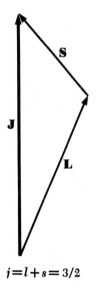

$j = l + s = 3/2$

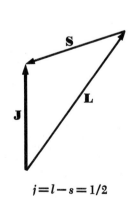

$j = l - s = 1/2$

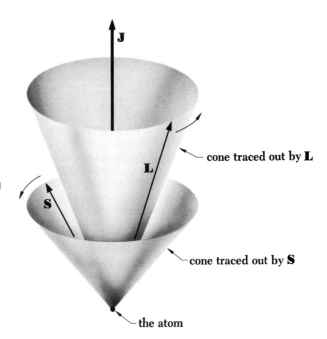

FIGURE 7-9 The orbital and spin angular-momentum vectors L and S precess about J according to the vector model of the atom.

cone traced out by **L**

cone traced out by **S**

the atom

magnetic moments are $\sim 10^{-3}$ the magnitude of electronic moments, and they lead to the *hyperfine structure* of spectral lines with typical spacings between components of $\sim 10^{-2}$ Å as compared with typical fine-structure spacings of several angstroms.

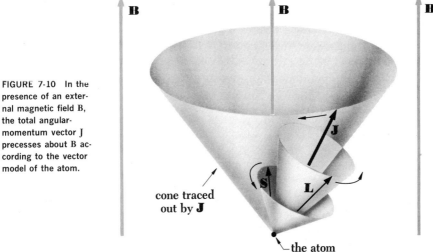

FIGURE 7-10 In the presence of an external magnetic field B, the total angular-momentum vector J precesses about B according to the vector model of the atom.

cone traced out by **J**

the atom

*7.8 LS COUPLING

When more than one electron contributes orbital and spin angular momenta to the total angular momentum $\mathbf{J}$ of an atom, $\mathbf{J}$ is still the vector sum of these individual momenta. Because the electrons involved interact with one another, the manner in which their individual momenta $\mathbf{L}_i$ and $\mathbf{S}_i$ add together to form $\mathbf{J}$ follows certain definite patterns depending upon the circumstances. The usual pattern for all but the heaviest atoms is that the orbital angular momenta $\mathbf{L}_i$ of the various electrons are coupled together electrostatically into a single resultant $\mathbf{L}$ and the spin angular momenta $\mathbf{S}_i$ are coupled together independently into another single resultant $\mathbf{S}$; we shall examine the reasons for this behavior later in this section. The momenta $\mathbf{L}$ and $\mathbf{S}$ then interact magnetically via the spin-orbit effect to form a total angular momentum $\mathbf{J}$. This scheme, called *LS coupling*, may be summarized as follows:

7.27
$$\mathbf{L} = \Sigma\,\mathbf{L}_i$$
$$\mathbf{S} = \Sigma\,\mathbf{S}_i \qquad\qquad \textbf{LS coupling}$$
$$\mathbf{J} = \mathbf{L} + \mathbf{S}$$

As usual, L, S, J, L_z, S_z, and J_z are quantized, with the respective quantum numbers being L, S, J, $\mathsf{M_L}$, $\mathsf{M_S}$, and $\mathsf{M_J}$. Hence

7.28 $\qquad L = \sqrt{\mathsf{L(L + 1)}}\,\hbar$

7.29 $\qquad L_z = \mathsf{M_L}\,\hbar$

7.30 $\qquad S = \sqrt{\mathsf{S(S + 1)}}\,\hbar$

7.31 $\qquad S_z = \mathsf{M_S}\hbar$

7.32 $\qquad J = \sqrt{\mathsf{J(J + 1)}}\,\hbar$

7.33 $\qquad J_z = \mathsf{M_J}\,\hbar$

Both L and $\mathsf{M_L}$ are always integers or 0, while the other quantum numbers are half-integral if an odd number of electrons is involved and integral or 0 if an even number of electrons is involved.

As an example, let us consider two electrons, one with $l_1 = 1$ and the other with $l_2 = 2$. There are three ways in which $\mathbf{L}_1$ and $\mathbf{L}_2$ can be combined into a single vector $\mathbf{L}$ that is quantized according to Eq. 7.28, as shown in Fig. 7-11. These correspond to $\mathsf{L} = 1, 2,$ and 3, since all values of L are possible from $l_1 + l_2$ to $l_1 - l_2$. The spin quantum number s is always $+\frac{1}{2}$, so that there are two possibilities for the sum $\mathbf{S}_1 + \mathbf{S}_2$, as in Fig. 7-11, that correspond to $\mathsf{S} = 0$ and $\mathsf{S} = 1$. We note that $\mathbf{L}_1$ and $\mathbf{L}_2$ can never be exactly parallel to $\mathbf{L}$, nor $\mathbf{S}_1$ and $\mathbf{S}_2$ to $\mathbf{S}$, except when the vector sum is 0. The quantum number J can have all values between $\mathsf{L} + \mathsf{S}$ and $\mathsf{L} - \mathsf{S}$, so here J can be 0, 1, 2, 3, or 4.

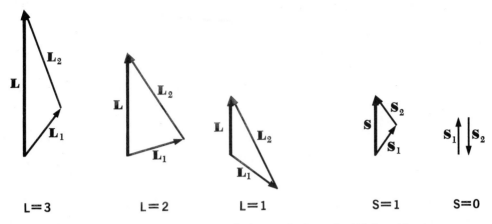

FIGURE 7-11 When $l_1 = 1$, $s_1 =$ ½, and $l_2 = 2$, $s_2 =$ ½, there are three ways in which $\mathbf{L}_1$ and $\mathbf{L}_2$ can combine to form $\mathbf{L}$ and two ways in which $\mathbf{S}_1$ and $\mathbf{S}_2$ can combine to form $\mathbf{S}$.

The *LS* scheme owes its existence to the relative strengths of the electrostatic forces that couple the individual orbital angular momenta into a resultant $\mathbf{L}$ and the individual spin angular momenta into a resultant $\mathbf{S}$. The origins of these forces are interesting. The coupling between orbital angular momenta can be understood by reference to Fig. 6-12, which shows how the electron probability density $|\psi|^2$ varies in space for different quantum states in hydrogen. The corresponding patterns for electrons in more complex atoms will be somewhat different, of course, but it remains true in general that $|\psi|^2$ is not spherically symmetric except for *s* states. (In the latter case $l = 0$ and the electron has no orbital angular momentum to contribute anyway.) Because of the asymmetrical distributions of their charge densities, the electrostatic forces between the electrons in an atom vary with the relative orientations of their angular-momentum vectors, and only certain relative orientations are stable. These stable configurations correspond to a total orbital angular momentum that is quantized according to the formula $L = \sqrt{L(L + 1)}\,\hbar$.

The coupling between the various $\mathbf{L}_i$ is usually such that the configuration of lowest energy is the one for which L is a maximum. This effect is easy to understand if we imagine two electrons in the same Bohr orbit. Because the electrons repel each other electrostatically, they tend to revolve around the nucleus in the same direction, which maximizes L. If they revolved in opposite directions to minimize L, the electrons would pass each other more frequently, leading to a higher energy for the system.

The origin of the strong coupling between electron spins is harder to visualize because it is a purely quantum-mechanical effect with no classical analog. (The direct interaction between the intrinsic electron magnetic moments, it should be noted, is insignificant and not responsible for the coupling between electron spin angular momenta.) The basic idea is that the complete wave function $\psi(1, 2, \ldots, n)$ of a system of n electrons is the product of a wave function $u(1, 2, \ldots, n)$ that describes the coordinates of the electrons and a spin function $s(1, 2, \ldots, n)$ that describes the orientations of their spins. As we saw in Sec. 7.3, the complete function $\psi(1, 2, \ldots, n)$ must be antisymmetric, which means that $u(1, 2, \ldots, n)$ is not independent of $s(1, 2, \ldots, n)$. A change in the relative orientations of the electron spin angular-momentum vectors must therefore be accompanied by a change in the spatial electronic configuration of the atom, which means a change in the atom's electrostatic potential energy. To go from one total spin angular momentum $\mathbf{S}$ to a different one involves altering the structure of the atom, and therefore a strong electrostatic force, besides altering the directions of the spin angular momenta $\mathbf{S}_1, \mathbf{S}_2, \ldots, \mathbf{S}_n$, which requires only a weak magnetic force. This situation is what is described when it is said that the spin momenta $\mathbf{S}_i$ are strongly coupled together electrostatically.

The $\mathbf{S}_i$ always combine into a ground-state configuration in which S is a maximum. This is an example of Hund's rule; as mentioned earlier, electrons with parallel spins have different m_l values and are described by different wave functions, which means that there is a greater average separation in space of the electrons and accordingly a lower total energy.

Although we shall not try to justify this conclusion, the combination of $\mathbf{L}$ and $\mathbf{S}$ that makes J a *minimum* results in the lowest energy.

*7.9   jj COUPLING

The electrostatic forces that couple the $\mathbf{L}_i$ into a single vector $\mathbf{L}$ and the $\mathbf{S}_i$ into another vector $\mathbf{S}$ are stronger than the magnetic spin-orbit forces that couple $\mathbf{L}$ and $\mathbf{S}$ to form $\mathbf{J}$ in light atoms, and dominate the situation even when a moderate external magnetic field is applied. (In the latter case the precession of $\mathbf{J}$ around $\mathbf{B}$ is accordingly slower than the precession of $\mathbf{L}$ and $\mathbf{S}$ around $\mathbf{J}$.) However, in heavy atoms the nuclear charge becomes great enough to produce spin-orbit interactions comparable in magnitude to the electrostatic ones between the $\mathbf{L}_i$ and between the $\mathbf{S}_i$, and the LS coupling scheme begins to break down. A similar breakdown occurs in strong external magnetic fields (typically ~ 10 T), which produces the Paschen-Back effect in atomic spectra. In the limit of the failure of LS coupling, the total angular momenta $\mathbf{J}_i$ of the individual electrons add together directly to form the angular momentum $\mathbf{J}$ of the entire atom, a situation referred to as *jj coupling* since each $\mathbf{J}_i$ is described by a quantum number j in the manner described earlier. Hence

7.34
$$\mathbf{J}_i = \mathbf{L}_i + \mathbf{S}_i$$
$$\mathbf{J} = \Sigma \mathbf{J}_i$$

jj coupling

In Sec. 6.4 we saw that individual orbital angular-momentum states are customarily described by a lowercase letter, with s corresponding to $l = 0$, p to $l = 1$, d to $l = 2$, and so on. A similar scheme using capital letters is used to designate the entire electronic state of an atom according to its total orbital angular-momentum quantum number L as follows:

$$L = 0 \quad 1 \quad 2 \quad 3 \quad 4 \quad 5 \quad 6 \dots$$
$$ S \quad P \quad D \quad F \quad G \quad H \quad I \dots$$

A superscript number before the letter (^2P for instance) is used to indicate the *multiplicity* of the state, which is the number of different possible orientations of **L** and **S** and hence the number of different possible values of J. The multiplicity is equal to $2S + 1$ in the usual situation where $L > S$, since J ranges from $L + S$ through 0 to $L - S$. Thus when $S = 0$, the multiplicity is 1 (a *singlet* state) and $J = L$; when $S = \frac{1}{2}$, the multiplicity is 2 (a *doublet* state) and $J = L \pm \frac{1}{2}$; when $S = 1$, the multiplicity is 3 (a *triplet* state) and $J = L + 1$, L, or $L - 1$; and so on. (In a configuration in which $S > L$, the multiplicity is given by $2L + 1$.) The total angular-momentum quantum number J is used as a subscript after the letter, so that a $^2P_{3/2}$ state (read as "doublet P three-halves") refers to an electronic configuration in which $S = \frac{1}{2}$, $L = 1$, and $J = \frac{3}{2}$. For historical reasons, these designations are called *term symbols*.

In the event that the angular momentum of the atom arises from a single outer electron, the principal quantum number n of this electron is used as a prefix: thus the ground state of the sodium atom is described by $3^2S_{1/2}$, since its electronic configuration has an electron with $n = 3$, $l = 0$, and $s = \frac{1}{2}$ (and hence $j = \frac{1}{2}$) outside closed $n = 1$ and $n = 2$ shells. For consistency it is conventional to denote the above state by $3^2S_{1/2}$ with the superscript 2 indicating a doublet, even though there is only a single possibility for J since $L = 0$.

*7.10 ONE-ELECTRON SPECTRA

We are now in a position to understand the chief features of the spectra of the various elements. Before we examine some representative examples, it should be mentioned that further complications exist which have not been considered here, for instance those that originate in relativistic effects and in the coupling between electrons and vacuum fluctuations in the electromagnetic field (see Sec. 6.10). These additional factors split certain energy states into closely spaced substates and therefore represent other sources of fine structure in spectral lines.

Figure 7-12 shows the various states of the hydrogen atom classified by their total quantum number n and orbital angular-momentum quantum number l. The selection rule for allowed transitions here is $\Delta l = \pm 1$, which is illustrated by

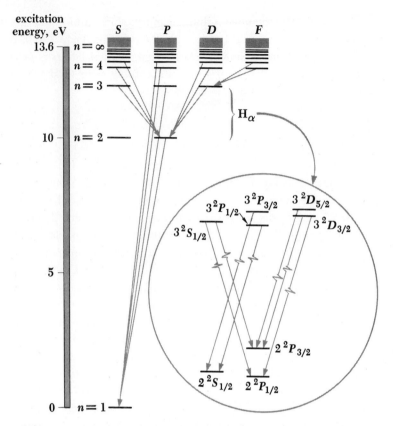

FIGURE 7-12 Energy-level diagram for hydrogen showing the origins of some of the more prominent spectral lines. The detailed structures of the $n = 2$ and $n = 3$ levels and the transitions that lead to the various components of the H_α line are pictured in the inset.

the transitions shown. To indicate some of the detail that is omitted in a simple diagram of this kind, the detailed structures of the $n = 2$ and $n = 3$ levels are pictured; not only are all substates of the same n and different j separated in energy, but the same is true of states of the same n and j but with different l. The latter effect is most marked for states of small n and l, and was first established in 1947 in the "Lamb shift" of the $2^2S_{1/2}$ state relative to the $2^2P_{1/2}$ state. The various separations conspire to split the H_α spectral line ($n = 3 \rightarrow n = 2$) into seven closely spaced components.

The sodium atom has a single $3s$ electron outside closed inner shells, and so, if we assume that the 10 electrons in its inner core completely shield $+10e$ of nuclear charge, the outer electron is acted upon by an effective nuclear charge of $+e$ just as in the hydrogen atom. Hence we expect, as a first approximation, that the energy levels of sodium will be the same as those of hydrogen except that the lowest one will correspond to $n = 3$ instead of $n = 1$ because of the exclusion principle. Figure 7-13 is the energy-level diagram for sodium and, by comparison with the hydrogen levels also shown, there is indeed agreement for the states of highest l, that is, for the states of highest angular momentum.

To understand the reason for the discrepancies at lower values of l, we need only refer to Fig. 6-11 to see how the probability for finding the electron in a hydrogen atom varies with distance from the nucleus. The smaller the value of l for a given n, the closer the electron gets to the nucleus on occasion. Although the sodium wave functions are not identical with those of hydrogen, their general behavior is similar, and accordingly we expect the outer electron

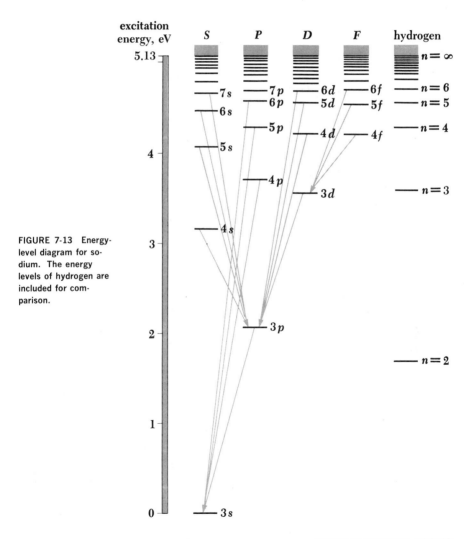

FIGURE 7-13 Energy-level diagram for sodium. The energy levels of hydrogen are included for comparison.

in a sodium atom to penetrate the core of inner electrons most often when it is in an s state, less often when it is in a p state, still less often when it is in a d state, and so on. The less shielded an outer electron is from the full nuclear charge, the greater the average force acting on it, and the smaller (that is, the more negative) its total energy. For this reason the states of small l in sodium are displaced downward from their equivalents in hydrogen, as in Fig. 7-13, and there are pronounced differences in energy between states of the same n but different l.

*7.11 TWO-ELECTRON SPECTRA

A single electron is responsible for the energy levels of both hydrogen and sodium. However, there are two $1s$ electrons in the ground state of helium, and it is interesting to consider the effect of LS coupling on the properties and behavior of the helium atom. To do this we first note the selection rules for allowed transitions under LS coupling:

7.35 $\quad\quad \Delta L = 0, \pm 1$

7.36 $\quad\quad \Delta J = 0, \pm 1$ **LS selection rules**

7.37 $\quad\quad \Delta S = 0$

When only a single electron is involved, $\Delta L = 0$ is prohibited and $\Delta L = \Delta l = \pm 1$ is the only possibility. Furthermore, J must change when the initial state has $J = 0$, so that $J = 0 \rightarrow J = 0$ is prohibited.

The helium energy-level diagram is shown in Fig. 7-14. The various levels represent configurations in which one electron is in its ground state and the other is in an excited state, but because the angular momenta of the two electrons are coupled, it is proper to consider the levels as characteristic of the entire atom. Three differences between this diagram and the corresponding ones for hydrogen and sodium are conspicuous.

First, there is the division into singlet and triplet states, which are, respectively, states in which the spins of the two electrons are antiparallel (to give $S = 0$) and parallel (to give $S = 1$). Because of the selection rule $\Delta S = 0$, no allowed transitions can occur between singlet states and triplet states, and the helium spectrum arises from transitions in one set or the other. Helium atoms in singlet states (antiparallel spins) constitute *parahelium* and those in triplet states (parallel spins) constitute *orthohelium*. An orthohelium atom can lose excitation energy in a collision and become one of parahelium, while a parahelium atom can gain excitation energy in a collision and become one of orthohelium; ordinary liquid or gaseous helium is therefore a mixture of both. The lowest triplet states are

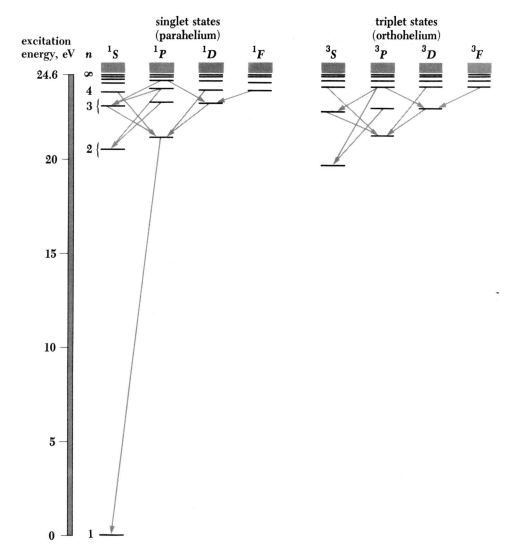

FIGURE 7-14 Energy-level diagram for helium showing the division into singlet (parahelium) and triplet (orthohelium) states. There is no 1^3S state.

called *metastable* because, in the absence of collisions, an atom in one of them can retain its excitation energy for a relatively long time (a second or more) before radiating.

The second obvious peculiarity in Fig. 7-14 is the absence of the 1^3S state. The lowest triplet state is 2^3S, although the lowest singlet state is 1^1S. The 1^3S state is missing as a consequence of the exclusion principle, since in this state the two electrons would have parallel spins and therefore identical sets of quantum numbers. Third, the energy difference between the ground state and the lowest excited state is relatively large, which reflects the tight binding of closed-shell electrons discussed earlier in this chapter. The ionization energy of helium—the work that must be done to remove an electron from a helium atom—is 24.6 eV, the highest of any element.

The last energy-level diagram we shall consider is that of mercury, which has two electrons outside an inner core of 78 electrons in closed shells or subshells (Table 7.2). We expect a division into singlet and triplet states as in helium, but because the atom is so heavy we might also expect signs of a breakdown in the LS coupling of angular momenta. As Fig. 7-15 reveals, both of these expectations are realized, and several prominent lines in the mercury spectrum arise from transitions that violate the $\Delta S = 0$ selection rule. The transition

FIGURE 7-15   Energy-level diagram for mercury. In each excited level one outer electron is in the ground state, and the designation of the levels in the diagram corresponds to the state of the other electron.

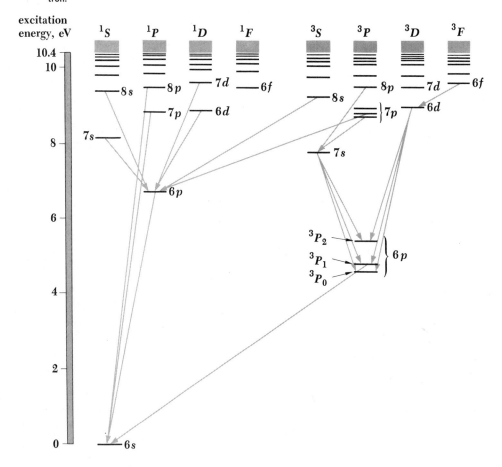

$^3P_1 \rightarrow {}^1S_0$ is an example, and is responsible for the strong 2,537-Å line in the ultraviolet. To be sure, this does not mean that the transition probability is necessarily very high, since the three $^3P_1$ states are the lowest of the triplet set and therefore tend to be highly populated in excited mercury vapor. The $^3P_0 \rightarrow {}^1S_0$ and $^3P_2 \rightarrow {}^1S_0$ transitions, respectively, violate the rules that forbid transitions from $J = 0$ to $J = 0$ and that limit ΔJ to 0 or ± 1, as well as violating $\Delta S = 0$, and hence are considerably less likely to occur than the $^3P_1 \rightarrow {}^1S_0$ transition. The $^3P_0$ and $^3P_2$ states are therefore metastable and, in the absence of collisions, an atom can persist in either of them for a relatively long time. The strong spin-orbit interaction in mercury that leads to the partial failure of LS coupling is also responsible for the wide spacing of the elements of the ^3P triplets.

7.12 X-RAY SPECTRA

In Chap. 2 we learned that the X-ray spectra of targets bombarded by fast electrons exhibit narrow spikes at wavelengths characteristic of the target material in addition to a continuous distribution of wavelengths down to a minimum wavelength inversely proportional to the electron energy. The continuous X-ray spectrum is the result of the inverse photoelectric effect, with electron kinetic energy being transformed into photon energy $h\nu$. The discrete spectrum, on the other hand, has its origin in electronic transitions within atoms that have been disturbed by the incident electrons.

Transitions involving the outer electrons of an atom usually involve only a few electron volts of energy, and even removing an outer electron requires at most 24.6 eV (for helium). These transitions accordingly are associated with photons whose wavelengths lie in or near the visible part of the electromagnetic spectrum, as is evident from the diagram in the back endpapers of this book. The inner electrons of heavier elements are a quite different matter, because these electrons experience all or much of the full nuclear charge without nearly complete shielding by intervening electron shells and in consequence are very tightly bound. In sodium, for example, only 5.13 eV is needed to remove the outermost 3s electron, while the corresponding figures for the inner ones are 31 eV for each 2p electron, 63 eV for each 2s electron, and 1,041 eV for each 1s electron. Transitions that involve the inner electrons in an atom are what give rise to discrete X-ray spectra because of the high photon energies involved.

Figure 7-16 shows the energy levels (not to scale) of a heavy atom classed by total quantum number n; energy differences between angular-momentum states within a shell are minor compared with the energy differences between shells. Let us consider what happens when an energetic electron strikes the atom

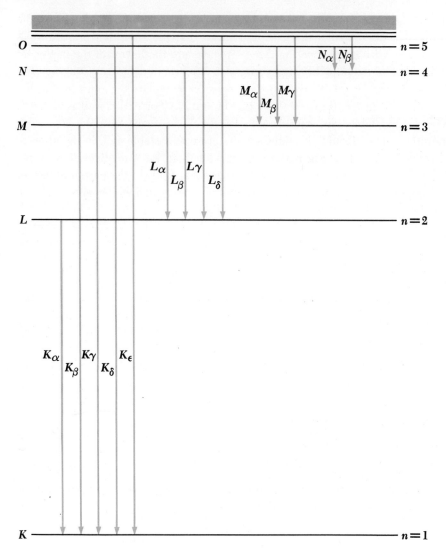

FIGURE 7-16 The origin of X-ray spectra.

and knocks out one of the K-shell electrons. (The K electron could also be elevated to one of the unfilled upper quantum states of the atom, but the difference between the energy needed to do this and that needed to remove the electron completely is insignificant, only 0.2 percent in sodium and still less in heavier atoms.) An atom with a missing K electron gives up most of its considerable excitation energy in the form of an X-ray photon when an electron from an outer shell drops into the "hole" in the K shell. As indicated in Fig. 7-16, the K *series* of lines in the X-ray spectrum of an element consists of wavelengths

236   THE ATOM

arising in transitions from the L, M, N, . . . levels to the K level. Similarly the longer-wavelength L series originates when an L electron is knocked out of the atom, the M series when an M electron is knocked out, and so on. The two spikes in the X-ray spectrum of molybdenum in Fig. 2-8 are the K_α and K_β lines of its K series.

An atom with a missing inner electron can also lose excitation energy by the *Auger effect* without emitting an X-ray photon. In the Auger effect an outer-shell electron is ejected from the atom at the same time that another outer-shell electron drops to the incomplete inner shell; the ejected electron carries off the atom's excitation energy instead of a photon doing this. In a sense the Auger effect represents an internal photoelectric effect, although the photon never actually comes into being within the atom. The Auger process is competitive with X-ray emission in most atoms, but the resulting electrons are usually absorbed in the target material while the X rays readily emerge to be detected.

Problems

1. If atoms could contain electrons with principal quantum numbers up to and including $n = 6$, how many elements would there be?

2. The ionization energies of the elements of atomic numbers 20 through 29 are very nearly equal. Why should this be so when considerable variations exist in the ionization energies of other consecutive sequences of elements?

3. The atomic radius of an element can be determined from measurements made on crystals of which it is a constituent. The results are shown in Fig. 10-3. Account for the general trend of the variation of radius with atomic number.

4. Many years ago it was pointed out that the atomic numbers of the rare gases are given by the following scheme:

$$Z(\text{He}) = 2(1^2) = 2$$
$$Z(\text{Ne}) = 2(1^2 + 2^2) = 10$$
$$Z(\text{Ar}) = 2(1^2 + 2^2 + 2^2) = 18$$
$$Z(\text{Kr}) = 2(1^2 + 2^2 + 2^2 + 3^2) = 36$$
$$Z(\text{Xe}) = 2(1^2 + 2^2 + 2^2 + 3^2 + 3^2) = 54$$
$$Z(\text{Rn}) = 2(1^2 + 2^2 + 2^2 + 3^2 + 3^2 + 4^2) = 86$$

Explain the origin of this scheme in terms of atomic theory.

5. A beam of electrons enters a uniform magnetic field of flux density 1.2 T. Find the energy difference between electrons whose spins are parallel and antiparallel to the field.

6. How does the agreement between observations of the normal Zeeman effect and the theory of this effect tend to confirm the existence of electrons as independent entities within atoms?

7. A sample of a certain element is placed in a magnetic field of flux density 0.3 T. How far apart are the Zeeman components of a spectral line of wavelength 4,500 Å?

8. Why does the normal Zeeman effect occur only in atoms with an even number of electrons?

°9. Find the S, L, and J values that correspond to each of the following states: $^1S_0$, $^3P_2$, $^2D_{3/2}$, $^5F_5$, $^6H_{5/2}$.

°10. The carbon atom has two $2s$ electrons and two $2p$ electrons outside a filled inner shell. Its ground state is $^3P_0$. What are the term symbols of the other allowed states, if any? Why would you think the $^3P_0$ state is the ground state?

°11. The lithium atom has one $2s$ electron outside a filled inner shell. Its ground state is $^2S_{1/2}$. What are the term symbols of the other allowed states, if any? Why would you think the $^2S_{1/2}$ state is the ground state?

°12. The magnesium atom has two $3s$ electrons outside filled inner shells. Find the term symbol of its ground state.

°13. The aluminum atom has two $3s$ electrons and one $3p$ electron outside filled inner shells. Find the term symbol of its ground state.

°14. The magnetic moment μ_J of an atom in which LS coupling holds has the magnitude

$$\mu_J = \sqrt{J(J + 1)}g_J\mu_B$$

where $\mu_B = e\hbar/2m$ is the Bohr magneton and

$$g_J = 1 + \frac{J(J + 1) - L(L + 1) + S(S + 1)}{2J(J + 1)}$$

is the *Landé g factor*. (*a*) Derive this result with the help of the law of cosines starting from the fact that, averaged over time, only the components of μ_L and μ_S parallel to **J** contribute to μ_J. (*b*) Consider an atom that obeys LS coupling that is in a weak magnetic field **B** in which the coupling is preserved. How many substates are there for a given value of J? What is the energy difference between different substates?

°15. The ground state of chlorine is $^2P_{3/2}$. Find its magnetic moment (see previous problem). Into how many substates will the ground state split in a weak magnetic field?

*16. Show that, if the angle between the directions of **L** and **S** in Fig. 7-8 is θ,

$$\cos \theta = \frac{j(j+1) - l(l+1) - s(s+1)}{2\sqrt{l(l+1)s(s+1)}}$$

17. The spin-orbit effect splits the $3P \rightarrow 3S$ transition in sodium (which gives rise to the yellow light of sodium-vapor highway lamps) into two lines, 5,890 Å corresponding to $3P_{3/2} \rightarrow 3S_{1/2}$ and 5,896 Å corresponding to $3P_{1/2} \rightarrow 3S_{1/2}$. Use these wavelengths to calculate the effective magnetic induction experienced by the outer electron in the sodium atom as a result of its orbital motion.

18. Show that the frequency of the K_α X-ray line of an element of atomic number Z is given by

$$\nu = \frac{3cR(Z-1)^2}{4}$$

where R is the Rydberg constant. Assume that each L electron in an atom may be regarded as the single electron in a hydrogenic atom whose effective nuclear charge is reduced by the presence of whatever K electrons are present. [The proportionality between ν and $(Z-1)^2$ was used by Moseley in 1913 to establish the atomic numbers of the elements from their X-ray spectra. This proportionality is referred to as *Moseley's law*.]

19. What element has a K_α X-ray line of wavelength 1.785 Å? Of wavelength 0.712 Å?

20. Explain why the X-ray spectra of elements of nearby atomic numbers are qualitatively very similar, whereas the optical spectra of these elements may differ considerably.

PROPERTIES OF MATTER 3

THE PHYSICS OF MOLECULES 8

What is the nature of the forces that bond atoms together to form molecules? This question, of fundamental importance to the chemist, is hardly less important to the physicist, whose theory of the atom cannot be correct unless it provides a satisfactory answer. The ability of the quantum theory of the atom not only to explain chemical bonding but to do so partly in terms of an effect that has no classical analog is further testimony to the power of this approach.

8.1 MOLECULAR FORMATION

A molecule is a stable arrangement of two or more atoms. By "stable" is meant that a molecule must be given energy from an outside source in order to break up into its constituent atoms. In other words, a molecule exists because the energy of the joint system is less than that of the system of separate noninteracting atoms. If the interactions among a certain group of atoms reduce their total energy, a molecule can be formed; if the interactions increase their total energy, the atoms repel one another.

Let us consider what happens when two atoms are brought closer and closer together. Three extreme situations may occur:

1. A *covalent bond* is formed. One or more pairs of electrons are shared by the two atoms. As these electrons circulate between the atoms, they spend more time between the atoms than elsewhere, which produces an attractive force. An example is H_2, the hydrogen molecule, whose electrons belong jointly to the two protons (Fig. 8-1a).

2. An *ionic bond* is formed. One or more electrons from one atom may transfer to the other, and the resulting positive and negative ions attract each other. An example is NaCl, where the bond exists between Na^+ and Cl^- ions and not between Na and Cl atoms (Fig. 8-1b).

3. No bond is formed. When the electron structures of two atoms overlap, they constitute a single system, and according to the exclusion principle no two

electrons in such a system can exist in the same quantum state. If some of the interacting electrons are thereby forced into higher energy states than they occupied in the separate atoms, the system may have much more energy than before and be unstable. To visualize this effect, we may regard the electrons as fleeing as far away from one another as possible to avoid forming a single system, which leads to a repulsive force between the nuclei. (Even when the exclusion principle can be obeyed with no increase in energy, there will be an electrostatic repulsive force between the various electrons; this is a much less significant factor than the exclusion principle in influencing bond formation, however.)

Ionic bonds usually do not result in the formation of molecules. A molecule is an electrically neutral aggregate of atoms that is held together strongly enough to be experimentally observable as a particle. Thus the individual units that constitute gaseous hydrogen each consist of two hydrogen atoms, and we are entitled to regard them as molecules. On the other hand, the crystals of rock salt (NaCl) are aggregates of sodium and chlorine ions which, although invariably arranged in a certain definite structure (Fig. 8-2), do not pair off into discrete molecules consisting of one Na⁺ ion and one Cl⁻ ion; rock salt crystals may in fact be of almost any size. There are always equal numbers of Na⁺ and Cl⁻ ions in rock salt, so that the formula NaCl correctly represents its composition. However, these ions form molecules rather than crystals only in the gaseous state.

FIGURE 8-1 (a) Covalent bonding. The shared electrons spend more time on the average between their parent nuclei and therefore lead to an attractive force. (b) Ionic bonding. Sodium and chlorine combine chemically by the transfer of electrons from sodium atoms to chlorine atoms; the resulting ions attract electrostatically.

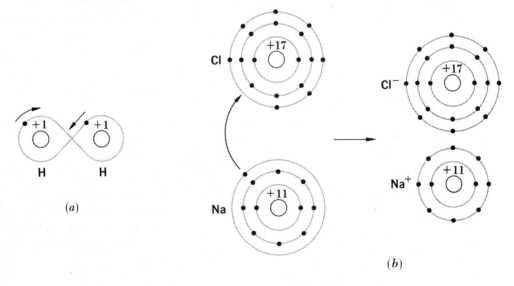

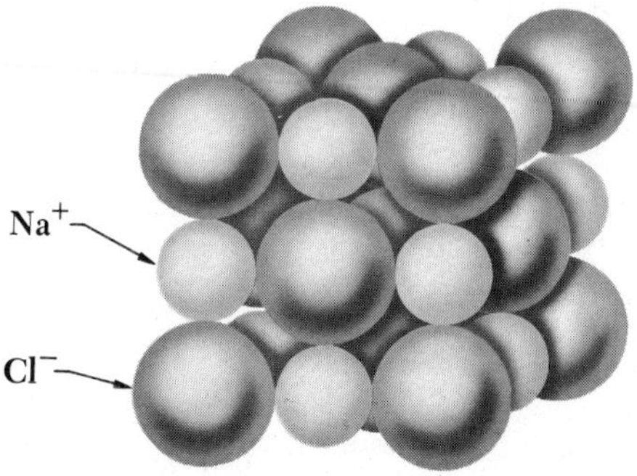

FIGURE 8-2   Scale model of NaCl crystal.

Na$^+$

Cl$^-$

In H_2 the bond is purely covalent and in NaCl it is purely ionic, but in many other molecules an intermediate type of bond occurs in which the atoms share electrons to an unequal extent. An example is the HCl molecule, where the Cl atom attracts the shared electrons more strongly than the H atom. A strong argument can be made for thinking of the ionic bond as no more than an extreme case of the covalent bond.

8.2   ELECTRON SHARING

The simplest possible molecular system is H_2^+, the hydrogen molecular ion, in which a single electron bonds two protons. Before we consider the bond in H_2^+ in detail, let us look in a general way into how it is possible for two protons to share an electron and why such sharing should lead to a lower total energy and hence to a stable system.

In Chap. 5 we discussed the phenomenon of quantum-mechanical barrier penetration: a particle can "leak" out of a box even though it does not have enough energy to break through the wall because the particle's wave function extends beyond it. Only if the wall is infinitely strong is the wave function wholly inside the box. The electric field around a proton is in effect a box for an electron, and two nearby protons correspond to a pair of boxes with a wall between them (Fig. 8-3). There is no mechanism in classical physics by which the electron in a hydrogen atom can transfer spontaneously to a neighboring proton more distant than its parent proton. In quantum physics, however, such a mechanism does exist. There is a certain probability that an electron trapped in one box will tunnel through the wall and get into the other box, and once there it has the same probability for tunneling back. This situation can be described by saying that the electron is shared by the protons.

To be sure, the likelihood that an electron will pass through the region of high potential energy—the "wall"—between two protons depends strongly upon how far apart the protons are. If the proton-proton distance is 1 Å, the electron

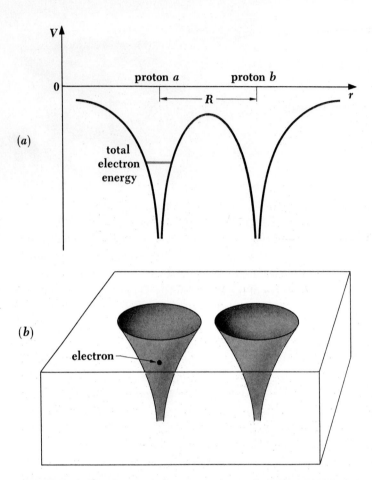

FIGURE 8-3 (a) Potential energy of an electron in the electric field of two nearby protons. The total energy of a ground-state electron in the hydrogen atom is indicated. (b) Two nearby protons correspond quantum-mechanically to a pair of boxes separated by a wall.

may be regarded as going from one proton to the other about every 10^{-15} s, which means that we can legitimately consider the electron as being shared by both. If the proton-proton distance is 10 Å, however, the electron shifts across an average of only about once per second, which is practically an infinite time on an atomic scale. Since the effective radius of the $1s$ wave function in hydrogen is 0.53 Å, we conclude that electron sharing can take place only between atoms whose wave functions overlap appreciably.

Granting that two protons can share an electron, there is a simple argument that shows why the energy of such a system could be less than that of a separate hydrogen atom and proton. According to the uncertainty principle, the smaller the region to which we restrict a particle, the greater must be its momentum and hence kinetic energy. An electron shared by two protons is less confined than one belonging to a single proton, which means that it has less kinetic energy. The total energy of the electron in H_2^+ is therefore less than that of the electron in $H + H^+$, and provided the magnitude of the proton-proton repulsion in H_2^+ is not too great, H_2^+ ought to be stable.

The preceding arguments are quantum-mechanical ones, while we normally tend to consider the interactions between charged particles in terms of electrostatic forces. There is a very important theorem, independently proved by Feynman and by Hellmann and named after them, which states in essence that both types of approach always yield identical results. According to the Feynman-Hellmann theorem, if the electron probability distribution in a molecule is known, the calculation of the system energy can proceed classically and will lead to the same conclusions as a purely quantum-mechanical calculation. The Feynman-Hellmann theorem is not an obvious one, because treating a molecule in terms of electrostatic forces does not explicitly take into account electron kinetic energy, while a quantum treatment involves the total electron energy; nevertheless, once the electron wave function ψ has been determined, either way of proceeding may be used.

8.3 THE H_2^+ MOLECULAR ION

What we would like to know is the wave function ψ of the electron in H_2^+, since from ψ we can calculate the energy of the system as a function of the separation R of the protons. If $E(R)$ has a minimum, we will know that a bond can exist, and we can also determine the bond energy and the equilibrium spacing of the protons.

Instead of solving Schrödinger's equation for ψ, which is a lengthy and complicated procedure, we shall use an intuitive approach. Let us begin by trying to predict what ψ is when R, the distance between the protons, is large compared with a_0, the radius of the smallest Bohr orbit in the hydrogen atom. In this event ψ near each proton must closely resemble the $1s$ wave function of the hydrogen atom, as pictured in Fig. 8-4 where the $1s$ wave function around proton a is called ψ_a and that around proton b is called ψ_b.

We also know what ψ looks like when R is 0, that is, when the protons are imagined to be fused together. Here the situation is that of the He^+ ion, since the electron is now in the presence of a single nucleus whose charge is $+2e$. The $1s$ wave function of He^+ has the same form as that of H but with a greater amplitude at the origin, as in Fig. 8-4e. Evidently ψ is going to be something like the wave function sketched in Fig. 8-4d when R is comparable with a_0. There is an enhanced likelihood of finding the electron in the region between the protons, which we have spoken of in terms of sharing of the electron by

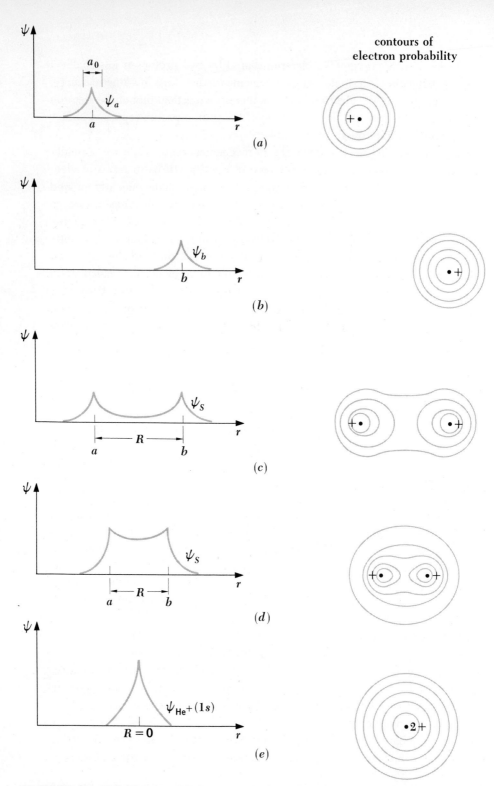

contours of
electron probability

FIGURE 8-4 The combination of two hydrogen-atom $1s$ wave functions to form the symmetric $H_2 +$ wave function ψ_S.

the protons. Thus there is on the average an excess of negative charge between the protons, and this attracts them together. We have still to establish whether the magnitude of this attraction is enough to overcome the mutual repulsion of the protons.

The combination of ψ_a and ψ_b in Fig. 8-4 is symmetric, since exchanging a and b does not affect ψ (see Sec. 7.3). However, it is also conceivable that we could have an *antisymmetric* combination of ψ_a and ψ_b, as in Fig. 8-5. Here there is a node between a and b where $\psi = 0$, which implies a diminished likelihood of finding the electron between the protons. Now there is on the average a deficiency of negative charge between the protons, and in consequence a repulsive force. With only repulsive forces acting, bonding cannot occur.

An interesting question concerns the behavior of the antisymmetric H_2^+ wave function ψ_A as $R \rightarrow 0$. Obviously ψ_A does not become the $1s$ wave function of He^+ when $R = 0$. However, ψ_A *does* approach the $2p$ wave function of He^+ (Fig. 8-5e), which has a node at the origin. Since the $2p$ state of He^+ is an excited state while the $1s$ state is the ground state, H_2^+ in the antisymmetric state ought to have more energy than when it is in the symmetric state, which agrees with our inference from the shapes of the wave functions ψ_A and ψ_S that in the former case there is a repulsive force and in the latter an attractive one.

A line of reasoning similar to the preceding one enables us to estimate how the total energy of the H_2^+ system varies with R. We first consider the symmetrical state. When R is large, the electron energy E_S must be the -13.6-eV energy of the hydrogen atom, while the electrostatic potential energy V_p of the protons,

8.1 $$V_p = \frac{e^2}{4\pi\varepsilon_0 R}$$

falls to 0 as $R \rightarrow \infty$. (V_p is a positive quantity, corresponding to a repulsive force.) When $R = 0$, the electron energy must equal that of the He^+ ion, which is Z^2 or 4 times that of the H atom. (See Prob. 25 of Chap. 4; the same result is obtained from the quantum theory of one-electron atoms.) Hence $E_S = -54.4$ eV when $R = 0$. Also, when $R \rightarrow 0$, $V_p \rightarrow \infty$ as $1/R$. Both E_S and V_p are sketched in Fig. 8-6 as functions of R; the shape of the curve for E_S can only be approximated without a detailed calculation, but we do have its value for both $R = 0$ and $R = \infty$ and, of course, V_p obeys Eq. 8.1.

The total energy $E_S{}^{\text{total}}$ of the system is the sum of the electron energy E_S and the potential energy V_p of the protons. Evidently $E_S{}^{\text{total}}$ has a minimum, which corresponds to a stable molecular state. This result is confirmed by the experimental data on H_2^+ which indicate a bond energy of 2.65 eV and an equilibrium separation R of 1.06 Å. By "bond energy" is meant the energy needed to break H_2^+ into $H + H^+$; the *total* energy of H_2^+ is the -13.6 eV of the hydrogen atom plus the -2.65-eV bond energy, or -16.3 eV in all.

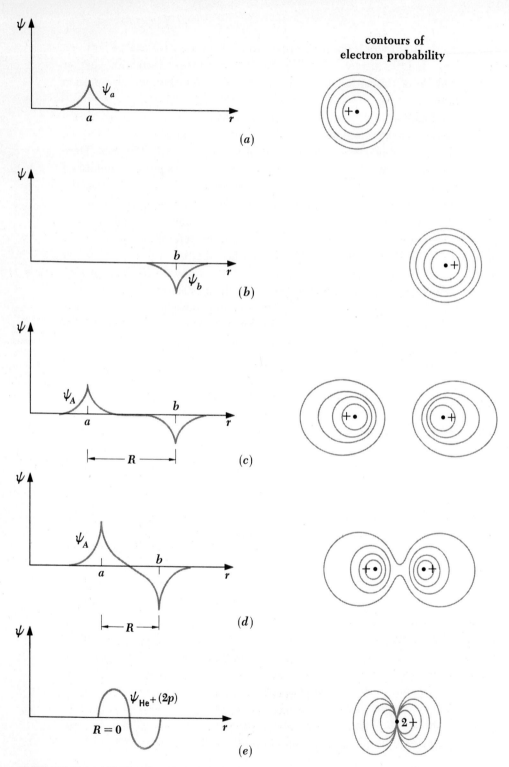

contours of
electron probability

ψ_a

a

r

(a)

b

ψ_b

r

(b)

ψ_A

a

b

r

R

(c)

ψ_A

a

b

r

R

(d)

$\psi_{\text{He}^+}(2p)$

$R = 0$

r

(e)

FIGURE 8-5 The combination of two hydrogen-atom $1s$ wave functions to form the antisymmetric H_2^+ wave function ψ_A.

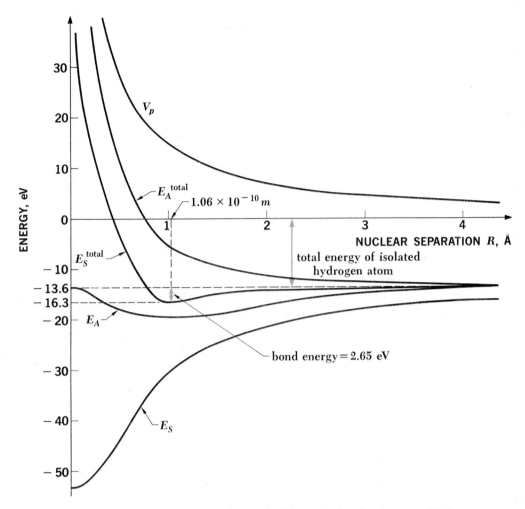

FIGURE 8-6 Electronic, proton repulsion, and total energy in H_2^+ as a function of nuclear separation R for the symmetric and antisymmetric states. The antisymmetric state has no minimum in its total energy.

In the case of the antisymmetric state, the analysis proceeds in the same way except that the electron energy E_A when $R = 0$ is that of the $2p$ state of He^+. This energy is proportional to Z^2/n^2; hence with $Z = 2$ and $n = 2$ it is just equal to the -13.6 eV of the ground-state hydrogen atom. Since $E_A \to -13.6$ eV also as $R \to \infty$, we might think that the electron energy is constant, but actually there is a small dip at intermediate distances. However, the dip is not nearly enough to yield a minimum in the total energy curve for the antisymmetric state, as indicated in Fig. 8-6, and so in this state no bond is formed.

8.4 THE H₂ MOLECULE

The H_2 molecule contains two electrons instead of the single electron of H_2^+. According to the exclusion principle, both electrons can share the same *orbital* (that is, be described by the same wave function ψ_{nlm_l}) provided their spins are antiparallel. With two electrons to contribute to the bond, H_2 ought to be more stable than H_2^+—at first glance, twice as stable, with a bond energy of 5.3 eV compared with 2.65 eV for H_2^+. However, the H_2 orbitals are not quite the same as those of H_2^+ because of the electrostatic repulsion between the two electrons in H_2, a factor absent in the case of H_2^+. The latter repulsion weakens the bond in H_2, so that the actual bond energy is 4.5 eV instead of 5.3 eV. For the same reason, the bond length in H_2 is 0.74 Å, which is somewhat larger than the use of unmodified H_2^+ wave functions would indicate. The general conclusion in the case of H_2^+ that the symmetric wave function ψ_S leads to a bound state and the antisymmetric wave function ψ_A to an unbound one remains valid for H_2.

In Sec. 7.3 the exclusion principle was formulated in terms of the symmetry and antisymmetry of wave functions, and it was concluded that systems of electrons are always described by antisymmetric wave functions (that is, by wave functions that reverse sign upon the exchange of any pair of electrons). However, we have just said that the bound state in H_2 corresponds to both electrons being described by a symmetrical wave function ψ_S, which seems to contradict the above conclusion.

A closer look shows that there is really no contradiction here. The *complete* wave function $\Psi(1,2)$ of a system of two electrons is the product of a spatial wave function $\psi(1,2)$ which describes the coordinates of the electrons and a spin function $s(1,2)$ which describes the orientations of their spins. The exclusion principle requires that the complete wave function

$$\Psi(1,2) = \psi(1,2)s(1,2)$$

be antisymmetric to an exchange of both coordinates and spins, not $\psi(1,2)$ by itself, and what we have been calling a molecular orbital is the same as $\psi(1,2)$. An antisymmetric complete wave function Ψ_A can result from the combination of a symmetric coordinate wave function ψ_S and an antisymmetric spin function s_A or from the combination of an antisymmetric coordinate wave function ψ_A and a symmetric spin function s_S. That is, only

$$\Psi = \psi_S s_A$$

and

$$\Psi = \psi_A s_S$$

are acceptable. If the spins of the two electrons are parallel, their spin function is symmetric since it does not change sign when the electrons are exchanged. Hence the coordinate wave function ψ for two electrons whose spins are parallel must be antisymmetric; we may express this by writing

$$\psi\uparrow\uparrow = \psi_A$$

On the other hand, if the spins of the two electrons are antiparallel, their spin function is antisymmetric since it reverses sign when the electrons are exchanged. Hence the coordinate wave function ψ for two electrons whose spins are anti-parallel must be symmetric, and we may express this by writing

$$\psi\uparrow\downarrow = \psi_S$$

Schrödinger's equation for the H_2 molecule has no exact solution. In fact, only for H_2^+ is an exact solution possible, and all other molecular systems must be treated approximately. The results of a detailed analysis of the H_2 molecule are shown in Fig. 8-7 for the case when the electrons have their spins parallel and the case when their spins are antiparallel. The difference between the two curves is due to the exclusion principle, which prevents two electrons in the same quantum state in a system from having the same spin and therefore leads to a dominating repulsion when the spins are parallel.

FIGURE 8-7   The variation of the energy of the system H + H with their distances apart when the electron spins are parallel and antiparallel.

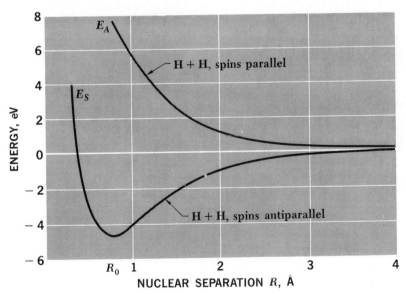

8.5 MOLECULAR ORBITALS

Covalent bonding in molecules other than H_2, diatomic as well as polyatomic, is usually a more complicated story. It would be yet more complicated but for the fact that any alteration in the electronic structure of an atom due to the proximity of another atom is confined to its outermost (or *valence*) electron shell. There are two reasons for this. First, the inner electrons are much more tightly bound and hence less responsive to external influences, partly because they are closer to their parent nucleus and partly because they are shielded from the nuclear charge by fewer intervening electrons. Second, the repulsive interatomic forces in a molecule become predominant while the inner shells of its atoms are still relatively far apart. Direct evidence in support of the idea that only the valence electrons are involved in chemical bonding is available from the X-ray spectra that arise from transitions to inner-shell electron states; it is found that these spectra are virtually independent of how the atoms are combined in molecules or solids.

In discussing chemical bonding it is helpful to be able to visualize the distributions in space of the various atomic orbitals, which qualitatively resemble those of hydrogen. The pictures in Fig. 6-11 are limited to two dimensions and hence are not suitable for this purpose. It is more appropriate here to draw boundary surfaces of constant $|\psi|^2$ in each case that outline the regions within which the total probability of finding the electron has some definite value, say 90 or 95 percent. Further, the sign of the wave function ψ can be indicated in each lobe of such a drawing, even though what is being represented is $|\psi|^2$. Figure 8-8 contains boundary-surface diagrams for s, p, and d orbitals. These diagrams show $|\Theta\Phi|^2$ in each case; for the corresponding radial probability densities $|R|^2$, Fig. 6-10 can be consulted. The total probability density $|\psi|^2$ is, of course, equal to the product of $|\Theta\Phi|^2$ and $|R|^2$.

In a number of cases the orbitals shown in Fig. 8-8 are derived from linear combinations of two atomic wave functions representing states of the same energy; such combinations are also solutions of Schrödinger's equation. For example, a p_x orbital is the result of adding together the $l = 1$ wave functions for $m_l = +1$ and $m_l = -1$:

$$\psi_{p_x} = \frac{1}{\sqrt{2}}(\psi_{p+1} + \psi_{p-1})$$

(The factor $1/\sqrt{2}$ is required to normalize ψ_{p_x}.) Similarly the p_y orbital is given by

$$\psi_{p_y} = \frac{-i}{\sqrt{2}}(\psi_{p+1} - \psi_{p-1})$$

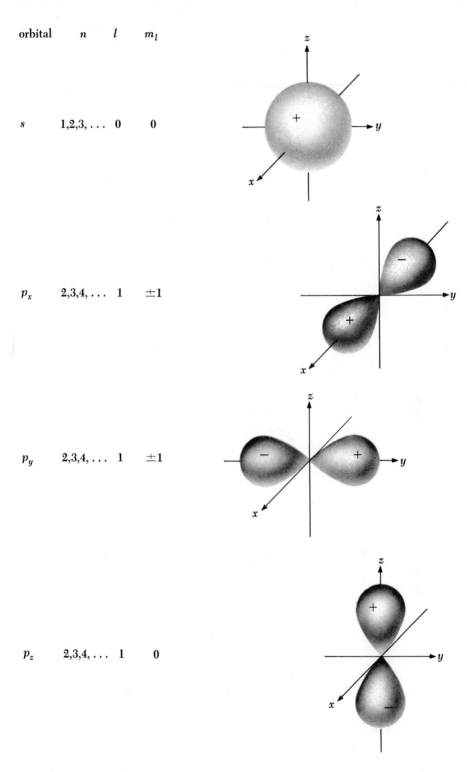

orbital	n	l	m_l
s	$1,2,3,\ldots$	0	0
p_x	$2,3,4,\ldots$	1	± 1
p_y	$2,3,4,\ldots$	1	± 1
p_z	$2,3,4,\ldots$	1	0

FIGURE 8-8 Boundary surface diagrams for s, p, and d atomic orbitals. The $+$ and $-$ signs refer to the sign of the wave function in each region.

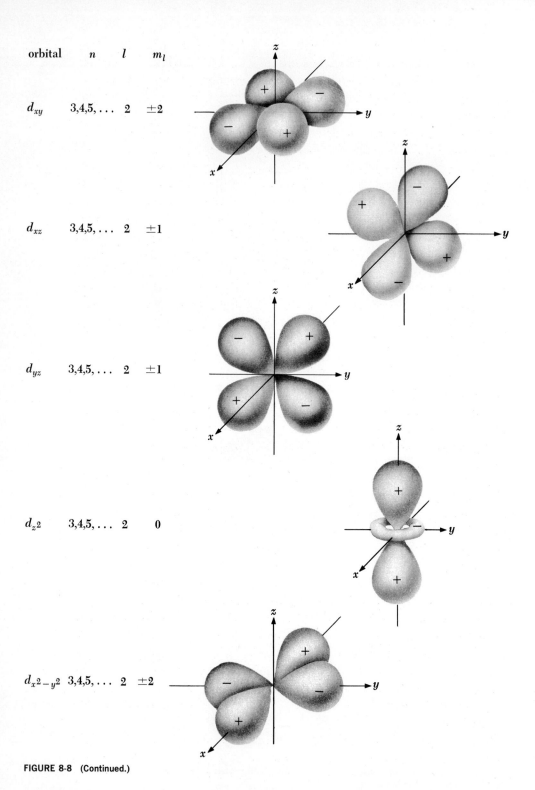

orbital	n	l	m_l
d_{xy}	3,4,5,...	2	± 2
d_{xz}	3,4,5,...	2	± 1
d_{yz}	3,4,5,...	2	± 1
d_{z^2}	3,4,5,...	2	0
$d_{x^2-y^2}$	3,4,5,...	2	± 2

FIGURE 8-8 (Continued.)

The p_z orbital, however, is identical with the $l = 1$, $m_l = 0$ wave function. The wave functions which are combined to form the d_{xz}, d_{yz}, d_{xy}, and $d_{x^2-y^2}$ orbitals are indicated in Fig. 8-8. The interactions between two atoms that yield a covalent bond between them have, as a consequence, different probability-density distributions for the electrons that participate in the bond than those which are characteristic of the atoms when alone in space, and these new distributions are easiest to understand in terms of the orbitals shown in Fig. 8-8.

When two atoms come together, their orbitals overlap and the result will be either an increased electron probability density between them that signifies a *bonding molecular orbital* or a decreased concentration that signifies an *antibonding molecular orbital*. In the previous section we saw how the 1s orbitals of two hydrogen atoms could combine to form either the bonding orbital ψ_S or the antibonding orbital ψ_A. In the terminology of molecular physics, ψ_S is referred to as a $1s\,\sigma$ orbital and ψ_A as a $1s\,\sigma°$ orbital. The "1s" identifies the atomic orbitals that are imagined to combine to form the molecular orbital. The Greek letter σ signifies that the molecular state has no angular momentum about the bond axis (which is taken to be the z axis). This component of the angular momentum of a molecule is quantized, and is restricted to the values $\lambda\hbar$ where $\lambda = 0, 1, 2, \ldots$. Molecular states for which $\lambda = 0$ are denoted by σ, those for which $\lambda = 1$ by π, those for which $\lambda = 2$ by δ, and so on in alphabetical order. Finally, an antibonding orbital is labeled with an asterisk, as in $1s\,\sigma°$ for the antibonding H_2 orbital ψ_A.

Figure 8-9 contains boundary-surface diagrams that show the formation of σ and π molecular orbitals from s and p atomic orbitals in homonuclear diatomic molecules. Evidently σ orbitals show rotational symmetry about the bond axis, while π orbitals change sign upon a 180° rotation about the bond axis. Since the lobes of p_z orbitals are on the bond axis, these atomic orbitals form σ molecular orbitals. The p_x and p_y orbitals both form π molecular orbitals.

A heteronuclear diatomic molecule consists of two unlike atoms. In general, the atomic orbitals are not the same in such molecules, so that the bonding electrons in them are not equally shared by both atoms. LiH is the simplest heteronuclear molecule and is a good example of this effect. The normal configuration of the H atom is 1s and that of the Li atom is $1s^2 2s$, which means in each case that there is a single valence electron. The 1s orbital of H and the 2s orbital of Li form a σ bonding orbital in LiH that is occupied by the two valence electrons (Fig. 8-10). In both atoms the effective nuclear charge acting on a valence electron is $+e$ (in Li the core of two 1s electrons shields $+2e$ of the total nuclear charge of $+3e$), but in Li a valence electron is on the average several times farther from the nucleus than in H. (The respective ionization energies reflect this difference, with the ionization energy of H being 13.6 eV while it is 5.4 eV in Li.) Hence the electrons in the σ bonding orbital of LiH

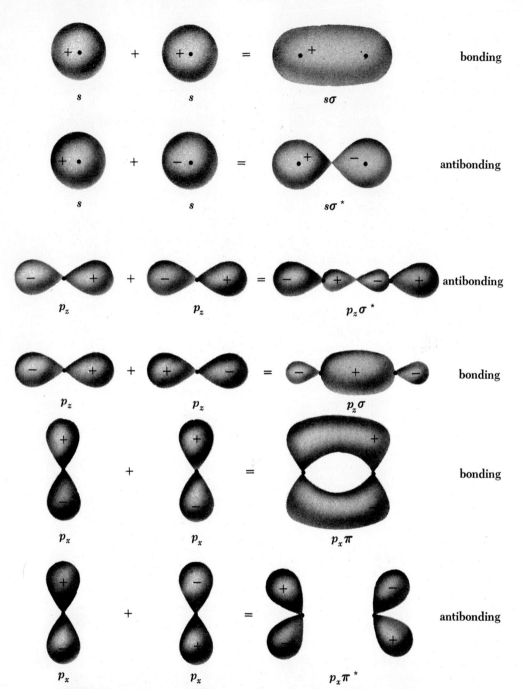

FIGURE 8-9 Boundary surface diagrams showing the formation of molecular orbitals from s and p atomic orbitals in homonuclear diatomic molecules. The z axis is along the internuclear axis of the molecule in each case, and the plane of the paper is the xz plane. The $p_y\pi$ and $p_y\pi^*$ orbitals are the same as the $p_x\pi$ and $p_x\pi^*$ orbitals except that they are rotated through 90°.

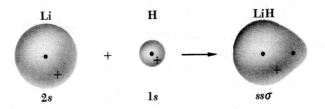

FIGURE 8-10 The bonding electrons in LiH occupy a σ molecular orbital formed from the $1s$ orbital of the H atom and the $2s$ orbital of the Li atom.

Li   H   LiH

$2s$ $1s$ $ss\sigma$

favor the H nucleus, and there is a partial separation of charge in the molecule.

If there were a complete separation of charge in LiH, as there is in NaCl, the molecule would consist of an Li^+ ion and a H^- ion, and the bond would be purely ionic. Instead the bond is only partially ionic, with the two bonding electrons spending perhaps 80 percent of the time in the neighborhood of the H nucleus and 20 percent in the neighborhood of the Li nucleus. In contrast, the bonding electrons in a homonuclear molecule such as H_2 or O_2 spend 50 percent of the time near each nucleus. Molecules whose bonds are neither purely covalent nor purely ionic are sometimes called *polar covalent*, since they possess electric dipole moments. The relative tendency of an atom to attract shared electrons when it is part of an atom is called its *electronegativity*. In the LiH molecule, for instance, H is more electronegative than Li.

In a heteronuclear molecule the atomic orbitals that are imagined to combine to form a molecular orbital may be of different character in each atom. An example is HF, where the $1s$ atomic orbital of H joins with the $2p_z$ orbital of F. There are two possibilities, as in Fig. 8-11, a bonding $sp\sigma$ molecular orbital and an antibonding $sp\sigma°$ orbital. Since the $1s$ orbital of H and the $2p_z$ orbital of F each contain one electron (Table 8.1), the $sp\sigma$ orbital in HF is occupied by two electrons, and we may regard HF as being held together by a single covalent bond. The electron structure of the HF molecule is shown in Fig. 8-12.

FIGURE 8-11   Bonding and antibonding molecular orbitals in HF.

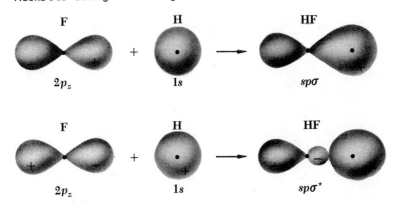

F   H   HF

$2p_z$ $1s$ $sp\sigma$

F   H   HF

$2p_z$ $1s$ $sp\sigma^*$

THE PHYSICS OF MOLECULES   259

Table 8.1.

ATOMIC STRUCTURES OF FIRST- AND SECOND-PERIOD ELEMENTS. Arrows indicate spin directions of electrons. According to Hund's rule (Sec. 7.6) electrons in the same subshell of an atom (same
orbital quantum number l) have as many of their spins parallel as possible.

Element		Atomic number	Atomic structure	$1s$	$2s$	$2p_x$	$2p_y$	$2p_z$
						Occupancy of orbitals		
Hydrogen	H	1	$1s$	↑				
Helium	He	2	$1s^2$	↑↓				
Lithium	Li	3	$1s^2 2s$	↑↓	↑			
Beryllium	Be	4	$1s^2 2s^2$	↑↓	↑↓			
Boron	B	5	$1s^2 2s^2 2p$	↑↓	↑↓	↑		
Carbon	C	6	$1s^2 2s^2 2p^2$	↑↓	↑↓	↑	↑	
Nitrogen	N	7	$1s^2 2s^2 2p^3$	↑↓	↑↓	↑	↑	↑
Oxygen	O	8	$1s^2 2s^2 2p^4$	↑↓	↑↓	↑↓	↑	↑
Fluorine	F	9	$1s^2 2s^2 2p^5$	↑↓	↑↓	↑↓	↑↓	↑
Neon	Ne	10	$1s^2 2s^2 2p^6$	↑↓	↑↓	↑↓	↑↓	↑↓

The configurations of the three p atomic orbitals together with their ability
to join with s orbitals to form bonding molecular orbitals make it possible to
understand the geometries of many polyatomic molecules. The water molecule
H_2O is an example. Offhand we might expect a linear molecule, H—O—H, since
oxygen is more electronegative than hydrogen and each H atom in H_2O accordingly exhibits a small positive charge. The resulting repulsion between the H
atoms should keep them as far apart as possible, namely on opposite sides of
the O atom. In reality, however, the water molecule has a structure closer to
O—H with an angle of 104.5° between the two O—H bonds.
|
H

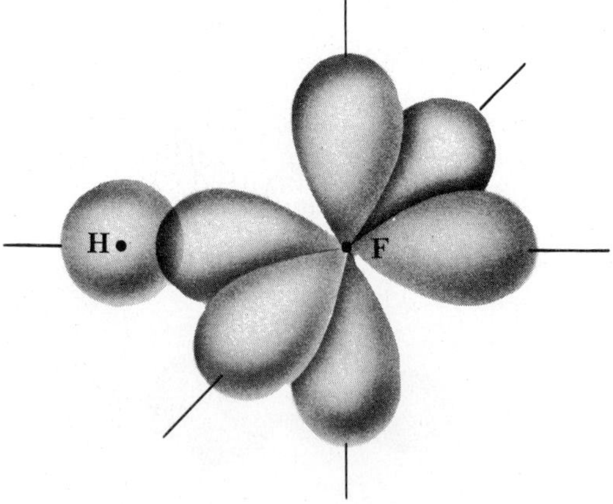

FIGURE 8-12   Valence atomic orbitals in
HF. The atomic orbitals shown as overlapping form a σ bonding molecular
orbital.

The bent shape of the water molecule is easy to explain. From Table 8.1 we find that the $2p_y$ and $2p_z$ orbitals in O are only singly occupied, so that each can join the $1s$ orbital of an H atom to form a $sp\sigma$ bonding orbital (Fig. 8-13). The y and z axes are 90° apart, and the larger 104.5° angle that is actually found may plausibly be attributed to the mutual repulsion of the H atoms. In support of the latter idea is the fact that the bond angles in the otherwise similar molecules H_2S and H_2Se are 92° and 90°, respectively, which we may ascribe to the greater separation of the H atoms around the larger atoms S (atomic number $Z = 16$) and Se ($Z = 34$).

A similar argument explains the pyramidal shape of the ammonia molecule NH_3. From Table 8.1 we find that the $2p_x$, $2p_y$, and $2p_z$ atomic orbitals in N are singly occupied, which means that each of them can form a $sp\sigma$ bonding orbital with the $1s$ orbital of an H atom. The bonding molecular orbitals in NH_3 should therefore be centered along the x, y, and z axes (Fig. 8-14) with N—H bonds 90° apart. As in H_2O, the actual bond angles in NH_3 are somewhat greater than 90°, in this case 107.5°, owing to repulsions among the H atoms. The similar hydrides of the larger atoms P ($Z = 15$) and As ($Z = 33$) exhibit the smaller bond angles of 94° and 90°, respectively, again in consequence of the reduced mutual repulsions among the more distant H atoms.

8.6 HYBRID ORBITALS

The straightforward way in which the shapes of the H_2O and NH_3 molecules are explained is a conspicuous failure in the case of methane, CH_4. A carbon atom has two electrons in its $2s$ orbital and one each in its $2p_x$ and $2p_y$ orbitals.

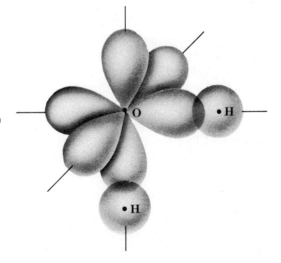

FIGURE 8-13 Valence atomic orbitals in H_2O. The bond angle is actually 104.5°.

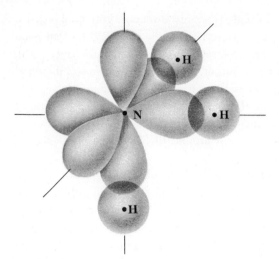

FIGURE 8-14   Valence atomic orbitals in NH_3. The bond angles are actually 107.5°.

Thus we would expect the hydride of carbon to be CH_2, with two $sp\sigma$ bonding orbitals and a bond angle of 90° or a little more. Yet CH_4 exists, and, furthermore, it is perfectly symmetrical in structure with tetrahedral molecules whose C—H bonds are exactly equivalent to one another.

We cannot consider carbon as an isolated exception, a sort of freak atom in which a fortuitous combination of circumstances leads to CH_4 instead of CH_2, because the same phenomenon occurs in other atoms as well. A boron atom, for example, has the configuration $1s^22s^22p$, and it forms BF_3 and BCl_3 instead of BF and BCl.

Clearly, what is happening in carbon and boron is that the $2s$ orbitals, despite being occupied by electron pairs and having less energy—more stability—than the $2p$ orbitals, somehow enter into the formation of molecular orbitals and thereby permit the $2s$ electrons to contribute to the bonds formed by C and B with other atoms. A C atom has four $n = 2$ electrons in all and forms CH_4, while a B atom has three $n = 2$ electrons in all and forms BF_3, respectively sharing four and three electrons with their partners.

The easiest way to explain the existence of CH_4 is to assume that one of the two $2s$ electrons in C is "promoted" to the vacant $2p_z$ orbital, so that there is now one electron each in the $2s$, $2p_x$, $2p_y$, and $2p_z$ orbitals and four bonds can be formed. To raise an electron from a $2s$ to a $2p$ state means increasing the energy of the C atom, but it is reasonable to suppose that the formation of four bonds (to yield CH_4) in place of two (to yield CH_2) lowers the energy of the resulting molecule more than enough to compensate for this. The foregoing picture suggests that three of the bonds in CH_4 are $sp\sigma$ bonds and one of them is a $ss\sigma$ bond involving the $1s$ orbital of H and now singly occupied $2s$ orbital of C; experimentally, however, all four bonds are found to be identical.

The correct explanation for CH_4 is based on a phenomenon called *hybridization* which can occur when the 2s and 2p states of an atom in a molecule are close together in energy. In this case the atom can contribute a linear combination of *both* its 2s and 2p atomic orbitals to *each* molecular orbital if in this way the resulting bonds are more stable than otherwise. That such composite atomic orbitals can occur follows from the nature of Schrödinger's equation, which is a partial differential equation. The 2s and 2p wave functions of an atom are both solutions of the same equation if the corresponding energies are the same, and a linear combination of solutions of a partial differential equation is always itself a solution. In an isolated atom, an electron in a 2s orbital has less energy (is more tightly bound) than an electron in a 2p orbital, and there is accordingly no tendency for hybrid atomic orbitals to occur. On the other hand, when an atom in a certain molecule contributes superposed s and p orbitals to the molecular orbitals, the resulting bonds may be stronger than the bonds that the s and p orbitals by themselves would lead to, even though the p parts of the hybrids had higher energies in the isolated atom. Hybrid orbitals thus occur when the bonding energy they give rise to is greater than that which pure orbitals would produce, which happens in practice when the s and p levels of an atom are close together.

In CH_4, then, carbon has four equivalent hybrid orbitals which participate in bonding. These four orbitals are hybrids of one 2s and three 2p orbitals, and we may consider each one as a combination of $\frac{1}{4}s$ and $\frac{3}{4}p$. This particular combination is therefore called a sp^3 hybrid. Its configuration can be visualized in terms of boundary-surface diagrams as shown in Fig. 8-15. Evidently a sp^3 hybrid orbital is strongly concentrated in a single direction, which accounts for its ability to produce an exceptionally strong bond—strong enough to compensate for the need to promote a 2s electron to a 2p state.

FIGURE 8-15 In sp^3 hybridization, an s orbital and three p orbitals in the same atom combine to form four sp^3 hybrid orbitals.

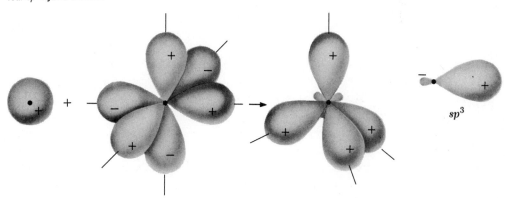

It must be kept in mind that hybrid orbitals do not exist in an isolated atom, even when it is in an excited state, but arise while that atom is interacting with others to form a molecule.

Figure 8-16 is a representation of the CH_4 molecule. Also shown is a model of this molecule that consists of a C atom in the center of a unit cube which has H atoms at alternate corners. A triangle with the C atom at one vertex and any two H atoms at the other vertexes has sides $\sqrt{3}/2$, $\sqrt{3}/2$, and $\sqrt{2}$ in length. If the angle between the C—H bonds is θ, from the law of cosines ($a^2 = b^2 + c^2 - 2bc \cos \theta$) we have

$$\cos \theta = -\frac{a^2 - b^2 - c^2}{2bc}$$

$$= -\frac{2 - \frac{3}{4} - \frac{3}{4}}{2 \times \frac{3}{4}} = -\frac{1}{3}$$

$$\theta = 109.5°$$

which is what is determined experimentally.

The bond angles of 104.5° in H_2O and 107.5° in NH_3 are evidently closer to the tetrahedral angle of 109.5° that occurs in sp^3 hybrid bonding than to the 90° expected if only p orbitals in the O and N atoms are involved. This fact provides a way to explain how the repulsions among the H atoms in these molecules that were spoken of earlier can be incorporated into the molecular-orbital description of bonding. In NH_3 there are three doubly occupied bonding orbitals, leaving one unshared pair of electrons that we earlier supposed to be

FIGURE 8-16 The tetrahedral methane (CH_4) molecule. The overlapping sp^3 hybrid orbitals of the C atom and 1s orbitals of the four H atoms form bonding molecular orbitals.

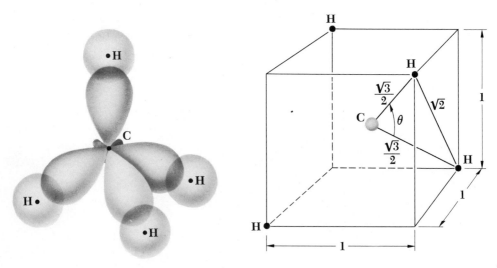

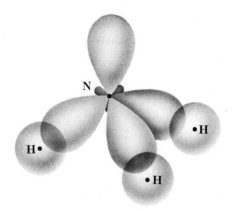

FIGURE 8-17 Valence atomic orbitals in the ammonia (NH_3) molecule based on the assumption of sp^3 hybridization in the N orbitals. One of the sp^3 orbitals is occupied by two N electrons and does not contribute to bonding.

in the $2s$ orbital of N. If there were sp^3 hybrid orbitals furnished by N instead of p orbitals, the more separated bonds would mean a lower energy for the system. Opposing this gain in molecular stability is the need to promote the pair of nonbonding $2s$ electrons to the higher-energy sp^3 hybrid state without any contribution by them to the bonding process (unlike the case of CH_4 where all four sp^3 orbitals participate in bonds). Hence we may regard the 107.5° bond angle in NH_3 as the result of a compromise between the two extremes of four sp^3 hybrid orbitals in N with one of them nonbonding and three $2p$ bonding orbitals and one $2s$ nonbonding (but low-energy) orbital. Figure 8-17 is a representation of the NH_3 molecule on the basis of sp^3 hybridization, which may be compared with Fig. 8-14 which was drawn on the basis of p orbitals.

In H_2O, since there are two bonding orbitals and two nonbonding orbitals in O, the tendency to form hybrid sp^3 orbitals is less than in NH_3 where there are three bonding orbitals and only one nonbonding orbital. The smaller bond angle in H_2O is in agreement with this conclusion. The structure of the H_2O molecule is further discussed in Sec. 10.4.

8.7 CARBON-CARBON BONDS

Two other types of hybrid orbital in addition to sp^3 can occur in carbon atoms. In sp^2 hybridization, one valence electron is in a pure p orbital and the other three are in hybrid orbitals that are $\frac{1}{3}s$ and $\frac{2}{3}p$ in character. In sp hybridization, two valence electrons are in pure p orbitals and the other two are in hybrid orbitals that are $\frac{1}{2}s$ and $\frac{1}{2}p$ in character.

Ethylene, C_2H_4, is an example of sp^2 hybridization in which the two carbon atoms are joined by two bonds. Figure 8-18 contains a boundary-surface diagram

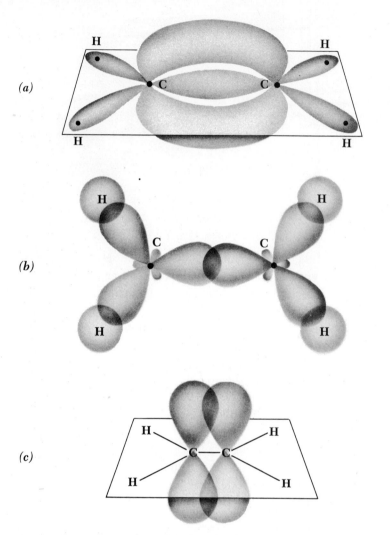

FIGURE 8-18 (a) The ethylene (C_2H_4) molecule. All the atoms lie in a plane perpendicular to the plane of the paper. (b) Top view, showing the sp^2 hybrid orbitals that form σ bonds between the C atoms and between each C atom and two H atoms. (c) Side view, showing the pure p_x orbitals that form a π bond between the C atoms.

showing the three sp^2 hybrid orbitals, which are 120° apart in the plane of the paper, and the pure p_x orbital in each C atom. Two of the sp^2 orbitals in each C atom overlap s orbitals in H atoms to form σ bonding orbitals, and the third sp^2 orbital in each C atom forms a σ bonding orbital with the same orbital in the other C atom. The p_x orbitals of the C atoms form a π bond with each other, so that one of the bonds between the carbon atoms is a σ bond and the other is a π bond. The conventional structural formula of ethylene is accordingly

$$\underset{H}{\overset{H}{\diagdown}}C=C\underset{H}{\overset{H}{\diagup}}$$

Ethylene

Acetylene, C_2H_2, is an example of sp hybridization in which the two carbon atoms are joined by three bonds. One sp hybrid orbital in each C atom forms a σ bond with an H atom, and the second forms a σ bond with the other atom. The $2p_x$ and $2p_y$ orbitals in each C atom form π bonds, so that one of the three bonds between the carbon atoms is a σ_z bond and the others are π_x and π_y bonds (Fig. 8-19). The conventional structural formula of acetylene is

$$H-C\equiv C-H$$

Acetylene

In both ethylene and acetylene the electrons in the π orbitals are "exposed" on the outside of the molecules. These compounds are much more reactive chemically than molecules with only single σ bonds between carbon atoms, such as ethane,

$$H-\underset{\overset{|}{H}}{\overset{\overset{|}{H}}{C}}-\underset{\overset{|}{H}}{\overset{\overset{|}{H}}{C}}-H$$

Ethane

in which all the bonds are formed from sp^3 hybrid orbitals in the carbon atoms. Carbon compounds with double and triple bonds are said to be *unsaturated* because they can add other atoms to their molecules in such reactions as

$$\underset{H}{\overset{H}{\diagdown}}C=C\underset{H}{\overset{H}{\diagup}}\ +\ HCl\ \longrightarrow\ H-\underset{\overset{|}{H}}{\overset{\overset{|}{H}}{C}}-\underset{\overset{|}{Cl}}{\overset{\overset{|}{H}}{C}}-H$$

FIGURE 8-19 The acetylene (C_2H_2) molecule. There are three bonds between the C atoms, one σ bond between sp hybrid orbitals and two π bonds between pure p_x and p_y orbitals.

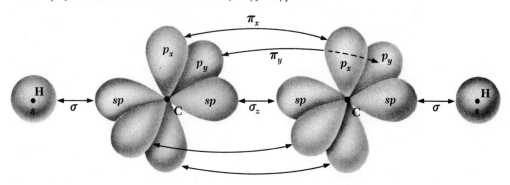

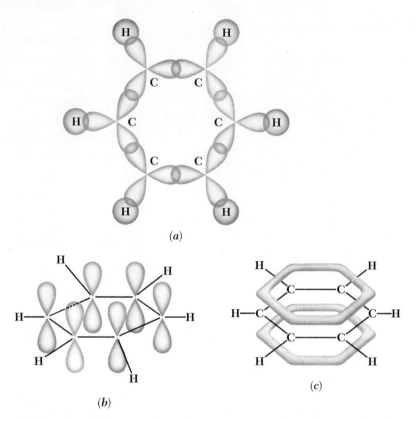

(a)

(b)

(c)

FIGURE 8-20 The benzene molecule. (a) The overlaps between the sp^2 hybrid orbitals in the C atoms with each other and with the s orbitals of the H atoms lead to σ bonds. (b) Each C atom has a pure p_x orbital occupied by one electron. (c) The bonding π molecular orbitals formed by the six p_x atomic orbitals constitute a continuous electron probability distribution around the molecule that contains six delocalized electrons.

In a *saturated* compound such as methane or ethane only single bonds are present.

In benzene, C_6H_6, the six C atoms are arranged in a flat hexagonal ring. Because the carbon-carbon bonds in the benzene ring are 120° apart, we conclude that the basic structure of the molecule is the result of bonding by sp^2 hybrid orbitals. Of the three sp^2 orbitals per C atom, one forms a σ bonding orbital with the 1s orbital of an H atom and the other two form σ bonding orbitals with the corresponding sp^2 orbitals of the C atoms on either side (Fig. 8-20). This leaves one $2p_x$ orbital per C atom, which has lobes above and below the plane of the ring. The total of six $2p_x$ orbitals in the molecule combine to produce bonding π orbitals which take the form of a continuous electron probability distribution above and below the plane of the ring. The six electrons belong to the molecule as a whole and not to any particular pair of atoms; these electrons are *delocalized*.

8.8 ROTATIONAL ENERGY LEVELS

Molecular energy states arise from the rotation of a molecule as a whole and from the vibrations of its constituent atoms relative to one another as well as from changes in its electronic configuration. Rotational states are separated by quite small energy intervals (10^{-3} eV is typical), and the spectra that arise from transitions between these states are in the microwave region with wavelengths of 0.1 mm to 1 cm. Vibrational states are separated by somewhat larger energy intervals (0.1 eV is typical), and vibrational spectra are in the infrared region with wavelengths of 10,000 Å to 0.1 mm. Molecular electronic states have higher energies, with typical separations between the energy levels of valence electrons of several electron volts and spectra in the visible and ultraviolet regions. A detailed picture of a particular molecule can often be obtained from its spectra, including bond lengths, force constants, and bond angles. For simplicity the treatment here will be restricted to diatomic molecules, but the main ideas apply to more complicated ones as well.

The lowest energy levels of a diatomic molecule arise from rotation about its center of mass. We may picture such a molecule as consisting of atoms of masses m_1 and m_2 a distance R apart, as in Fig. 8-21. The moment of inertia of this molecule about an axis passing through its center of mass and perpendicular to a line joining the atoms is

8.2
$$I = m_1 r_1^2 + m_2 r_2^2$$

where r_1 and r_2 are the distances of atoms 1 and 2 respectively from the center of mass. Since

8.3
$$m_1 r_1 = m_2 r_2$$

by definition, the moment of inertia may be written

$$I = \left(\frac{m_1 m_2}{m_1 + m_2}\right)(r_1 + r_2)^2$$

8.4
$$= m'R^2$$

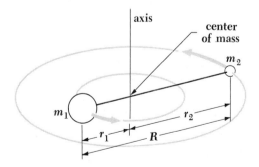

FIGURE 8-21 A diatomic molecule can rotate about its center of mass.

where

8.5 $m' = \dfrac{m_1 m_2}{m_1 + m_2}$ **Reduced mass**

is the reduced mass of the molecule as mentioned in Sec. 4.9. Equation 8.4 states that the rotation of a diatomic molecule is equivalent to the rotation of a single particle of mass m' about an axis located a distance of R away.

The angular momentum **L** of the molecule has the magnitude

8.6 $L = I\omega$

where ω is its angular velocity. Angular momentum is always quantized in nature, as we know. If we denote the *rotational quantum number* by J, we have here

8.7 $L = \sqrt{J(J + 1)}\, \hbar$ $J = 0, 1, 2, 3, \ldots$

The energy of a rotating molecule is $\frac{1}{2}I\omega^2$, and so its energy levels are specified by

$$E_J = \tfrac{1}{2}I\omega^2$$

$$= \frac{L^2}{2I}$$

8.8 $\qquad = \dfrac{J(J + 1)\hbar^2}{2I}$ **Rotational energy levels**

Let us see what sorts of energies and angular velocities are involved in molecular rotation. The carbon monoxide (CO) molecule has a bond length R of 1.13 Å and the masses of the ^{12}C and ^{16}O atoms are respectively 1.99×10^{-26} kg and 2.66×10^{-26} kg. The reduced mass m' of the CO molecule is therefore

$$m' = \frac{m_1 m_2}{m_1 + m_2}$$

$$= \frac{1.99 \times 2.66}{1.99 + 2.66} \times 10^{-26} \text{ kg}$$

$$= 1.14 \times 10^{-26} \text{ kg}$$

and its moment of inertia I is

$$I = m'R^2$$
$$= 1.14 \times 10^{-26} \text{ kg} \times (1.13 \times 10^{-10} \text{ m})^2$$
$$= 1.46 \times 10^{-46} \text{ kg-m}^2$$

The lowest rotational energy level corresponds to $J = 1$, and for this level in CO

$$E_{J=1} = \frac{J(J+1)\hbar^2}{2I} = \frac{\hbar^2}{I}$$

$$= \frac{(1.054 \times 10^{-34} \text{ J-s})^2}{1.46 \times 10^{-46} \text{ kg-m}^2}$$

$$= 7.61 \times 10^{-23} \text{ J}$$

$$= 5.07 \times 10^{-4} \text{ eV}$$

This is not a great deal of energy, and at room temperature, when $kT \approx 2.6 \times 10^{-2}$ eV, nearly all the molecules in a sample of CO are in excited rotational states. The angular velocity of the CO molecule when $J = 1$ is

$$\omega = \sqrt{\frac{2E}{I}}$$

$$= \sqrt{\frac{2 \times 7.61 \times 10^{-23} \text{ J}}{1.46 \times 10^{-46} \text{ kg-m}^2}}$$

$$= 3.23 \times 10^{11} \text{ rad/s}$$

Thus far we have been considering only rotation about an axis perpendicular to the axis of symmetry of a diatomic molecule, as in Fig. 8-21—end-over-end rotations. What about rotations about the axis of symmetry itself? The reason the latter can be neglected is that the mass of an atom is almost entirely concentrated in its nucleus, whose radius is only $\sim 10^{-4}$ of the radius of the atom itself. The principal contribution to the moment of inertia of a diatomic molecule about its symmetry axis therefore comes from its electrons, which are concentrated in a region whose radius about the axis is roughly half the bond length R but whose total mass is only about $\frac{1}{4,000}$ of the total molecular mass. Since the allowed rotational energy levels are proportional to $1/I$, rotation about the symmetry axis must involve energies $\sim 10^4$ times the E_J values for end-over-end rotations. Hence energies of at least several eV would be involved in any rotation about the symmetry axis of a diatomic molecule. Since bond energies are of this order of magnitude too, the molecule would be likely to dissociate in any environment in which such a rotation could be excited.

Rotational spectra arise from transitions between rotational energy states. Only molecules that have electric dipole moments can absorb or emit electromagnetic photons in such transitions, which means that nonpolar diatomic molecules such as H_2 and symmetric polyatomic molecules such as CO_2 (O=C=O) and CH_4 (Fig. 8-16) do not exhibit rotational spectra. (Transitions between rotational states in molecules like H_2, CO_2, and CH_4 can take place during collisions, however.) Furthermore, even in molecules that possess permanent dipole moments, not all transitions between rotational states involve radiation. As in the case of atomic spectra (Sec. 6.10), certain *selection rules* summarize the conditions for a radiative transition between rotational states to be possible. For a rigid diatomic molecule the selection rule for rotational transitions is

8.9 $$\Delta J = \pm 1$$

In practice, rotational spectra are always obtained in absorption, so that each transition that is found involves a change from some initial state of quantum number J to the next higher state of quantum number $J + 1$. In the case of a rigid molecule, the frequency of the photon absorbed is

$$\nu_{J \to J+1} = \frac{\Delta E}{h} = \frac{E_{J+1} - E_J}{h}$$

8.10 $$= \frac{\hbar}{2\pi I}(J + 1) \qquad \text{Rotational spectra}$$

where I is the moment of inertia for end-over-end rotations. The spectrum of a rigid molecule therefore consists of equally spaced lines, as in Fig. 8-22. The frequency of each line can be measured, and the transition it corresponds to can often be ascertained from the sequence of lines; from these data the moment of inertia of the molecule can be readily calculated. (Alternatively, the frequencies of any two successive lines may be used to determine I if the spectrometer used does not record the lowest-frequency lines in a particular spectral sequence.) In CO, for instance, the $J = 0 \to J = 1$ absorption line occurs at a frequency of 1.153×10^{11} Hz. Hence

$$I_{CO} = \frac{\hbar}{2\pi \nu}(J + 1)$$

$$= \frac{1.054 \times 10^{-34} \text{ J-s}}{2\pi \times 1.153 \times 10^{11} \text{ s}^{-1}}$$

$$= 1.46 \times 10^{-46} \text{ kg-m}^2$$

Since the reduced mass of the CO molecule is 1.14×10^{26} kg, the bond length R_{CO} is $\sqrt{I/m'} = 1.13$ Å. This is the way in which the bond length for CO quoted earlier in this section was determined.

8.9 VIBRATIONAL ENERGY LEVELS

When sufficiently excited, a molecule can vibrate as well as rotate. As before, we shall only consider diatomic molecules. Figure 8-23 shows how the potential energy of a molecule varies with the internuclear distance R. In the neighborhood

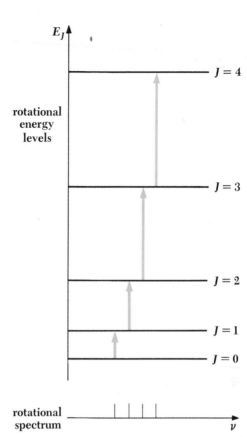

FIGURE 8-22 Energy levels and spectrum of molecular rotation.

of the minimum of this curve, which corresponds to the normal configuration of the molecule, the shape of the curve is very nearly a parabola. In this region, then,

8.11 $$V = V_0 + \tfrac{1}{2}k(R - R_0)^2$$

where R_0 is the equilibrium separation of the atoms. The interatomic force that gives rise to this potential energy may be found by differentiating V:

$$F = -\frac{dV}{dR}$$

8.12 $$= -k(R - R_0)$$

The force is just the restoring force that a stretched or compressed spring exerts—a Hooke's law force—and, as with a spring, a molecule suitably excited can undergo simple harmonic oscillations.

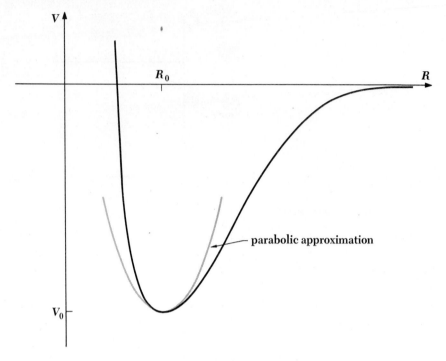

FIGURE 8-23   The potential energy of a diatomic molecule as a function of internuclear distance.

Classically, the frequency of a vibrating body of mass m connected to a spring of force constant k is

8.13
$$\nu_0 = \frac{1}{2\pi}\sqrt{\frac{k}{m}}$$

What we have in the case of a diatomic molecule is the somewhat different situation of two bodies of masses m_1 and m_2 joined by a spring, as in Fig. 8.24. In the absence of external forces the linear momentum of the system remains constant, and the oscillations of the bodies therefore cannot affect the motion of their center of mass. For this reason m_1 and m_2 vibrate back and forth relative to their center of mass in opposite directions, and both reach the extremes of their respective motions at the same times. The frequency of oscillation of such a two-body oscillator is given by Eq. 8.13 with the reduced mass m' of Eq. 8.5

force constant k

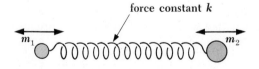

FIGURE 8-24   A two-body oscillator.

substituted for m:

8.14 $$\nu_0 = \frac{1}{2\pi} \sqrt{\frac{k}{m'}}$$ **Two-body oscillator**

When the harmonic-oscillator problem is solved quantum mechanically, as was done in Chap. 5, the energy of the oscillator is found to be restricted to the values

8.15 $$E_v = (v + \tfrac{1}{2})\, h\nu_0$$

where v, the *vibrational quantum number*, may have the values

$$v = 0, 1, 2, 3, \ldots$$

The lowest vibrational state $(v = 0)$ has the finite energy $\tfrac{1}{2}h\nu_0$, not the classical value of 0; as discussed in Chap. 5, this result is in accord with the uncertainty principle, because if the oscillating particle were stationary, the uncertainty in its position would be $\Delta x = 0$ and its momentum uncertainty would then have to be infinite—and a particle with $E = 0$ cannot have an infinitely uncertain momentum. In view of Eq. 8.14 the vibrational energy levels of a diatomic molecule are specified by

8.16 $$E_v = (v + \tfrac{1}{2})\, \hbar \sqrt{\frac{k}{m'}}$$ **Vibrational energy levels**

Let us calculate the frequency of vibration of the CO molecule and the spacing between its vibrational energy levels. The force constant k of the bond in CO is 187 N/m (which is 10 lb/in.—not an exceptional figure for an ordinary spring) and, as we found in Sec. 8.8, the reduced mass of the CO molecule is $m' = 1.14 \times 10^{-26}$ kg. The frequency of vibration is therefore

$$\nu_0 = \frac{1}{2\pi} \sqrt{\frac{k}{m'}}$$

$$= \frac{1}{2\pi} \sqrt{\frac{187 \text{ N/m}}{1.14 \times 10^{-26} \text{ kg}}}$$

$$= 2.04 \times 10^{13} \text{ Hz}$$

The separation ΔE between the vibrational energy levels in CO is

$$\Delta E = E_{v+1} - E_v = h\nu_0$$
$$= 6.63 \times 10^{-34} \text{ J-s} \times 2.04 \times 10^{13} \text{ s}^{-1}$$
$$= 8.44 \times 10^{-2} \text{ eV}$$

which is considerably more than the spacing between its rotational energy levels. Because $\Delta E > kT$ for vibrational states in a sample at room temperature, most of the molecules in such a sample exist in the $v = 0$ state with only their zero-point energies. This situation is very different from that characteristic of rotational states, where the much smaller energies mean that the majority of the molecules in a room-temperature sample are excited to higher states.

The higher vibrational states of a molecule do not obey Eq. 8.15 because the parabolic approximation to its potential-energy curve becomes less and less valid with increasing energy. As a result, the spacing between adjacent energy levels of high v is less than the spacing between adjacent levels of low v, which is shown in Fig. 8-25. This diagram also shows the fine structure in the vibrational levels caused by the simultaneous excitation of rotational levels.

The selection rule for transitions between vibrational states is $\Delta v = \pm 1$ in the harmonic oscillator approximation. This rule is easy to understand. An oscillating dipole whose frequency is v_0 can only absorb or emit electromagnetic radiation of the same frequency, and all quanta of frequency v_0 have the energy

FIGURE 8-25 The potential energy of a diatomic molecule as a function of interatomic distance, showing vibrational and rotational energy levels.

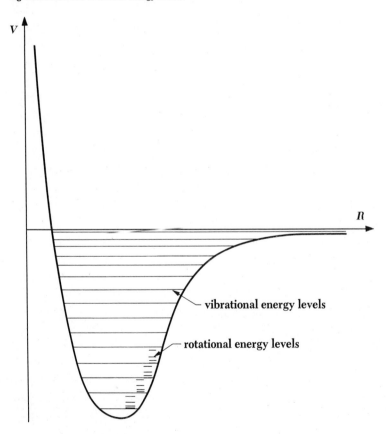

vibrational energy levels

rotational energy levels

hv_0. The oscillating dipole accordingly can only absorb $\Delta E = hv_0$ at a time, in which case its energy increases from $(v + \frac{1}{2})hv_0$ to $(v + \frac{1}{2} + 1)hv_0$, and it can only emit $\Delta E = hv_0$ at a time, in which case its energy decreases from $(v + \frac{1}{2})hv_0$ to $(v + \frac{1}{2} - 1)hv_0$. Hence the selection rule $\Delta v = \pm 1$.

Pure vibrational spectra are observed only in liquids where interactions between adjacent molecules inhibit rotation. Because the excitation energies involved in molecular rotation are considerably smaller than those involved in vibration, the freely moving molecules in a gas or vapor nearly always are rotating, regardless of their vibrational state. The spectra of such molecules do not show isolated lines corresponding to each vibrational transition, but instead a large number of closely spaced lines due to transitions between the various rotational states of one vibrational level and the rotational states of the other. In spectra obtained using a spectrometer with inadequate resolution, the lines appear as a broad streak called a vibration-rotation band.

To a first approximation, the vibrations and rotations of a molecule take place independently of each other, and we can also ignore the effects of centrifugal distortion and anharmonicity. Under these circumstances the energy levels of a diatomic molecule are specified by

8.17
$$E_{v,J} = (v + \frac{1}{2}) \hbar \sqrt{\frac{k}{m'}} + J(J + 1)\frac{\hbar^2}{2I}$$

Figure 8.26 shows the $J = 0$, 1, 2, 3, and 4 levels of a diatomic molecule for the $v = 0$ and $v = 1$ vibrational states, together with the spectral lines in absorption that are consistent with the selection rules $\Delta v = +1$ and $\Delta J = \pm 1$. The $v = 0 \to v = 1$ transitions fall into two categories, the *P branch* in which $\Delta J = -1$ (that is, $J \to J - 1$) and the *R branch* in which $\Delta J = +1$ ($J \to J + 1$). From Eq. 8.17 the frequencies of the spectral lines in each branch are given by

$$v_P = (E_{1,J-1} - E_{0,J})/h$$
$$= \frac{1}{2\pi} \sqrt{\frac{k}{m'}} + [(J - 1)J - J(J + 1)]\frac{\hbar}{4\pi I}$$

8.18
$$= v_0 - J\frac{\hbar}{2\pi I} \qquad J = 1, 2, 3, \ldots \qquad \text{P branch}$$

and

$$v_R = (E_{1,J+1} - E_{0,J})/h$$
$$= \frac{1}{2\pi} \sqrt{\frac{k}{m'}} + [(J + 1)(J + 2) - J(J + 1)]\frac{\hbar}{4\pi I}$$

8.19
$$= v_0 + (J + 1)\frac{\hbar}{2\pi I} \qquad J = 0, 1, 2, \ldots \qquad \text{R branch}$$

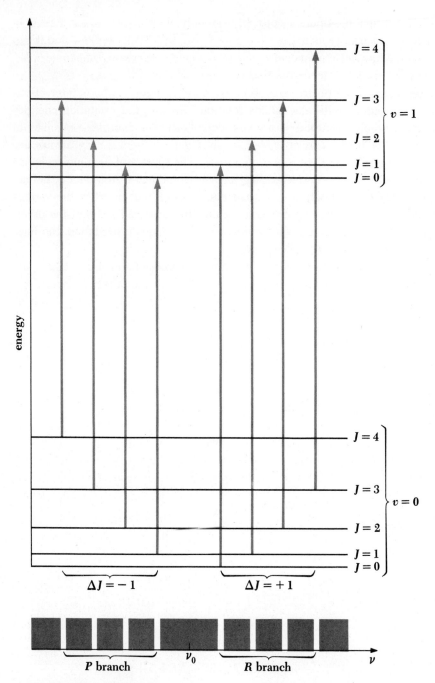

FIGURE 8-26   The rotational structure of the $v = 0 \rightarrow v = 1$ transitions in a diatomic molecule. There is no line at $\nu = \nu_0$ (the Q branch) because of the selection rule $\Delta J = \pm 1$.

There is no line at $\nu = \nu_0$ because transitions for which $\Delta J = 0$ are forbidden in diatomic molecules. The spacing between the lines in both the P and the R branch is $\Delta\nu = \hbar/2\pi I$; hence the moment of inertia of a molecule can be ascertained from its infrared vibration-rotation spectrum as well as from its microwave pure-rotation spectrum. Figure 8-27 shows the $v = 0 \rightarrow v = 1$ vibration-rotation absorption band in CO.

A molecule that consists of many atoms may have a large number of different normal modes of vibration. Some of these modes involve the entire molecule, but others involve only groups of atoms whose vibrations occur more or less independently of the rest of the molecule. Thus the —OH group has a characteristic vibrational frequency of 1.1×10^{14} Hz and the —NH$_2$ group has a frequency of 1.0×10^{14} Hz. The characteristic vibrational frequency of a carbon-carbon group depends upon the number of bonds between the C atoms: the $\overset{\diagdown}{}C—C\overset{\diagup}{}$ group vibrates at about 3.3×10^{13} Hz, the $\overset{\diagdown}{}C{=}C\overset{\diagup}{}$ group vibrates at about 5.0×10^{13} Hz, and the —C≡C— group vibrates at about 6.7×10^{13} Hz (Figs. 8-28 and 8-29). (As we would expect, the greater the number of carbon-carbon bonds, the larger the value of the force constant k and the higher the frequency.) In each case the frequency does not depend on the particular molecule or the location in the molecule of the group. This independence makes vibrational spectra a valuable tool in determining molecular

FIGURE 8-27 The $v = 0 \rightarrow v = 1$ vibration-rotation absorption band in CO under high resolution. The lines are identified by the value of J in the initial rotational state.

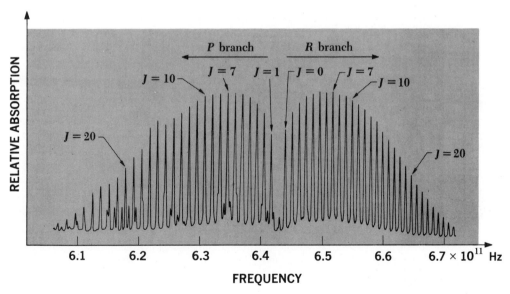

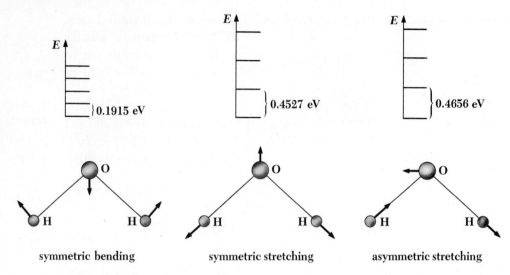

symmetric bending   symmetric stretching   asymmetric stretching

FIGURE 8-28   The normal modes of vibration of the H_2O molecule and the energy levels of each mode.

structures. An example is thioacetic acid, whose structure might conceivably be either CH_3CO—SH or CH_3CS—OH. The infrared absorption spectrum of thioacetic acid contains lines at frequencies equal to the vibrational frequencies of the $\diagdown C{=}O$ and —SH groups, but no lines corresponding to the $\diagdown C{=}S$ or —OH groups, so the former alternative is the correct one.

FIGURE 8-29   The normal modes of vibration of the CO_2 molecule and the energy levels of each mode. The symmetric bending mode can occur in two perpendicular planes.

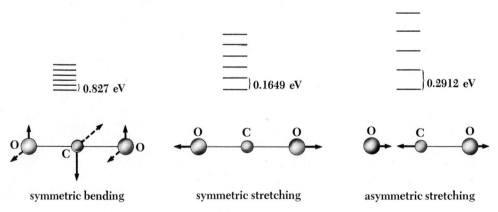

symmetric bending   symmetric stretching   asymmetric stretching

8.10 ELECTRONIC SPECTRA OF MOLECULES

The energies of rotation and vibration in a molecule are due to the motion of its atomic nuclei, since the nuclei contain essentially all of the molecule's mass. The molecular electrons also can be excited to higher energy levels than those corresponding to the ground state of the molecule, though the spacing of these levels is much greater than the spacing of rotational or vibrational levels. Electronic transitions involve radiation in the visible or ultraviolet parts of the spectrum, with each transition appearing as a series of closely spaced lines, called a band, due to the presence of different rotational and vibrational states in each electronic state (see Fig. 4-12). All molecules exhibit electronic spectra, since a dipole moment change always accompanies a change in the electronic configuration of a molecule. Therefore homonuclear molecules, such as H_2 and N_2, which have neither rotational nor vibrational spectra because they lack permanent dipole moments, nevertheless have electronic spectra which possess rotational and vibrational fine structures that permit their moments of inertia and bond force constants to be ascertained.

Electronic excitation in a polyatomic molecule often leads to a change in its shape, which can be determined from the rotational fine structure in its band spectrum. The origin of such changes lies in the different characters of the wave functions of electrons in different states, which lead to correspondingly different types of bond. For example, a possible electronic transition in a molecule whose bonds involve sp hybrid orbitals is to a higher-energy state in which the bonds involve pure p orbitals. From the sketches earlier in this chapter we can see that, in a molecule such as BeH_2, the bond angle in the case of sp hybridization is 180° and the molecule is linear (H—Be—H), while the bond angle in the case of pure p orbitals is 90° and the molecule is bent (H—Be).

$$\begin{matrix} & \\ & | \\ & H \end{matrix}$$

There are various ways in which a molecule in an excited electronic state can lose energy and return to its ground state. The molecule may, of course, simply emit a photon of the same frequency as that of the photon it absorbed, thereby returning to the ground state in a single step. Another possibility is *fluorescence;* the molecule may give up some of its vibrational energy in collisions with other molecules, so that the downward radiative transition originates from a lower vibrational level in the upper electronic state (Fig. 8-30). Fluorescent radiation is therefore of lower frequency than that of the absorbed radiation.

In molecular spectra, as in atomic spectra, radiative transitions between electronic states of different total spin are prohibited (see Sec. 7-11). Figure 8-31

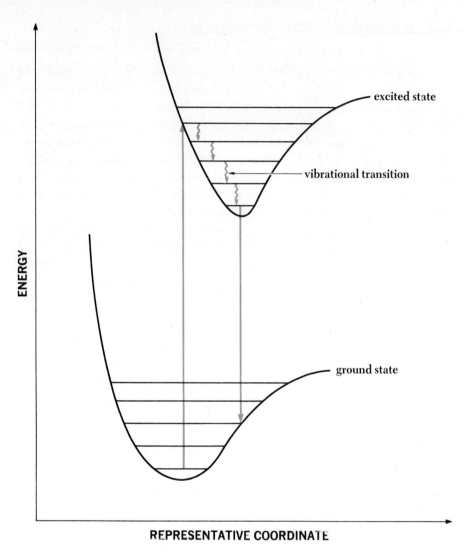

ENERGY

REPRESENTATIVE COORDINATE

excited state

vibrational transition

ground state

FIGURE 8-30 The origin of fluorescence.

shows a situation in which the molecule in its singlet (S = 0) ground state absorbs a photon and is elevated to a singlet excited state. In collisions the molecule can undergo radiationless transitions to a lower vibrational level that may happen to have about the same energy as one of the levels in the triplet (S = 1) excited state, and there is then a certain probability for a shift to the triplet state to occur. Further collisions in the triplet state bring the molecule's energy below that of the crossover point, so that it is now trapped in the triplet state and ultimately reaches the $v = 0$ level. A radiative transition from a triplet to a singlet state is "forbidden" by the selection rules, which really means not that it is impossible to occur but that it has only a minute likelihood of doing so. Such transitions accordingly have very long half lives, and the resulting *phosphorescent radiation* may be emitted minutes or even hours after the initial absorption.

FIGURE 8-31 The origin of phosphorescence. The final transition is delayed because it violates the selection rules for electronic transitions.

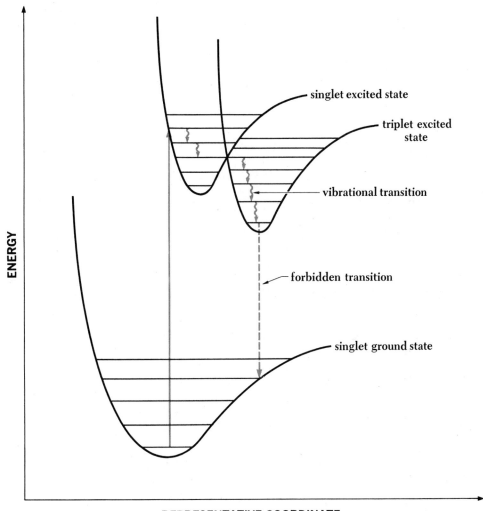

Problems

1. At what temperature would the average kinetic energy of the molecules in a hydrogen sample be equal to their binding energy?

2. Although the molecule He_2 is unstable and does not occur, the molecular ion He_2^+ is stable and has a bond energy about equal to that of H_2^+. Explain this observation.

3. Which would you expect to have the highest bond energy, F_2, F_2^+, or F_2^-? The lowest bond energy?

4. The ionization energy of H_2 is 15.7 eV and that of H is 13.6. Why are they so different?

5. The $J = 0 \rightarrow J = 1$ rotational absorption line occurs at 1.153×10^{11} Hz in $^{12}C^{16}O$ and at 1.102×10^{11} Hz in $^{?}C^{16}O$. Find the mass number of the unknown carbon isotope.

6. Calculate the energies of the four lowest rotational energy states of the H_2 and D_2 molecules, where D represents the deuterium atom ^2_1H.

7. The rotational spectrum of HCl contains the following wavelengths:

12.03×10^{-5} m
9.60×10^{-5} m
8.04×10^{-5} m
6.89×10^{-5} m
6.04×10^{-5} m

If the isotopes involved are ^1_1H and $^{35}_{17}Cl$, find the distance between the hydrogen and chlorine nuclei in an HCl molecule. (The mass of ^{35}Cl is 5.81×10^{-26} kg.)

8. Calculate the classical frequency of rotation of a rigid body whose energy is given by Eq. 8.8 for states of $J = J$ and $J = J + 1$, and show that the frequency of the spectral line associated with a transition between these states is intermediate between the rotational frequencies of the states.

9. A $^{200}Hg^{35}Cl$ molecule emits a 4.4-cm photon when it undergoes a rotational transition from $J = 1$ to $J = 0$. Find the interatomic distance in this molecule. (The masses of ^{200}Hg and ^{35}Cl are, respectively, 3.32×10^{-25} kg and 5.81×10^{-26} kg.)

10. Assume that the H_2 molecule behaves exactly like a harmonic oscillator with a force constant of 573 N/m and find the vibrational quantum number corresponding to its 4.5-eV dissociation energy.

11.   The bond between the hydrogen and chlorine atoms in a $^1H^{35}Cl$ molecule has a force constant of 516 N/m. Is it likely that a HCl molecule will be vibrating in its first excited vibrational state at room temperature?

12.   The hydrogen isotope deuterium has an atomic mass approximately twice that of ordinary hydrogen. Does H_2 or HD have the greater zero-point energy? How does this affect the binding energies of the two molecules?

13.   The force constant of the $^1H^{19}F$ molecule is 966 N/m. Find the frequency of vibration of the molecule.

14.   The observed molar specific heat of hydrogen gas at constant volume is plotted in Fig. 8-32 versus absolute temperature. (The temperature scale is logarithmic.) Since each degree of freedom (that is, each mode of energy possession) in a gas molecule contributes ~1 kcal/kmol K to the specific heat of the gas, this curve is interpreted as indicating that only translational motion, with three degrees of freedom, is possible for hydrogen molecules at very low temperatures. At higher temperatures the specific heat rises to ~5 kcal/kmol K, indicating that two more degrees of freedom are available, and at still higher temperatures the specific heat is ~7 kcal/kmol K, indicating two further degrees of freedom. The additional pairs of degrees of freedom represent, respectively, rotation, which can occur about two independent axes perpendicular to the axis of symmetry of the H_2 molecule, and non-zero-point vibration, where the degrees of freedom correspond to kinetic and potential modes of energy possession by the molecule. (a) Verify this interpretation of Fig. 8-32 by calculating the temperatures at which kT is equal to the minimum rotational energy and to the

FIGURE 8-32   Molar specific heat of hydrogen at constant volume.

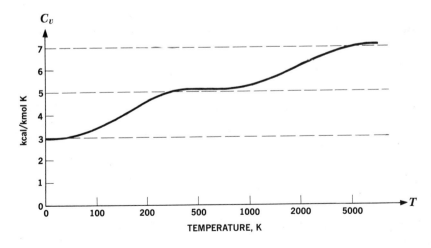

minimum vibrational energy a H_2 molecule can have. Assume that the force constant of the bond in H_2 is 573 N/m and that the H atoms are 7.42×10^{-11} m apart. (At these temperatures, approximately half the molecules are rotating or vibrating, respectively, though in each case some are in higher states than $J = 1$ or $v = 1$.) (b) To justify considering only two degrees of rotational freedom in the H_2 molecule, calculate the temperature at which kT is equal to the minimum rotational energy a H_2 molecule can have for rotation about its axis of symmetry. (c) How many rotations does a H_2 molecule with $J = 1$ and $v = 1$ make per vibration?

STATISTICAL MECHANICS 9

The branch of physics known as *statistical mechanics* attempts to relate the macroscopic properties of an assembly of particles to the microscopic properties of the particles themselves. Statistical mechanics, as its name implies, is not concerned with the actual motions or interactions of individual particles, but investigates instead their *most probable* behavior. While statistical mechanics cannot help us determine the life history of a particular particle, it *is* able to inform us of the likelihood that a particle (exactly which one we cannot know in advance) has a certain position and momentum at a certain instant. Because so many phenomena in the physical world involve assemblies of particles, the value of a statistical rather than deterministic approach is clear. Owing to the generality of its arguments, statistical mechanics can be applied with equal facility to classical problems (such as that of molecules in a gas) and quantum-mechanical problems (such as those of free electrons in a metal or photons in a box), and it is one of the most powerful tools of the theoretical physicist.

9.1 STATISTICAL DISTRIBUTION LAWS

We shall use statistical mechanics to determine the most probable way in which a fixed total amount of energy is distributed among the various members of an assembly of identical particles; that is, how many particles are likely to have the energy ε_1, how many to have the energy ε_2, and so on. The particles are assumed to interact with one another (or with the walls of their container) to an extent sufficient to establish thermal equilibrium in the assembly but not sufficient to result in any correlation between the motions of individual particles. We shall consider assemblies of three kinds of particles:

1. Identical particles of any spin that are sufficiently widely separated to be distinguished. The molecules of a gas are particles of this kind, and the *Maxwell-Boltzmann distribution law* holds for them.

2. Identical particles of 0 or integral spin that cannot be distinguished one from another. Such particles do not obey the exclusion principle, and the *Bose-Einstein distribution law* holds for them. Photons are Bose particles, or *bosons,* and we shall use the Bose-Einstein distribution law to explain the spectrum of radiation from a black body.

3. Identical particles of spin ½ that cannot be distinguished one from another. Such particles obey the exclusion principle, and the *Fermi-Dirac distribution law* holds for them. Electrons are Fermi particles, or *fermions,* and we shall use the Fermi-Dirac distribution law to explain the behavior of the free electrons in a metal.

9.2 PHASE SPACE

The state of a system of particles is completely specified classically at a particular instant if the position and momentum of each of its constituent particles are known. Since position and momentum are vectors with three components apiece, we must know six quantities,

$$x, y, z, p_x, p_y, p_z$$

for each particle.

The position of a particle is a point having the coordinates x, y, z in ordinary three-dimensional space. It is convenient to generalize this conception by imagining a six-dimensional space in which a point has the six coordinates x, y, z, p_x, p_y, p_z. This combined position and momentum space is called *phase space.* The notion of phase space is introduced to enable us to develop statistical mechanics in a geometrical framework, thereby permitting a simpler and more straightforward method of analysis than an equivalent one wholly abstract in character. A point in phase space corresponds to a particular position *and* momentum, while a point in ordinary space corresponds to a particular position only. Thus every particle is completely specified by a point in phase space, and the state of a system of particles corresponds to a certain distribution of points in phase space.

The uncertainty principle compels us to elaborate what we mean by a "point" in phase space. Let us divide phase space into tiny six-dimensional cells whose sides are $dx, dy, dz, dp_x, dp_y, dp_z$. As we reduce the size of the cells, we approach more and more closely to the limit of a point in phase space. However, the volume of each of these cells is

$$\tau = dx\, dy\, dz\, dp_x\, dp_y\, dp_z$$

and, according to the uncertainty principle,

$$dx\,dp_x \geqslant \hbar$$
$$dy\,dp_y \geqslant \hbar$$
$$dz\,dp_z \geqslant \hbar$$

Hence we see that

$$\tau \geqslant \hbar^3$$

A "point" in phase space is actually a cell whose minimum volume is of the order of $\hbar^3$. We must think of a particle in phase space as being located somewhere in such a cell centered at some location x, y, z, p_x, p_y, p_z instead of being precisely at the point itself.

A more detailed analysis shows that each cell in phase space actually has the volume h^3, which does not contradict the uncertainty-principle argument since $h^3 > \hbar^3$. In general, each cell in a phase space consisting of k coordinates and k momenta occupies a volume of h^k. It is the task of statistical mechanics to determine the state of a system by investigating how the particles constituting the system distribute themselves among the cells in phase space.

While the notion of a point of infinitesimal size in phase space can have no physical significance, since it violates the uncertainty principle, the notion of a point of infinitesimal size in either position space or momentum space alone is perfectly acceptable: we *can* in principle determine the position of a particle with as much precision as we like merely by accepting an unlimited uncertainty in our knowledge of its momentum, and vice versa.

*9.3 MAXWELL-BOLTZMANN DISTRIBUTION

Let us consider an assembly of N molecules whose energies are limited to $\varepsilon_1, \varepsilon_2, \ldots, \varepsilon_i, \ldots$. These energies may represent either discrete quantum states or average energies within a sequence of energy intervals, and more than one cell in phase space may correspond to a given energy. What we would like to know is the most probable distribution of molecules among the various possible energies.

A fundamental premise of statistical mechanics is that the greater the number W of different ways in which the molecules can be arranged among the cells in phase space to yield a particular distribution of molecules among the different energy levels, the more probable is the distribution. The most probable distribution is therefore the one for which W is a maximum. Our first step, then, is to find a general expression for W. We assume that each cell in phase space is equally likely to be occupied; this assumption is plausible, but the ultimate justification for it (as in the case of Schrödinger's equation) is that the conclusions arrived at with its help agree with experimental results.

If there are g_i cells with the energy ε_i, the number of ways in which one molecule can have the energy ε_i is g_i. The total number of ways in which two molecules can have the energy ε_i is g_i^2, and the total number of ways that n_i molecules can have the energy ε_i is $(g_i)^{n_i}$. Hence the number of ways in which all N molecules can be distributed among the various energies is the product of factors of the form $(g_i)^{n_i}$, namely,

9.1 $$(g_1)^{n_1}(g_2)^{n_2}(g_3)^{n_3} \dots$$

subject to the condition that

9.2 $$\Sigma n_i = n_1 + n_2 + n_3 + \dots = N$$

Equation 9.1 does not equal W, however, since we must take into account the possible permutations of the molecules among the different energy levels. The total number of permutations possible for N molecules is $N!$; in other words, N molecules can be arranged in $N!$ different sequences. As an example, we might have four molecules, a, b, c, and d. The value of $4!$ is

$$4! = 4 \times 3 \times 2 \times 1 = 24$$

and there are indeed 24 ways of arranging them:

abcd	bacd	cabd	dabc
abdc	badc	cadb	dacb
acbd	bcad	cbad	dbac
acdb	bcda	cbda	dbca
adbc	bdac	cdab	dcab
adcb	bdca	cdba	dcba

When more than one molecule is in an energy level, however, permuting them among themselves has no significance in this situation. For instance, if molecules a, b, and c happen to be in level j, it does not matter here whether we enumerate them as abc, acb, bca, bac, cab, or cba; these six distributions are equivalent, since all we care about is the fact that $n_j = 3$. Thus the n_i molecules in the ith level contribute $n_i!$ irrelevant permutations. If there are n_1 molecules in level 1, n_2 molecules in level 2, and so on, there are $n_1! n_2! n_3! \dots$ irrelevant permutations. What we want is the total number of possible permutations $N!$ divided by the total number of irrelevant ones, or

9.3 $$\frac{N!}{n_1! n_2! n_3! \dots}$$

The total number of ways in which the N molecules can be distributed among the possible energy levels is the product of Eqs. 9.1 and 9.3:

9.4
$$W = \frac{N!}{n_1! n_2! n_3! \dots} (g_1)^{n_1} (g_2)^{n_2} (g_3)^{n_3} \dots$$

What we now must do is determine just which distribution of the molecules is most probable, that is, which distribution yields the largest value of W. Our first step is to obtain a suitable analytic approximation for the factorial of a large number. We note that, since

$$n! = n(n-1)(n-2) \dots (4)\,(3)\,(2)$$

the natural logarithm of $n!$ is

$$\ln n! = \ln 2 + \ln 3 + \ln 4 + \dots + \ln(n-1) + \ln n$$

Figure 9-1 is a plot of $\ln n$ versus n. The area under the stepped curve is $\ln n!$ When n is very large, the stepped curve and the smooth curve of $\ln n$ become indistinguishable, and we can find $\ln n!$ by merely integrating $\ln n$ from $n = 1$ to $n = n$:

$$\ln n! = \int_1^n \ln n \, dn$$
$$= n \ln n - n + 1$$

Because we are assuming that $n \gg 1$, we may neglect the 1 in the above result, and so we obtain

9.5 $\ln n! = n \ln n - n \qquad n \gg 1$ **Stirling's formula**

Equation 9.5 is known as *Stirling's formula*.
The natural logarithm of Eq. 9.4 is

$$\ln W = \ln N! - \Sigma \ln n_i! + \Sigma \, n_i \ln g_i$$

FIGURE 9-1 The area under the stepped curve is ln $n!$ When n is very large, the smooth curve is a good approximation of the stepped curve, and ln $n!$ can be found by integrating ln n from $n = 1$ to $n = n$.

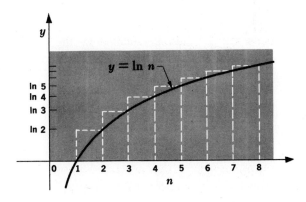

Stirling's formula enables us to write this expression as

$$\ln W = N \ln N - N - \Sigma\, n_i \ln n_i + \Sigma\, n_i + \Sigma\, n_i \ln g_i$$

Since $\Sigma\, n_i = N$,

9.6 $\qquad \ln W = N \ln N - \Sigma\, n_i \ln n_i + \Sigma\, n_i \ln g_i$

While we have an equation for $\ln W$ rather than for W itself, this is no handicap since

$$(\ln W)_{\max} = \ln W_{\max}$$

The condition for a distribution to be the most probable one is that small changes δn_i in any of the n_i's not affect the value of W. (If the n_i's were continuous variables instead of being restricted to integral values, we could express this condition in the usual way as $\partial W/\partial n_i = 0$.) If the change in $\ln W$ corresponding to a change in n_i of δn_i is $\delta \ln W$, from Eq. 9.6 we see that

9.7 $\qquad \delta \ln W_{\max} = -\Sigma\, n_i \delta \ln n_i - \Sigma \ln n_i \delta n_i + \Sigma \ln g_i \delta n_i = 0$

since $N \ln N$ is constant. Now

$$\delta \ln n_i = \frac{1}{n_i}\, \delta n_i$$

and so

$$\Sigma\, n_i \delta \ln n_i = \Sigma\, \delta n_i$$

Because the total number of molecules is constant, the sum $\Sigma\, \delta n_i$ of all the changes in the number of molecules in each energy level must be 0, which means that

$$\Sigma\, n_i \delta \ln n_i = 0$$

Hence Eq. 9.7 becomes

9.8 $\qquad -\Sigma \ln n_i \delta n_i + \Sigma \ln g_i \delta n_i = 0$

While Eq. 9.8 must be fulfilled by the most probable distribution of the molecules among the energy levels, it does not by itself completely specify this distribution. We must also take into account the conservation of particles

(9.2) $\qquad \Sigma n_i = n_1 + n_2 + n_3 + \cdots = N$

and the conservation of energy

9.9 $\qquad \Sigma n_i \varepsilon_i = n_1 \varepsilon_1 + n_2 \varepsilon_2 + n_3 \varepsilon_3 + \cdots = E$

where E is the total energy of the assembly of molecules. In consequence the variations $\delta n_1, \delta n_2, \ldots$ in the number of molecules in each energy level are not independent of one another but must obey the relationships

9.10 $\qquad \Sigma\, \delta n_i = \delta n_1 + \delta n_2 + \delta n_3 + \cdots = 0$

9.11 $\qquad \Sigma\, \varepsilon_i \delta n_i = \varepsilon_i \delta n_i + \varepsilon_2 \delta n_2 + \varepsilon_3 \delta n_3 + \cdots = 0$

To incorporate the above conditions on the various δn_i into Eq. 9.8 we make use of Lagrange's method of undetermined multipliers, which is simply a convenient mathematical device. What we do is multiply Eq. 9.10 by $-\alpha$ and Eq. 9.11 by $-\beta$, where α and β are quantities independent of the n_i's, and add these expressions to Eq. 9.8. We obtain

9.12 $\qquad \Sigma(-\ln n_i + \ln g_i - \alpha - \beta \varepsilon_i)\delta n_i = 0$

In each of the separate equations added together to give Eq. 9.12, the variation δn_i is effectively an independent variable. In order for Eq. 9.12 to hold, then, the quantity in parentheses must be 0 for each value of i. Hence

$$-\ln n_i + \ln g_i - \alpha - \beta \varepsilon_i = 0$$

from which we obtain the Maxwell-Boltzmann distribution law:

9.13 $\qquad n_i = g_i e^{-\alpha} e^{-\beta \varepsilon_i}$
Maxwell-Boltzmann distribution law

This formula gives the number of molecules n_i that have the energy ε_i in terms of the number of cells in phase space g_i that have the energy ε_i and the constants α and β. We must now evaluate g_i, α, and β.

*9.4 EVALUATION OF CONSTANTS

Energy quantization is inconspicuous in the translational motion of the molecules in a gas, and the total number of molecules in a sample is usually very large. It is therefore more convenient to consider a continuous distribution of molecular energies rather than the discrete set $\varepsilon_1, \varepsilon_2, \varepsilon_3, \ldots$. If $n(\varepsilon)\, d\varepsilon$ is the number of molecules whose energies lie between ε and $\varepsilon + d\varepsilon$, Eq. 9.13 becomes

9.14 $\qquad n(\varepsilon)\, d\varepsilon = g(\varepsilon)e^{-\alpha}e^{-\beta \varepsilon}\, d\varepsilon$

In terms of molecular momentum, since

$$\varepsilon = \frac{p^2}{2m}$$

we have

9.15 $\qquad n(p)\, dp = g(p)e^{-\alpha}e^{-\beta p^2/2m}\, dp$

The quantity $g(p)$ is equal to the number of cells in phase space in which a molecule has a momentum between p and $p + dp$. Since each cell has the volume h^3,

$$g(p)\,dp = \frac{\iiiiint dx\,dy\,dz\,dp_x\,dp_y\,dp_z}{h^3}$$

where the numerator is the phase-space volume occupied by particles with the specified momenta. Here

$$\iiint dx\,dy\,dz = V$$

where V is the volume occupied by the gas in ordinary position space, and

$$\iint dp_x\,dp_y\,dp_z = 4\pi p^2\,dp$$

where $4\pi p^2\,dp$ is the volume of a spherical shell of radius p and thickness dp in momentum space. Hence

9.16 $$g(p)\,dp = \frac{4\pi V p^2\,dp}{h^3}$$

and

9.17 $$n(p)\,dp = \frac{4\pi V p^2 e^{-\alpha} e^{-\beta p^2/2m}}{h^3}\,dp$$

We are now able to find $e^{-\alpha}$. Since

$$\int_0^\infty n(p)\,dp = N$$

we find by integrating Eq. 9.17 that

$$N = \frac{4\pi e^{-\alpha} V}{h^3} \int_0^\infty p^2 e^{-\beta p^2/2m}\,dp$$

$$= \frac{e^{-\alpha} V}{h^3} \left(\frac{2\pi m}{\beta}\right)^{3/2}$$

where we have made use of the definite integral

$$\int_0^\infty x^2 e^{-ax^2}\,dx = \frac{1}{4}\sqrt{\frac{\pi}{a^3}}$$

Hence

$$e^{-\alpha} = \frac{Nh^3}{V}\left(\frac{\beta}{2\pi m}\right)^{3/2}$$

and

9.18 $$n(p)\,dp = 4\pi N \left(\frac{\beta}{2\pi m}\right)^{3/2} p^2 e^{-\beta p^2/2m}\,dp$$

To find β, we compute the total energy E of the assembly of molecules. Since

$$p^2 = 2m\varepsilon \quad \text{and} \quad dp = \frac{m\,d\varepsilon}{\sqrt{2m\varepsilon}}$$

we can write Eq. 9.18 in the form

9.19
$$n(\varepsilon)\,d\varepsilon = \frac{2N\beta^{3/2}}{\sqrt{\pi}}\,\sqrt{\varepsilon}\,e^{-\beta\varepsilon}\,d\varepsilon$$

The total energy is

$$E = \int_0^\infty \varepsilon n(\varepsilon)\,d\varepsilon$$

$$= \frac{2N\beta^{3/2}}{\sqrt{\pi}}\int_0^\infty \varepsilon^{3/2}e^{-\beta\varepsilon}\,d\varepsilon$$

9.20
$$= \frac{3}{2}\frac{N}{\beta}$$

where we have made use of the definite integral

$$\int_0^\infty x^{3/2}e^{-ax}\,dx = \frac{3}{4a^2}\sqrt{\frac{\pi}{a}}$$

According to the kinetic theory of gases, the total energy E of N molecules of an ideal gas (which is what we have been considering) at the absolute temperature T is

9.21
$$E = \frac{3}{2}NkT$$

where k is Boltzmann's constant

$$k = 1.380 \times 10^{-23}\text{ J/molecule-degree}$$

Equations 9.20 and 9.21 agree if

9.22
$$\beta = \frac{1}{kT}$$

9.5   MOLECULAR ENERGIES IN AN IDEAL GAS

Now that the parameters α and β have been evaluated, we can write the Boltzmann distribution law in its final form,

9.23
$$n(\varepsilon)\,d\varepsilon = \frac{2\pi N}{(\pi kT)^{3/2}}\,\sqrt{\varepsilon}\,e^{-\varepsilon/kT}\,d\varepsilon \qquad \textbf{Boltzmann distribution of energies}$$

This equation gives the number of molecules with energies between ε and $\varepsilon + d\varepsilon$ in a sample of an ideal gas that contains a total of N molecules and whose absolute temperature is T. The Boltzmann energy distribution is plotted in Fig. 9-2 in terms of kT. The curve is not symmetrical because the lower limit to ε is $\varepsilon = 0$ while there is, in principle, no upper limit (although the likelihood of energies many times greater than kT is small).

According to Eq. 9.20, the total energy E of an assembly of N molecules is

$$E = \frac{3}{2}\frac{N}{\beta}$$

The *average energy* $\overline{\varepsilon}$ per molecule is E/N, so that

$$\overline{\varepsilon} = \frac{3}{2}\frac{1}{\beta}$$

9.24
$$= \frac{3}{2}kT$$ **Average molecular energy**

At 300 K, which is approximately room temperature,

$$\overline{\varepsilon} = 6.21 \times 10^{-21}\ \text{J/molecule}$$
$$\approx \tfrac{1}{25}\ \text{eV/molecule}$$

FIGURE 9-2   Maxwell-Boltzmann energy distribution.

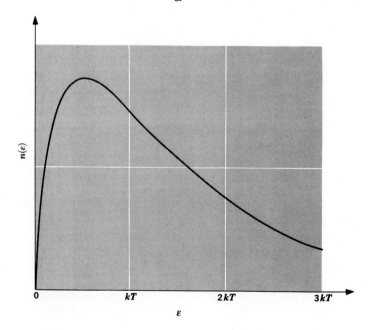

This average energy is the same for all molecules at 300 K, regardless of their mass. The Boltzmann distributions of molecular momenta and speeds can be obtained from Eq. 9.23 by noting that

$$\varepsilon = \frac{p^2}{2m} = \frac{1}{2}mv^2$$

$$d\varepsilon = \frac{p}{m}dp = mv\,dv$$

We find that

9.25 $$n(p)\,dp = \frac{\sqrt{2\pi}\,N}{(\pi mkT)^{3/2}}p^2 e^{-p^2/2mkT}\,dp$$ **Boltzmann distribution of momenta**

is the number of molecules having momenta between p and $p + dp$, and

9.26 $$n(v)\,dv = \frac{\sqrt{2\pi}\,Nm^{3/2}}{(\pi kT)^{3/2}}v^2 e^{-mv^2/2kT}\,dv$$ **Boltzmann distribution of speeds**

is the number of molecules having speeds between v and $v + dv$. The last formula, which was first obtained by Maxwell in 1859, is plotted in Fig. 9-3. The speed of a molecule with the average energy of $\frac{3}{2}kT$ is

9.27 $$v_{\mathrm{rms}} = \sqrt{\overline{v^2}} = \sqrt{\frac{3kT}{m}}$$ **Rms speed**

since $\frac{1}{2}m\overline{v^2} = \frac{3}{2}kT$. This speed is denoted v_{rms} because it is the square root of the average of the squared molecular speeds—the *root-mean-square* speed— and is not the same as the simple arithmetical average speed $\bar{v}$. The relationship between $\bar{v}$ and v_{rms} depends upon the distribution law that governs the molecular speeds being considered. For a Boltzmann distribution,

$$v_{\mathrm{rms}} = \sqrt{\frac{3\pi}{8}}\,\bar{v} \approx 1.09\bar{v}$$

so that the rms speed is about 9 percent greater than the arithmetical average speed.

Because the Boltzmann distribution of speeds is not symmetrical, the most probable speed v_p is smaller than either $\bar{v}$ or v_{rms}. To find v_p, we set equal to zero the derivative of $n(v)$ with respect to v and solve the resulting equation for v. We obtain

9.28 $$v_p = \sqrt{\frac{2kT}{m}}$$ **Most probable speed**

Molecular speeds in a gas vary considerably on either side of v_p. Figure 9-4 shows the distribution of molecular speeds in oxygen at 73 K ($-200°C$), in oxygen

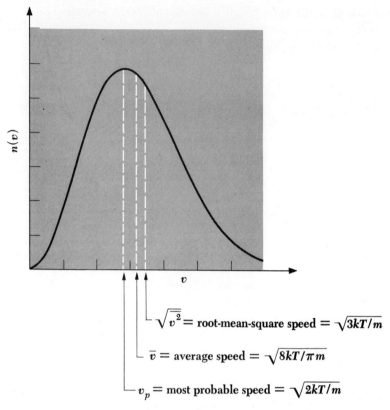

$\sqrt{\overline{v^2}}$ = root-mean-square speed = $\sqrt{3kT/m}$

$\overline{v}$ = average speed = $\sqrt{8kT/\pi m}$

v_p = most probable speed = $\sqrt{2kT/m}$

FIGURE 9-3 Maxwell-Boltzmann velocity distribution.

at 273 K (0°C), and in hydrogen at 273 K. The most probable molecular speed increases with temperature and decreases with molecular mass. Accordingly molecular speeds in oxygen at 73 K are on the whole less than at 273 K, and at 273 K molecular speeds in hydrogen are on the whole greater than in oxygen at the same temperature. (The average molecular *energy* is the same in both oxygen and hydrogen at 273 K, of course.)

9.6 ROTATIONAL SPECTRA

A continuous distribution of energies occurs only in the translational motions of molecules. As we saw in Chap. 8, molecular rotations and vibrations are quantized, with only certain specific energies E_i being possible. The Boltzmann distribution law for modes of energy possession of the latter sort may be written

9.29 $n_i = n_0 \, g_i \, e^{-E_i/kT}$

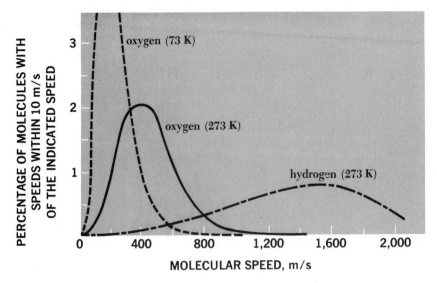

FIGURE 9-4 The distributions of molecular speeds in oxygen at 73 K, in oxygen at 273 K, and in hydrogen at 273 K.

which is simply Eq. 9.13 with n_0 replacing $e^{-\alpha}$ and E_i/kT replacing $\beta\varepsilon_i$. The factor $e^{-E_i/kT}$, often called the *Boltzmann factor*, expresses the relative probability that a quantum state of energy E_i be occupied at the temperature T. The factor g_i, the multiplicity (or *statistical weight*) of the level, is the number of quantum states that have the same energy E_i.

Let us apply Eq. 9.29 to the rotational energy levels of a molecule. (The relative populations of atomic energy levels can be treated in the same way.) As we know, more than one rotational state may correspond to a particular rotational quantum number J. The degeneracy arises because the component L_z in any specified direction of the angular momentum $\mathbf{L}$ may have any value in multiples of $\hbar$ from $J\hbar$ through 0 to $-J\hbar$, for a total of $2J + 1$ possible values. That is, there are $2J + 1$ possible orientations of $\mathbf{L}$ relative to the specified (z) direction, with each of these orientations constituting a separate quantum state. Hence an energy level whose rotational quantum number is J has a statistical weight of

$$g_J = 2J + 1$$

For a rigid diatomic molecule,

$$E_J = J(J + 1)\frac{\hbar^2}{2I}$$

and so the Boltzmann factor corresponding to the quantum number J is

$$e^{-J(J+1)\hbar^2/2IkT}$$

The Boltzmann distribution formula for the probabilities of occupancy of the rotational energy levels of a rigid diatomic molecule is therefore

9.30 $\qquad n_J = (2J + 1)\, n_0\, e^{-J(J+1)\hbar^2/2IkT}$

Here the quantity n_0 is the number of molecules in the $J = 0$ rotational state.

In Sec. 8.8 we found that the moment of inertia of the CO molecule is 1.46×10^{-46} kg-m². For a sample of carbon monoxide gas at room temperature (293 K, which is 20°C)

$$\frac{\hbar^2}{2IkT} = \frac{(1.054 \times 10^{-34}\ \text{J-s})^2}{2 \times 1.46 \times 10^{-46}\ \text{kg-m}^2 \times 1.38 \times 10^{-23}\ \text{J/K} \times 293\ \text{K}}$$
$$= 0.00941$$

and so

$$n_J = (2J + 1)\, n_0\, e^{-0.00941\, J(J+1)}$$

Figure 9-5 contains graphs of the statistical weight $2J + 1$, the Boltzmann factor $e^{-0.00941\, J(J+1)}$, and the relative population n_J/n_0 for CO at 20°C, all as functions of J. The $J = 7$ rotational energy level is evidently the most highly populated, and about as many molecules in a sample of CO at room temperature are in the $J = 19$ level as are in the $J = 0$ level.

The intensities of the rotational lines in a molecular spectrum are proportional to the relative populations of the various rotational energy levels. Figure 8-27 shows the vibration-rotation band of CO for the $v = 0 \to v = 1$ vibrational transition under high resolution; lines are identified according to the J value of the initial rotational level. The P and R branches both have their maxima at $J = 7$, as expected.

*9.7 BOSE-EINSTEIN DISTRIBUTION

The basic distinction between Maxwell-Boltzmann statistics and Bose-Einstein statistics is that the former governs identical particles which can be distinguished from one another in some way, while the latter governs identical particles which cannot be distinguished, though they can be counted. In Bose-Einstein statistics, as before, all quantum states are assumed to have equal probabilities of occupancy, so that g_i represents the number of states that have the same energy ε_i. Each quantum state corresponds to a cell in phase space, and our first step is to determine the number of ways in which n_i indistinguishable particles can be distributed in g_i cells.

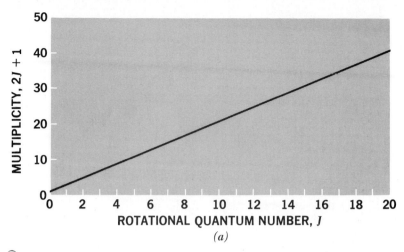

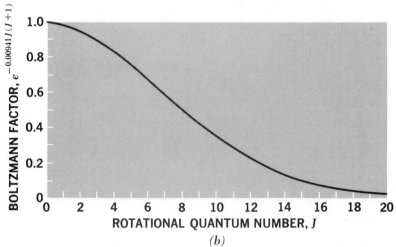

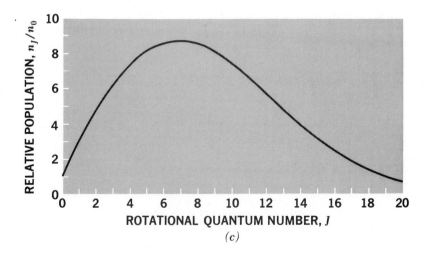

FIGURE 9-5 The multiplicities (a), Boltzmann factors (b), and relative populations (c) of the rotational energy levels of the CO molecule at 20°C.

To carry out the required enumeration, we consider a series of $n_i + g_i - 1$ objects placed in a line (Fig. 9-6). We note that $g_i - 1$ of the objects can be regarded as partitions separating a total of g_i intervals, with the entire series therefore representing n_i particles arranged in g_i cells. In the picture $g_i = 12$ and $n_i = 20$; 11 partitions separate the 20 particles into 12 cells. The first cell contains two particles, the second none, the third one particle, the fourth three particles, and so on. There are $(n_i + g_i - 1)!$ possible permutations among $n_i + g_i - 1$ objects, but of these the $n_i!$ permutations of the n_i particles among themselves and the $(g_i - 1)!$ permutations of the $g_i - 1$ partitions among themselves do not affect the distribution and are irrelevant. Hence there are

$$\frac{(n_i + g_i - 1)!}{n_i!(g_i - 1)!}$$

possible distinguishably different arrangements of the n_i indistinguishable particles among the g_i cells.

The number of ways W in which the N particles can be distributed is the product

9.31 $$W = \Pi \frac{(n_i + g_i - 1)!}{n_i!(g_i - 1)!}$$

of the numbers of distinct arrangements of particles among the states having each energy. We now assume that

$$(n_i + g_i) \gg 1$$

so that $(n_i + g_i - 1)$ can be replaced by $(n_i + g_i)$, and take the natural logarithm of both sides of Eq. 9.31 to give

$$\ln W = \Sigma \left[\ln (n_i + g_i)! - \ln n_i! - \ln (g_i - 1)! \right]$$

FIGURE 9-6 A series of n_i indistinguishable particles separated by $g_i - 1$ partitions into g_i cells.

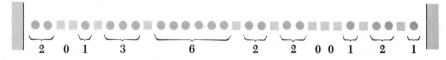

2   0 1   3   6   2   2   0 0 1   2   1

● particle
▨ partition

number of indistinguishable particles $= n_i = 20$
number of partitions $= g_i - 1 = 11$
number of cells $= g_i = 12$

PROPERTIES OF MATTER

Stirling's formula

$$\ln n! = n \ln n - n$$

permits us to rewrite $\ln W$ as

9.32 $\quad \ln W = \Sigma \left[(n_i + g_i) \ln (n_i + g_i) - n_i \ln n_i - \ln (g_i - 1)! - g_i \right]$

As before, the condition that this distribution be the most probable one is that small changes δn_i in any of the individual n_i's not affect the value of W. If a change in $\ln W$ of $\delta \ln W$ occurs when n_i changes by δn_i, the above condition may be written

$$\delta \ln W_{max} = 0$$

Hence, if the W of Eq. 9.32 represents a maximum,

9.33 $\quad \delta \ln W_{max} = \Sigma \left[\ln (n_i + g_i) - \ln n_i \right] \delta n_i = 0$

where we have made use of the fact that

$$\delta \ln n = \frac{1}{n} \delta n$$

As in Sec. 9.3 we incorporate the conservation of particles, expressed in the form

$$\Sigma \, \delta n_i = 0$$

and the conservation of energy, expressed in the form

$$\Sigma \, \varepsilon_i \, \delta n_i = 0$$

by multiplying the former equation by $-\alpha$ and the latter by $-\beta$ and adding to Eq. 9.33. The result is

$$\Sigma \left[\ln (n_i + g_i) - \ln n_i - \alpha - \beta \varepsilon_i \right] \delta n_i = 0$$

Since the δn_i's are independent, the quantity in brackets must vanish for each value of i. Hence

$$\ln \frac{n_i + g_i}{n_i} - \alpha - \beta \varepsilon_i = 0$$

$$1 + \frac{g_i}{n_i} = e^{\alpha} e^{\beta \varepsilon_i}$$

and

9.34 $\quad n_i = \dfrac{g_i}{e^{\alpha} e^{\beta \varepsilon_i} - 1}$

Substituting for β from Eq. 9.22,

(9.22) $$\beta = \frac{1}{kT}$$

we arrive at the *Bose-Einstein distribution law:*

9.35 $$n_i = \frac{g_i}{e^\alpha e^{\varepsilon_i/kT} - 1}$$ **Bose-Einstein distribution law**

9.8 BLACK-BODY RADIATION

Every substance emits electromagnetic radiation, the character of which depends upon the nature and temperature of the substance. We have already discussed the discrete spectra of excited gases which arise from electronic transitions within isolated atoms. At the other extreme, dense bodies such as solids radiate continuous spectra in which all frequencies are present; the atoms in a solid are so close together that their mutual interactions result in a multitude of adjacent quantum states indistinguishable from a continuous band of permitted energies.

The ability of a body to radiate is closely related to its ability to absorb radiation. This is to be expected, since a body at a constant temperature is in thermal equilibrium with its surroundings and must absorb energy from them at the same rate as it emits energy. It is convenient to consider as an ideal body one that absorbs *all* radiation incident upon it, regardless of frequency. Such a body is called a *black body*.

It is easy to show experimentally that a black body is a better emitter of radiation than anything else. The experiment, illustrated in Fig. 9-7, involves two identical pairs of dissimilar surfaces. No temperature difference is observed between surfaces I′ and II′. At a given temperature the surfaces I and I′ radiate at the rate of e_1 W/m², while II and II′ radiate at the different rate e_2. The surfaces I and I′ absorb some fraction a_1 of the radiation falling on them, while II and II′ absorb some other fraction a_2. Hence I′ absorbs energy from II at a rate proportional to $a_1 e_2$, and II′ absorbs energy from I at a rate proportional to $a_2 e_1$. Because I′ and II′ remain at the same temperature, it must be true that

$$a_1 e_2 = a_2 e_1$$

and

$$\frac{e_1}{a_1} = \frac{e_2}{a_2}$$

The ability of a body to emit radiation is proportional to its ability to absorb radiation. Let us suppose that I and I′ are black bodies, so that $a_1 = 1$, while

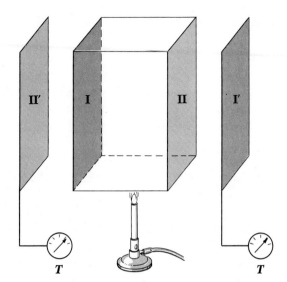

FIGURE 9-7 Surfaces I and I' are identical to each other and are different from the identical pair of surfaces II and II'.

II and II' are not, so that $a_2 < 1$. Hence

$$e_1 = \frac{e_2}{a_2}$$

and, since $a_2 < 1$, $e_1 > e_2$. A black body at a given temperature radiates energy at a faster rate than any other body.

The point of introducing the idealized black body in a discussion of thermal radiation is that we can now disregard the precise nature of whatever is radiating, since all black bodies behave identically. In the laboratory a black body can be approximated by a hollow object with a very small hole leading to its interior (Fig. 9-8). Any radiation striking the hole enters the cavity, where it is trapped

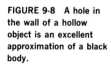

FIGURE 9-8 A hole in the wall of a hollow object is an excellent approximation of a black body.

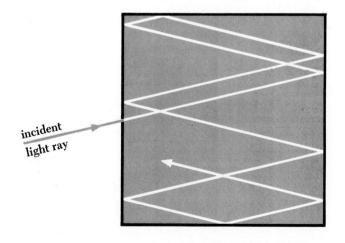

incident
light ray

by reflection back and forth until it is absorbed. The cavity walls are constantly emitting and absorbing radiation, and it is in the properties of this radiation (*black-body radiation*) that we are interested. Experimentally we can sample black-body radiation simply by inspecting what emerges from the hole. The results agree with our everyday experience; a black body radiates more when it is hot than when it is cold, and the spectrum of a hot black body has its peak at a higher frequency than the peak in the spectrum of a cooler one. We recall the familiar behavior of an iron bar as it is heated to progressively higher temperatures: at first it glows dull red, then bright orange-red, and eventually becomes "white hot." The spectrum of black-body radiation is shown in Fig. 9-9 for two temperatures.

The principles of classical physics are unable to account for the observed black-body spectrum. In fact, it was this particular failure of classical physics that led Max Planck in 1900 to suggest that light emission is a quantum phenomenon. We shall use quantum-statistical mechanics to derive the Planck radiation formula, which predicts the same spectrum as that found by experiment.

$T \lambda_m = \text{CONSTANT}$

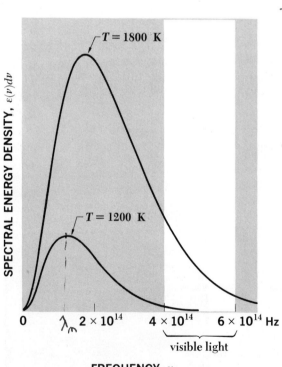

FIGURE 9-9 Black-body spectra. The spectral distribution of energy in the radiation depends only upon the temperature of the body.

Our theoretical model of a black body will be the same as the laboratory version, namely, a cavity in some opaque material. This cavity has some volume V, and it contains a large number of indistinguishable photons of various frequencies. Photons do not obey the exclusion principle, and so they are Bose particles that follow the Bose-Einstein distribution law. The number of states $g(p)$ in which a photon can have a momentum between p and $p + dp$ is equal to twice the number of cells in phase space within which such a photon may exist. The reason for the possible double occupancy of each cell is that photons of the same frequency can have two different directions of polarization (circularly clockwise and circularly counter-clockwise). Hence, using the argument that led to Eq. 9.16,

$$g(p)\, dp = \frac{8\pi V p^2\, dp}{h^3}$$

Since the momentum of a photon is $p = h\nu/c$,

$$p^2\, dp = \frac{h^3 \nu^2\, d\nu}{c^3}$$

and

9.36 $$g(\nu)\, d\nu = \frac{8\pi V}{c^3} \nu^2\, d\nu$$

We must now evaluate the Lagrangian multiplier α in Eq. 9.35. To do this, we note that the number of photons in the cavity need *not* be conserved. Unlike gas molecules or electrons, photons may be created and destroyed, and so, while the total radiant energy within the cavity must remain constant, the number of photons that incorporate this energy can change. For instance, two photons of energy $h\nu$ can be emitted simultaneously with the absorption of a single photon of energy $2h\nu$. Hence

$$\Sigma\, \delta n_i \neq 0$$

which we can express by letting $\alpha = 0$ since it multiplies $\Sigma\, \delta n_i = 0$.

Substituting Eq. 9.36 for g_i and $h\nu$ for ε_i, and letting $\alpha = 0$ in the Bose-Einstein distribution law (Eq. 9.35), we find that the number of photons with frequencies between ν and $\nu + d\nu$ in the radiation within a cavity of volume V whose walls are at the absolute temperature T is

9.37 $$n(\nu)\, d\nu = \frac{8\pi V}{c^3} \frac{\nu^2\, d\nu}{e^{h\nu/kT} - 1}$$

The corresponding spectral energy density $\varepsilon(\nu)\, d\nu$, which is the energy per unit volume in radiation between ν and $\nu + d\nu$ in frequency, is given by

$$\varepsilon(\nu)\, d\nu = \frac{h\nu n(\nu)\, d\nu}{V}$$

9.38 $$= \frac{8\pi h}{c^3} \frac{\nu^3\, d\nu}{e^{h\nu/kT} - 1}$$ **Planck radiation formula**

Equation 9.38 is the *Planck radiation formula*, which agrees with experiment.

Two interesting results can be obtained from the Planck radiation formula. To find the wavelength whose energy density is greatest, we express Eq. 9.38 in terms of wavelength and set

$$\frac{d\varepsilon(\lambda)}{d\lambda} = 0$$

and then solve for $\lambda = \lambda_{\text{max}}$. We obtain

$$\frac{hc}{kT\lambda_{\text{max}}} = 4.965$$

which is more conveniently expressed as

$$\lambda_{\text{max}} T = \frac{hc}{4.965k}$$

9.39 $$= 2.898 \times 10^{-3} \text{ m K}$$

Equation 9.39 is known as *Wien's displacement law*. It quantitatively expresses the empirical fact that the peak in the black-body spectrum shifts to progressively shorter wavelengths (higher frequencies) as the temperature is increased.

Another result we can obtain from Eq. 9.38 is the total energy density ε within the cavity. This is the integral of the energy density over all frequencies,

$$\varepsilon = \int_0^\infty \varepsilon(\nu)\, d\nu$$

$$= \frac{8\pi^5 k^4}{15 c^3 h^3} T^4$$

$$= aT^4$$

where a is a universal constant. The total energy density is proportional to the fourth power of the absolute temperature of the cavity walls. We therefore expect that the energy e radiated by a black body per second per unit area is also proportional to T^4, a conclusion embodied in the *Stefan-Boltzmann law*:

9.40 $$e = \sigma T^4$$

The value of Stefan's constant σ is

$$\sigma = 5.67 \times 10^{-8} \text{ W/m}^2 \text{ K}^4$$

Both Wien's displacement law and the Stefan-Boltzmann law are evident in qualitative fashion in Fig. 9-9; the maxima in the various curves shift to higher frequencies and the total areas underneath them increase rapidly with rising temperature.

*9.9 FERMI-DIRAC DISTRIBUTION

Fermi-Dirac statistics apply to indistinguishable particles which are governed by the exclusion principle. Our derivation of the Fermi-Dirac distribution law will therefore parallel that of the Bose-Einstein distribution law except that now each cell (that is, quantum state) can be occupied by at most one particle.

If there are g_i cells having the same energy ε_i and n_i particles, n_i cells are filled and $(g_i - n_i)$ are vacant. The g_i cells can be rearranged in $g_i!$ different ways, but the $n_i!$ permutations of the filled cells among themselves are irrelevant since the particles are indistinguishable and the $(g_i - n_i)!$ permutations of the vacant cells among themselves are irrelevant since the cells are not occupied. The number of distinguishable arrangements of the particles among the cells is therefore

$$\frac{g_i!}{n_i!(g_i - n_i)!}$$

The probability W of the entire distribution of particles is the product

9.41
$$W = \Pi \frac{g_i!}{n_i!(g_i - n_i)!}$$

Taking the natural logarithm of both sides,

$$\ln W = \Sigma \left[\ln g_i! - \ln n_i! - \ln (g_i - n_i)!\right]$$

which Stirling's formula

$$\ln n! = n \ln n - n$$

permits us to rewrite as

9.42
$$\ln W = \Sigma \left[g_i \ln g_i - n_i \ln n_i - (g_i - n_i) \ln (g_i - n_i)\right]$$

For this distribution to represent maximum probability, small changes δn_i in any of the individual n_i's must not alter W. Hence

9.43
$$\delta \ln W_{\max} = \Sigma \left[-\ln n_i + \ln (g_i - n_i)\right] \delta n_i = 0$$

As before, we take into account the conservation of particles and of energy by adding

$$-\alpha \Sigma \, \delta n_i = 0$$

...d

$$-\beta \Sigma \, \varepsilon_i \delta n_i = 0$$

to Eq. 9.43, with the result that

9.44 $\quad \Sigma \, [-\ln n_i + \ln (g_i - n_i) - \alpha - \beta \varepsilon_i] \, \delta n_i = 0$

Since the δn_i's are independent, the quantity in brackets must vanish for each value of i, and so

$$\ln \frac{g_i - n_i}{n_i} - \alpha - \beta \varepsilon_i = 0$$

$$\frac{g_i}{n_i} - 1 = e^\alpha e^{\beta \varepsilon_i}$$

9.45 $\quad n_i = \dfrac{g_i}{e^\alpha e^{\beta \varepsilon_i} + 1}$

Substituting

$$\beta = \frac{1}{kT}$$

yields the *Fermi-Dirac distribution law,*

9.46 $\quad n_i = \dfrac{g_i}{e^\alpha e^{\varepsilon_i/kT} + 1}$ $\qquad\qquad$ **Fermi-Dirac distribution law**

The most important application of the Fermi-Dirac distribution law is in the free-electron theory of metals, which we shall examine in the next chapter.

9.10 COMPARISON OF RESULTS

The three statistical distribution laws are as follows:

$$n_i = \frac{g_i}{e^\alpha e^{\varepsilon_i/kT}}$$ $\qquad\qquad$ **Maxwell-Boltzmann**
distinguishable
particles

$$n_i = \frac{g_i}{e^\alpha e^{\varepsilon_i/kT} - 1}$$ $\qquad\qquad$ **Bose-Einstein**
indistinguishable
particles

$$n_i = \frac{g_i}{e^\alpha e^{\varepsilon_i/kT} + 1}$$ $\qquad\qquad$ **Fermi-Dirac**
INDISTINGUISHABLE & EXCLUSION PRINCIPLE

In these formulas n_i is the number of particles whose energy is ε_i and g_i is the number of states that have the same energy ε_i. The quantity

$$f(\varepsilon_i) = \frac{n_i}{g_i}$$

called the *occupation index* of a state of energy ε_i, is therefore the average number of particles in each of the states of that energy. The occupation index does not depend upon how the energy levels of a system of particles are distributed, and for this reason it provides a convenient way of comparing the essential natures of the three distribution laws.

The Maxwell-Boltzmann occupation index is a pure exponential, dropping by the factor $1/e$ for each increase in ε_i of kT. While $f(\varepsilon_i)$ depends upon the parameter α, the *ratio* between the occupation indices $f(\varepsilon_i)$ and $f(\varepsilon_j)$ of the two energy levels ε_i and ε_j does not;

9.48
$$\frac{f(\varepsilon_i)}{f(\varepsilon_j)} = e^{(\varepsilon_j - \varepsilon_i)/kT}$$
Boltzmann factor

This formula is useful because when $f(\varepsilon) \ll 1$, the Bose-Einstein and Fermi-Dirac distributions resemble the Maxwell-Boltzmann distribution, and it then permits us to determine the relative degrees of occupancy of two quantum states in a simple way.

In the case of a photon gas, $\alpha = 0$, and the Bose-Einstein occupation index approaches the Maxwell-Boltzmann one when $\varepsilon_i \gg kT$, whereas when $\varepsilon_i \ll kT$ the -1 term in the denominator of the formula for the former occupation index causes it to exceed the latter. The Fermi-Dirac occupation index never goes above 1, signifying one particle per state at most, which is a consequence of the obedience of Fermi particles to the exclusion principle. At low temperatures virtually all the lower energy states are filled, with the occupation index dropping rapidly near a certain critical energy known as the *Fermi energy*. At high temperatures the occupation index is sufficiently small at all energies for the effects of the exclusion principle to be unimportant, and the Fermi-Dirac distribution becomes similar to the Maxwell-Boltzmann one.

9.11 THE LASER

Three kinds of transition involving electromagnetic radiation can occur between two energy levels in an atom, a lower one i and an upper one j (Fig. 9-10). If the atom is initially in state i, it can be raised to state j by absorbing a photon of light whose energy is $h\nu = E_j - E_i$. This process is called *induced absorption*. If the atom is initially in the upper state j, it can drop to state i by emitting a photon of energy $h\nu$; this is *spontaneous emission*.

There is also a third possibility, *induced emission*, in which an incident photon of energy $h\nu$ causes a transition from the upper state to the lower one. Induced

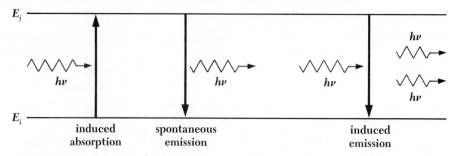

FIGURE 9-10 Transitions between two energy levels in an atom can occur by induced absorption, spontaneous emission, and induced emission.

emission involves no novel concepts. An analogy is a harmonic oscillator, for instance a pendulum, which has a sinusoidal force applied to it whose period is the same as its natural period of vibration. If the applied force is exactly in phase with the pendulum swings, the amplitude of the latter increases; this corresponds to induced absorption of energy. However, if the applied force is 180° out of phase with the pendulum swings, the amplitude of the latter *decreases;* this corresponds to induced emission of energy.

Since $h\nu$ is normally much greater than kT for atomic and molecular radiations, at thermal equilibrium the population of upper energy states in an atomic system is considerably smaller than that of the lowest state. Suppose we shine light of frequency ν upon a system in which the energy difference between the ground state and excited state is $h\nu$. With the upper state largely unoccupied, there will be little stimulated emission, and the chief events that occur will be absorption of incident photons by atoms in the ground state and the subsequent spontaneous random reradiation of photons of the same frequency. (A certain proportion of excited atoms will give up their energies in collisions.)

Certain atomic systems can sustain inverted energy populations, with an upper state occupied to a greater extent than the ground state. Figure 9-11 shows a three-level system in which the intermediate state 1 is metastable, which means that the transition from it to the ground state is forbidden by selection rules. The system can be "pumped" to the upper state 2 by radiation of frequency $\nu' = (E_2 - E_0)/h$. (Electron impacts are another way to raise the system to the upper state.) Atoms in state 2 have lifetimes of about 10^{-8} s against spontaneous emission via an allowed transition, so they fall to the metastable state 1 (or to the ground state) almost at once. Metastable states may have lifetimes of well over 1 s against spontaneous emission, and it is therefore possible to continue pumping until there is a higher population in state 1 than there is in state 0. If now we direct radiation of frequency $\nu = (E_1 - E_0)/h$ on the system, the induced emission of photons of this frequency will exceed their absorption since

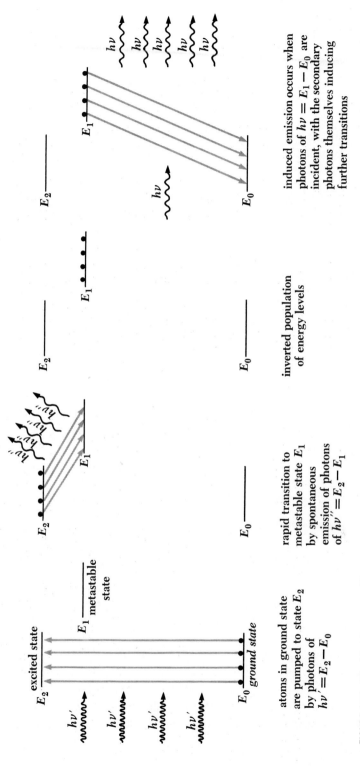

FIGURE 9-11 Principle of masers and lasers.

more atoms are in the higher state, and the net result will be an output of radiation of frequency ν that exceeds the input. This is the principle of the *maser* (*microwave amplification by stimulated emission of radiation*) and the *laser* (*light amplification by stimulated emission of radiation*).

The radiated waves from spontaneous emission are, as might be expected, incoherent, with random phase relationships in space and time since there is no coordination among the atoms involved. The radiated waves from induced emission, however, are in phase with the inducing waves, which makes it possible for a maser or laser to produce a completely coherent beam. A typical laser is a gas-filled tube or a transparent solid that has mirrors at both ends, one of them partially transmitting to allow some of the light produced to emerge. The pumping light of frequency ν' is directed at the active medium from the sides of the tube, while the back-and-forth traversals of the trapped light stimulate emissions of frequency ν that maintain the emerging beam collimated. A wide variety of masers and lasers have been devised; usually, the required inverted energy distribution is obtained less directly than by the straightforward mechanism described above.

Problems

1. Verify that the most probable speed of a molecule of an ideal gas is equal to $\sqrt{2kT/m}$.

2. Verify that the average speed of a molecule of an ideal gas is equal to $\sqrt{8kT/\pi m}$.

3. Find the average value of $1/v$ in a gas obeying Maxwell-Boltzmann statistics.

4. What proportion of the molecules of an ideal gas have components of velocity in any particular direction greater than twice the most probable speed?

5. A flux of 10^{12} neutrons/m² emerges each second from a port in a nuclear reactor. If these neutrons have a Maxwell-Boltzmann energy distribution corresponding to $T = 300$ K, calculate the density of neutrons in the beam.

6. The frequency of vibration of the H_2 molecule is 1.32×10^{14} Hz. (a) Find the relative populations of the $v = 0, 1, 2, 3,$ and 4 vibrational states at 5000 K. (b) Can the populations of the $v = 2$ and $v = 3$ states ever be equal? If so, at what temperature does this occur?

7. The moment of inertia of the H_2 molecule is 4.64×10^{-48} kg-m². (a) Find the relative populations of the $J = 0, 1, 2, 3,$ and 4 rotational states at 300 K.

(b) Can the populations of the $J = 2$ and $J = 3$ states ever be equal? If so, at what temperature does this occur?

8. The N_2O molecule is linear with an N—N bond length of 1.126 Å and an N—O bond length of 1.191 Å. The mass of the ^{18}O atom is 2.66×10^{-26} kg and that of the ^{14}N atom is 2.32×10^{-26} kg. (a) What is the quantum number of the most populated rotational energy level at 300 K? (b) Plot n_J/n_0 versus J at 300 K.

9. The temperature of the sun's chromosphere is approximately 5000 K. Find the relative numbers of hydrogen atoms in the chromosphere in the $n = 1$, 2, 3, and 4 energy levels. Be sure to take into account the multiplicity of each level.

10. If the tungsten filament of a light bulb is equivalent to a black body at 2900 K, find the percentage of the emitted radiant energy in the form of visible light with frequencies between 4×10^{14} and 7×10^{14} Hz.

11. Sunlight arrives at the earth at the rate of about 1,400 W/m^2 when the sun is directly overhead. The sun's radius is 6.96×10^8 m and the mean radius of the earth's orbit is 1.49×10^{11} m. From these data find the surface temperature of the sun on the assumption that it radiates like a black body. (The actual surface temperature of the sun is slightly less than this value.)

12. The problem of the black-body spectrum was examined at the end of the nineteenth century by Rayleigh and Jeans, using classical physics, since the notion of electromagnetic quanta was as yet unknown. They obtained the formula

$$\varepsilon(v) \, dv = \frac{8\pi v^2 kT \, dv}{c^3}$$

(a) Why is it impossible for a formula with this dependence on frequency to be correct? (b) Show that, in the limit of $v \to 0$, the Planck radiation law reduces to the Rayleigh-Jeans formula.

13. At the same temperature, will a gas of classical molecules, a gas of *bosons* (particles that obey Bose-Einstein statistics), or a gas of *fermions* (particles that obey Fermi-Dirac statistics) exert the greatest pressure? The least pressure? Why?

14. Derive the Stefan-Boltzmann law in the following way. Consider a Carnot engine that consists of a cylinder and piston whose inside surfaces are perfect reflectors and which uses electromagnetic radiation as its working substance. The operating cycle of this engine has four steps: an isothermal expansion at the temperature T during which the pressure remains constant at p; an adiabatic

expansion during which the temperature drops by dT and the pressure drops by dp; an isothermal compression at the temperature $T - dT$ and pressure $p - dp$; and an adiabatic compression to the original temperature, pressure, and volume. The pressure exerted by radiation of energy density u in a container with reflecting walls is $u/3$, and the efficiency of all Carnot engines is $dW/Q = 1 - (T - dT)/T$, where Q is the heat input during the isothermal expansion and dW is the work done by the engine during the entire cycle. Calculate the efficiency of this particular engine in terms of u and T with the help of a p-V diagram and show that $u = aT^4$, where a is a constant.

15. In a continuous helium-neon laser, He and Ne atoms are pumped to meta-stable states respectively 20.61 and 20.66 eV above their ground states by electron impact. Some of the excited He atoms transfer energy to Ne atoms in collisions, with the 0.05 eV additional energy provided by the kinetic energy of the atoms. An excited Ne atom emits a 6328-Å photon in the forbidden transition that leads to laser action. Then a 6680-Å photon is emitted in an allowed transition to another metastable state, and the remaining excitation energy is lost in collisions with the tube walls. Find the excitation energies of the two intermediate states in Ne. Why are He atoms needed?

THE SOLID STATE 10

A solid consists of atoms, ions, or molecules packed closely together, and their proximity is responsible for the characteristic properties of this state of matter. The covalent bonds involved in the formation of a molecule are also present in certain solids. In addition, *ionic, van der Waals,* and *metallic* bonds provide the cohesive forces in solids whose structural elements are, respectively, ions, molecules, and metal atoms. All these bonds involve electric forces, so that the chief distinctions among them lie in the distribution of electrons around the various particles whose regular arrangement forms a solid.

10.1 CRYSTALLINE AND AMORPHOUS SOLIDS

The majority of solids are crystalline, with the atoms, ions, or molecules of which they are composed falling into regular, repeated three-dimensional patterns. The presence of *long-range order* is thus the defining property of a crystal. Other solids lack long-range order in the arrangements of their constituent particles and may properly be regarded as supercooled liquids whose stiffness is due to an exceptionally high viscosity. Glass, pitch, and many plastics are examples of such amorphous ("without form") solids.

Amorphous solids do exhibit *short-range order* in their structures, however. The distinction between the two kinds of order is nicely exhibited in boron trioxide (B_2O_3), which can occur in both crystalline and amorphous forms. In each case every boron atom is surrounded by three oxygen atoms, which represents a short-range order. In a B_2O_3 crystal the oxygen atoms are present in hexagonal arrays, as in Fig. 10-1, which is a long-range ordering, while amorphous B_2O_3, a vitreous or "glassy" substance, lacks this additional regularity. A conspicuous example of short-range order in a liquid occurs in water just above the melting point, where the result is a lower density than at higher temperatures because H_2O molecules are less tightly packed when linked in crystals than when free to move.

The analogy between an amorphous solid and a liquid is worth pursuing as a means of better understanding both states of matter. Liquids are usually

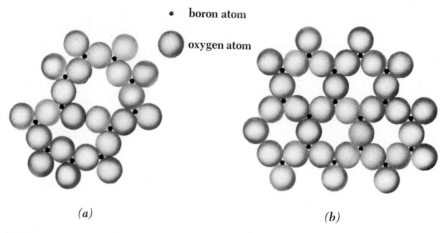

boron atom

oxygen atom

(a)   *(b)*

FIGURE 10-1 Two-dimensional representation of B_2O_3. *(a)* Amorphous B_2O_3 exhibits only short-range order. *(b)* Crystalline B_2O_3 exhibits long-range order as well.

regarded as resembling gases more closely than solids; after all, liquids and gases are both fluids, and at temperatures above the critical point the two become indistinguishable. However, from a microscopic point of view, liquids and solids also have much in common. The density of a given liquid is usually close to that of the corresponding solid, for instance, which suggests that the degree of packing is similar, an inference supported by the compressibilities of these states. Furthermore, X-ray diffraction indicates that many liquids have definite short-range structures at any instant, quite similar to those of amorphous solids except that the groupings of liquid molecules are continually shifting.

Since amorphous solids are essentially liquids, they have no sharp melting points. We can interpret this behavior on a microscopic basis by noting that, since an amorphous solid lacks long-range order, the bonds between its molecules vary in strength. When the solid is heated, the weakest bonds rupture at lower temperatures than the others, so that it softens gradually. In a crystalline solid the transition between long-range and short-range order (or no order at all) involves the breaking of bonds whose strengths are more or less identical, and melting occurs at a precisely defined temperature.

10.2 IONIC CRYSTALS

Covalent bonds come into being when atoms share pairs of electrons in such a way that attractive forces are produced. Ionic bonds come into being when atoms that have low ionization energies, and hence lose electrons readily, interact with other atoms that tend to acquire excess electrons. The former atoms give up electrons to the latter, and they thereupon become positive and negative ions respectively. In an ionic crystal these ions come together in an equilibrium configuration in which the attractive forces between positive and negative ions predominate over the repulsive forces between similar ions. As in the case of molecules, crystals of all types are prevented from collapsing under the influence of the cohesive forces present by the action of the exclusion principle, which requires the occupancy of higher energy states when electron shells of different atoms overlap and mesh together.

Table 10.1 contains the ionization energies of the elements, and Fig. 10-2 shows how these energies vary with atomic number. It is not hard to see why the ionization energies of the elements vary as they do. For instance, an atom of any of the alkali metals of group I has only a single s electron in its outer shell. The electrons in the inner shells partially shield the outer electron from the nuclear charge $+Ze$, so that the effective charge holding the outer electron to the atom is just $+e$ rather than $+Ze$. Relatively little work must be done to detach an electron from such an atom, and the alkali metals form positive ions readily. The larger the atom, the farther the outer electron is from the nucleus (Fig. 10-3) and the weaker is the electrostatic force on it; this is why the ionization energy generally decreases as we go down any group. The increase in ionization energy from left to right across any period is accounted for by the increase in nuclear charge while the number of inner shielding electrons stays constant. There are two electrons in the common inner shell of period 2 elements, and the effective nuclear charge acting on the outer electrons of these atoms is

FIGURE 10-2 The variation of ionization energy with atomic number.

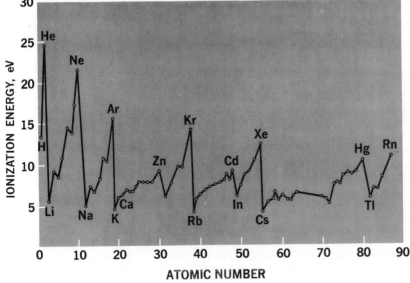

Table 10.1.

IONIZATION ENERGIES OF THE ELEMENTS. In electron volts.

1																	2
H																	He
13.6																	24.6

3	4											5	6	7	8	9	10
Li	Be											B	C	N	O	F	Ne
5.4	9.3											8.3	11.3	14.5	13.6	17.4	21.6

11	12											13	14	15	16	17	18
Na	Mg											Al	Si	P	S	Cl	Ar
5.1	7.6											6.0	8.1	11.0	10.4	13.0	15.8

19	20	21	22	23	24	25	26	27	28	29	30	31	32	33	34	35	36
K	Ca	Sc	Ti	V	Cr	Mn	Fe	Co	Ni	Cu	Zn	Ga	Ge	As	Se	Br	Kr
4.3	6.1	6.6	6.8	6.7	6.8	7.4	7.9	7.9	7.6	7.7	9.4	6.0	7.9	9.8	9.8	11.8	14.0

37	38	39	40	41	42	43	44	45	46	47	48	49	50	51	52	53	54
Rb	Sr	Y	Zr	Nb	Mo	Tc	Ru	Rh	Pd	Ag	Cd	In	Sn	Sb	Te	I	Xe
4.2	5.7	6.5	7.0	6.8	7.1	7.3	7.4	7.5	8.3	7.6	9.0	5.8	7.3	8.6	9.0	10.5	12.1

55	56	°	72	73	74	75	76	77	78	79	80	81	82	83	84	85	86
Cs	Ba		Hf	Ta	W	Re	Os	Ir	Pt	Au	Hg	Tl	Pb	Bi	Po	At	Rn
3.9	5.2		5.5	7.9	8.0	7.9	8.7	9.2	9.0	9.2	10.4	6.1	7.4	7.3	8.4	—	10.7

87	88	†															
Fr	Ra																
—	5.3																

°	57	58	59	60	61	62	63	64	65	66	67	68	69	70	71
	La	Ce	Pr	Nd	Pm	Sm	Eu	Gd	Tb	Dy	Ho	Er	Tm	Yb	Lu
	5.6	6.9	5.8	6.3	—	5.6	5.7	6.2	6.7	6.8	—	6.1	5.8	6.2	5.0

†	89	90	91	92	93	94	95	96	97	98	99	100	101	102	103
	Ac	Th	Pa	U	Np	Pu	Am	Cm	Bk	Cf	Es	Fm	Md	No	Lw
	—	7.0	—	6.1	—	5.1	6.0	—	—	—	—	—	—	—	—

therefore $+(z - 2)e$. The outer electron in a lithium atom is held to the atom by an effective charge of $+e$, while each outer electron in beryllium, boron, carbon, etc., atoms is held to its parent atom by effective charges of $+2e$, $+3e$, $+4e$, etc.

At the other extreme from alkali metal atoms, which tend to lose their outermost electrons, are halogen atoms, which tend to complete their outer p subshells by picking up an additional electron each. The *electron affinity* of an element is defined as the energy released when an electron is added to an atom of each element. The greater the electron affinity, the more tightly bound is the added

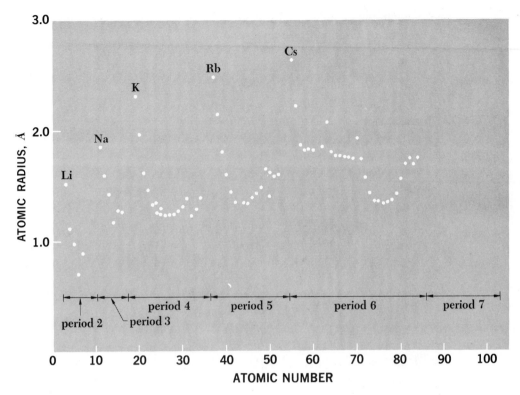

FIGURE 10-3 Atomic radii of the elements. Several have two radii, corresponding to different crystal structures.

electron. Table 10.2 shows the electron affinities of the halogens. In general, electron affinities decrease going down any group of the periodic table and increase going from left to right across any period. The experimental determination of electron affinities is quite difficult, and those for only a few elements are accurately known.

An ionic bond between two atoms can occur when one of them has a low ionization energy, and hence a tendency to become a positive ion, while the other one has a high electron affinity, and hence a tendency to become a negative

Table 10.2.

ELECTRON AFFINITIES OF THE HALOGENS.
In electron volts.

Fluorine	3.45
Chlorine	3.61
Bromine	3.36
Iodine	3.06

ion. Sodium, with an ionization energy of 5.14 eV, is an example of the former and chlorine, with an electron affinity of 3.61 eV, is an example of the latter. When a Na^+ ion and a Cl^- ion are in the same vicinity and are free to move, the attractive electrostatic force between them brings them together. The condition that a stable molecule of NaCl result is simply that the total energy of the system of the two ions be less than the total energy of a system of two atoms of the same elements; otherwise the surplus electron on the Cl^- ion would transfer to the Na^+ ion, and the neutral Na and Cl atoms would no longer be bound together. Let us see how this criterion is met by NaCl.

In general, in an ionic crystal each ion is surrounded by as many ions of the opposite sign as can fit closely, which leads to maximum stability. The relative sizes of the ions involved therefore govern the type of structure that occurs. Two common types of structure found in ionic crystals are shown in Figs. 10-4 and 10-5. In a sodium chloride crystal, the ions of either kind may be thought of as being located at the corners and at the centers of the faces of an assembly of cubes, with the Na^+ and Cl^- assemblies interleaved. Each ion thus has six nearest neighbors of the other kind, a consequence of the considerable difference in the sizes of the Na^+ and Cl^- ions. Such a structure is called *face-centered cubic*. A different arrangement is found in cesium chloride crystals, where each ion is located at the center of a cube at whose corners are ions of the other kind. Each ion has eight nearest neighbors of the other kind in such a *body-centered cubic* structure, which results when the participating ions are comparable in size.

The *cohesive energy* of an ionic crystal is the energy that would be liberated by the formation of the crystal from individual neutral atoms. Cohesive energy

FIGURE 10-4 (a) The face-centered cubic structure of a NaCl crystal. The coordination number (number of nearest neighbors about each ion) is 6. (b) Scale model of NaCl crystal.

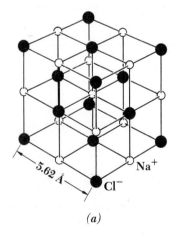

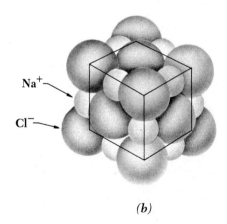

(a)   (b)

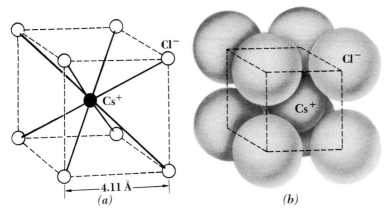

FIGURE 10-5 (a) The body-centered cubic structure of a CsCl crystal. The coordination number is 8. (b) Scale model of CsCl crystal.

is usually expressed in eV/atom, in eV/molecule, or in kcal/mol, where "molecule" and "mol" here refer to sets of atoms specified by the formula of the compound involved (for instance NaCl in the case of a sodium chloride crystal) even though molecules as such do not exist in the crystal.

The principal contribution to the cohesive energy of an ionic crystal is the electrostatic potential energy $V_{coulomb}$ of the ions. Let us consider an Na^+ ion in NaCl. Its nearest neighbors are six Cl^- ions, each one the distance r away. The potential energy of the Na^+ ion due to these six Cl^- ions is therefore

$$V_1 = -\frac{6e^2}{4\pi\varepsilon_0 r}$$

The next nearest neighbors are 12 Na^+ ions, each one the distance $\sqrt{2}\,r$ away since the diagonal of a square r long on a side is $\sqrt{2}\,r$. The potential energy of the Na^+ ion due to the 12 Na^+ ions is

$$V_2 = +\frac{12e^2}{4\pi\varepsilon_0 \sqrt{2}\,r}$$

When the summation is continued over all the $+$ and $-$ ions in a crystal of infinite size, the result is

$$V_{coulomb} = -\frac{e^2}{4\pi\varepsilon_0 r}\left(6 - \frac{12}{\sqrt{2}} + \cdots\right)$$

$$= -1.748\frac{e^2}{4\pi\varepsilon_0 r}$$

10.1 $$= -\alpha\frac{e^2}{4\pi\varepsilon_0 r}$$ **Coulomb energy of ionic crystal**

(This result holds for the potential energy of a Cl⁻ ion as well, of course.) The quantity α is called the *Madelung constant* of the crystal, and it has the same value for all crystals of the same structure. Similar calculations for other crystal varieties yield different Madelung constants: crystals whose structures are like that of cesium chloride (Fig. 10-5), for instance, have $\alpha = 1.763$, and those with structures like that of zinc blende (one form of ZnS) have $\alpha = 1.638$. Simple crystal structures have Madelung constants that lie between 1.6 and 1.8.

The potential energy contribution of the repulsive forces due to the action of the exclusion principle can be expressed to a fair degree of approximation in the form

10.2
$$V_{repulsive} = \frac{B}{r^n}$$

The sign of $V_{repulsive}$ is positive, which corresponds to a repulsive interaction, and the dependence on r^{-n} (where n is a large number) corresponds to a short-range force that increases rapidly with decreasing internuclear distance r. The total potential energy V of each ion due to its interactions with all the other ions is therefore

$$V = V_{coulomb} + V_{repulsive}$$

10.3
$$= -\frac{\alpha e^2}{4\pi\varepsilon_0 r} + \frac{B}{r^n}$$

At the equilibrium separation r_0 of the ions, V is a minimum by definition, and so $(dV/dr) = 0$ when $r = r_0$. Hence

$$\left(\frac{dV}{dr}\right)_{r=r_0} = \frac{\alpha e^2}{4\pi\varepsilon_0 r_0{}^2} - \frac{nB}{r_0{}^{n+1}} = 0$$

$$\frac{\alpha e^2}{4\pi\varepsilon_0 r_0{}^2} = \frac{nB}{r_0{}^{n+1}}$$

10.4
$$B = \frac{\alpha e^2}{4\pi\varepsilon_0 n} r_0{}^{n-1}$$

and the total potential energy is

10.5
$$V = -\frac{\alpha e^2}{4\pi\varepsilon_0 r_0}\left(1 - \frac{1}{n}\right)$$

It is possible to evaluate the exponent n from the observed compressibilities of ionic crystals. The average result is $n \approx 9$, which means that the repulsive force varies quite sharply with r: the ions are "hard" rather than "soft" and strongly resist being packed too tightly. At the equilibrium ion spacing, the mutual repulsion due to the exclusion principle (as distinct from the electrostatic repulsion between like ions) decreases the potential energy by about 11 percent. A really precise knowledge of n is evidently not essential; if $n = 10$ instead of $n = 9$, V would change by only 1 percent.

In an NaCl crystal, the equilibrium distance r_0 between ions is 2.81 Å. Since $\alpha = 1.748$ and $n = 9$, the potential energy of an ion of either sign is

$$V = -\frac{\alpha e^2}{4\pi\varepsilon_0 r_0}\left(1 - \frac{1}{n}\right)$$

$$= -\frac{9 \times 10^9 \text{ N-m}^2/\text{C}^2 \times 1.748 \times (1.60 \times 10^{-19} \text{ C})^2}{2.81 \times 10^{-10} \text{ m}}\left(1 - \frac{1}{9}\right)$$

$$= -1.27 \times 10^{-18} \text{ J}$$

$$= -7.97 \text{ eV}$$

Because we may not count each ion more than once, only half of this potential energy, or -3.99 eV, represents the contribution *per ion* to the cohesive energy of the crystal.

We must also take into account the energy needed to transfer an electron from a Na atom to a Cl atom to yield a Na⁺—Cl⁻ ion pair. This electron transfer energy is the difference between the $+5.14$-eV ionization energy of Na and the -3.61-eV electron affinity of Cl, or $+1.53$ eV. Each atom therefore contributes $+0.77$ eV to the cohesive energy from this source. The total cohesive energy per atom is thus

$$E_{\text{cohesive}} = (-3.99 + 0.77) \text{ eV/atom} = -3.22 \text{ eV/atom}$$

An empirical figure for the cohesive energy of an ionic crystal can be obtained from measurements of its heat of vaporization, dissociation energy, and electron exchange energy. The result for NaCl is 3.28 eV, in close agreement with the calculated value.

Most ionic solids are hard, owing to the strength of the bonds between their constituent ions, and have high melting points. They are usually brittle as well, since the slipping of atoms past one another that accounts for the ductility of metals is prevented by the ordering of positive and negative ions imposed by the nature of the bonds. Polar liquids such as water are able to dissolve many ionic crystals, but covalent liquids such as gasoline generally cannot.

10.3 COVALENT CRYSTALS

The cohesive forces in covalent crystals arise from the presence of electrons between adjacent atoms. Each atom participating in a covalent bond contributes an electron to the bond, and these electrons are shared by both atoms rather

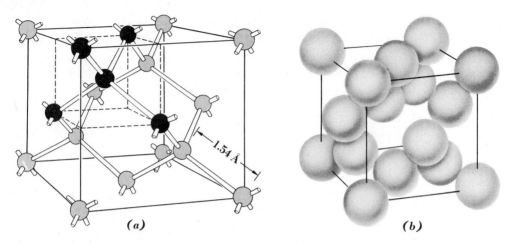

FIGURE 10-6 (a) The tetrahedral structure of diamond. The coordination number is 4. (b) Scale model of diamond crystal.

than being the virtually exclusive property of one of them as in an ionic bond. Diamond is an example of a crystal whose atoms are linked by covalent bonds. Figure 10-6 shows the structure of a diamond crystal; the tetrahedral arrangement is a consequence of the ability of each carbon atom to form covalent bonds with four other atoms (see Fig. 8-16).

Purely covalent crystals are relatively few in number. In addition to diamond, some examples are silicon, germanium, and silicon carbide; in SiC each atom is surrounded by four atoms of the other kind in the same tetrahedral structure as that of diamond. All covalent crystals are hard (diamond is the hardest substance known, and SiC is the industrial abrasive carborundum), have high melting points, and are insoluble in all ordinary liquids, behavior which reflects the strength of the covalent bonds. Cohesive energies of 3 to 5 eV/atom are typical of covalent crystals, which is the same order of magnitude as the cohesive energies in ionic crystals.

There are several ways to ascertain whether the bonds in a given nonmetallic, nonmolecular crystal are predominantly ionic or covalent. In general, a compound of an element from group I or II of the periodic table with one from group VI or VII exhibits ionic bonding in the solid state. Another guide is the coordination number of the crystal, which is the number of nearest neighbors about each constituent particle. A high coordination number suggests an ionic crystal, since it is hard to see how an atom can form purely covalent bonds with six other atoms (as in a face-centered cubic structure like that of NaCl) or with eight other atoms (as in a body-centered cubic structure like that of CsCl). A coordination number of 4, however, as in the diamond structure, is compatible with an exclusively covalent character. To be sure, as with molecules, it is not always possible to classify a particular crystal as being wholly ionic or covalent: AgCl, whose structure is the same as that of NaCl, and CuCl, whose structure resembles that of diamond, both have bonds of intermediate character, as do a great many other solids.

10.4   VAN DER WAALS FORCES

All atoms and molecules—even inert-gas atoms such as those of helium and argon—exhibit weak, short-range attractions for one another due to *van der Waals* forces. These forces are responsible for the condensation of gases into liquids and the freezing of liquids into solids despite the absence of ionic, covalent, or metallic bonding mechanisms. Such familiar aspects of the behavior of matter in bulk as friction, surface tension, viscosity, adhesion, cohesion, and so on also arise from van der Waals forces. The van der Waals attraction between two molecules r apart is proportional to r^{-7}, so that it is significant only for molecules very close together.

We begin by noting that many molecules (called *polar molecules*) possess permanent electric dipole moments. An example is the H_2O molecule, in which the concentration of electrons around the oxygen atom makes that end of the molecule more negative than the end where the hydrogen atoms are. Such molecules tend to align themselves so that ends of opposite sign are adjacent, as in Fig. 10-7, and in this orientation the molecules strongly attract each other.

A polar molecule is also able to attract molecules which do not normally have a permanent dipole moment. The process is illustrated in Fig. 10-8: the electric field of the polar molecule causes a separation of charge in the other molecule, with the induced moment the same in direction as that of the polar molecule.

FIGURE 10-7   Polar molecules attract each other.

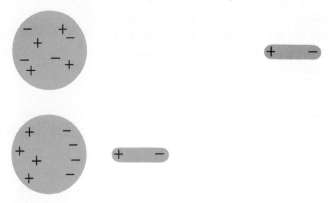

FIGURE 10-8 Polar molecules attract polarizable molecules.

The result is an attractive force. (The effect is the same as that involved in the attraction of an unmagnetized piece of iron by a magnet.)

More remarkably, two nonpolar molecules can attract each other by the above mechanism. Even though the electron distribution in a nonpolar molecule is symmetric *on the average*, the electrons themselves are in constant motion and at *any given instant* one part or another of the molecule has an excess of them. Instead of the fixed charge asymmetry of a polar molecule, a nonpolar molecule has a constantly shifting asymmetry. When two nonpolar molecules are close enough, their fluctuating charge distributions tend to shift together, adjacent ends always having opposite sign (Fig. 10-9) and so always causing an attractive force. This kind of force is named after the Dutch physicist van der Waals, who suggested its existence nearly a century ago to explain observed departures from the ideal-gas law; the explanation of the actual mechanism of the force, of course, is more recent.

Van der Waals forces are much weaker than those found in ionic and covalent bonds, and as a result molecular crystals generally have low melting and boiling points and little mechanical strength. Cohesive energies are low, only 0.08 eV/atom in solid argon (melting point $-189°C$), 0.01 eV/molecule in solid hydrogen (mp $-259°C$), and 0.1 eV/molecule in solid methane, CH_4 (mp $-183°C$).

An especially strong type of van der Waals bond called a *hydrogen bond* occurs between certain molecules containing hydrogen atoms. The electron distribution in such an atom is so distorted by the affinity of the "parent" atom for electrons that each hydrogen atom in essence has donated most of its negative charge to the parent atom, leaving behind a poorly shielded proton. The result is a molecule with a localized positive charge which can link up with the concentration of negative charge elsewhere in another molecule of the same kind. The key factor here is the small effective size of the poorly shielded proton, since electric forces vary as r^{-2}.

Water molecules are exceptionally prone to form hydrogen bonds because the four pairs of electrons around the O atom occupy sp^3 hybrid orbitals that project outward as though toward the vertexes of a tetrahedron (Fig. 10-10). Hydrogen atoms are at two of these vertexes, which accordingly exhibit localized positive charges, while the other two vertexes exhibit somewhat more diffuse negative charges. Each H_2O molecule can therefore form hydrogen bonds with *four* other H_2O molecules; in two of these bonds the central molecule provides the bridging protons, and in the other two the attached molecules provide them. In the liquid state, the hydrogen bonds between adjacent H_2O molecules are continually being broken and reformed owing to thermal agitation, but even so at any instant the molecules are combined in definite clusters. In the solid state, these clusters are large and stable and constitute ice crystals.

The characteristic hexagonal pattern (Fig. 10-11) of an ice crystal arises from the tetrahedral arrangement of the four hydrogen bonds each H_2O molecule can participate in. With only four nearest neighbors around each molecule, ice crystals have extremely open structures, which is the reason for the exceptionally low density of ice. Because the molecular clusters are smaller and less stable in the liquid state, water molecules on the average are packed more closely

FIGURE 10-9 On the average, nonpolar molecules have symmetric charge distributions, but at any instant the distributions are asymmetric. The fluctuations in the charge distributions of nearby molecules are coordinated as shown, which leads to an attractive force between them whose magnitude is proportional to $1/r^7$.

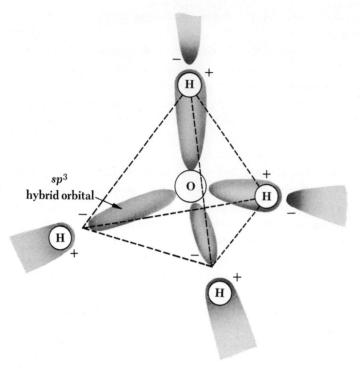

sp^3
hybrid orbital

FIGURE 10-10 In an H_2O molecule, the four pairs of valence electrons around the oxygen atom (six contributed by the O atom and one each by the H atoms) occupy four sp^3 hybrid orbitals that form a tetrahedral pattern. Each H_2O molecule can form hydrogen bonds with four other H_2O molecules.

FIGURE 10-11 Top view of an ice crystal, showing the open hexagonal arrangement of H_2O molecules. Each molecule has four nearest neighbors to which it is attached by hydrogen bonds.

together than are ice molecules, and water has the higher density: hence ice floats. The density of water increases from $0°C$ to a maximum at $4°C$ as large clusters of H_2O molecules are broken up into smaller ones that occupy less space in the aggregate; only past $4°C$ does the normal thermal expansion of a liquid manifest itself in a decreasing density with increasing temperature.

10.5 THE METALLIC BOND

The underlying theme of the modern theory of metals is that the valence electrons of the atoms comprising a metal are common to the entire aggregate, so that a kind of "gas" of free electrons pervades it. The interaction between this electron gas and the positive metal ions leads to a strong cohesive force. The presence of such free electrons accounts nicely for the high electrical and thermal conductivities, opacity, surface luster, and other unique properties of metals. To be sure, no electrons in any solid, even a metal, are able to move about its interior with total freedom. All of them are influenced to some extent by the other particles present, and when the theory of metals is refined to include these complications, there emerges a comprehensive picture that is in excellent accord with experiment.

Some insight into the ability of metal atoms to bond together to form crystals of unlimited size can be gained by viewing the metallic bond as an unsaturated covalent bond. Let us compare the bonding processes in hydrogen and in lithium, both members of group I of the periodic table. A H_2 molecule contains two $1s$ electrons with opposite spins, the maximum number of K electrons that can be present. The H_2 molecule is therefore saturated, since the exclusion principle requires that any additional electrons be in states of higher energy and the stable attachment of further H atoms cannot occur unless their electrons are in $1s$ states. Superficially lithium might seem obliged to behave in a similar way, having the electron configuration $1s^22s$. There are, however, six *unfilled* $2p$ states in every Li atom whose energies are only very slightly greater than those of the $2s$ states. When a Li atom comes near a Li_2 molecule, it readily becomes attached with a covalent bond without violating the exclusion principle, and the resulting Li_3 molecule is stable since all its valence electrons remain in L shells. There is no limit to the number of Li atoms that can join together in this way, since lithium forms body-centered cubic crystals (Fig. 10-5) in which each atom has eight nearest neighbors. With only one electron per atom available to enter into bonds, each bond involves one-fourth of an electron on the average instead of two electrons as in ordinary covalent bonds. Hence the bonds are far from being saturated; this is true of the bonds in other metals as well.

One consequence of the unsaturated nature of the metallic bond is the fact that the properties of a mixture of different metal atoms do not depend critically on the proportion of each kind of atom, provided their sizes are similar. Thus the characteristics of an alloy often vary smoothly with changes in its composition, in contrast to the specific atomic proportions found in ionic solids and in covalent solids such as SiC.

The most striking consequence of the unsaturated bonds in a metal is the ability of the valence electrons to wander freely from atom to atom. To understand this phenomenon intuitively, we can think of each valence electron as constantly moving from bond to bond. In solid Li, each electron participates in eight bonds, so that it only spends a short time between any pair of Li^+ ions. The electron cannot remember (so to speak) which of the two ions it really belongs to, and it is just as likely to move on to a bond that does not involve its parent ion at all. The valence electrons in a metal therefore behave in a manner quite similar to that of molecules in a gas.

As in the case of any other solid, metal atoms cohere because their collective energy is lower when they are bound together than when they exist as separate atoms. To understand why this reduction in energy occurs in a metallic crystal, we note that, because of the proximity of the ions, each valence electron is on the average closer to one nucleus or another than it would be if it belonged to an isolated atom. Hence the potential energy of the electron is less in the crystal than in the atom, and it is this decrease in potential energy that is responsible for the metallic bond.

There is another factor to be considered, however. Whereas the electron potential energy is reduced in a metallic crystal, the electron kinetic energy is increased. The free electrons in a metal constitute a single system of electrons, and the exclusion principle prohibits more than two of them (one with each spin) from occupying each energy level. It would seem at first glance that, again using lithium as an example, only eight valence electrons in an entire Li crystal could occupy $n = 2$ quantum states, with the rest being forced into higher and higher states of such great energy as to disrupt the entire structure. What actually happens is less dramatic. The valence energy levels of the various metal atoms are all slightly altered by their interactions, and an *energy band* comes into being that consists of as many closely spaced energy levels as the total number of valence energy levels in all the atoms in the crystal. The free electrons accordingly range in kinetic energy from 0 to some maximum u_F, called the *Fermi energy*; the Fermi energy in lithium, for example, is 4.72 eV, and the average kinetic energy of the free electrons in metallic lithium is 2.8 eV. Since electron kinetic energy is a positive quantity, its increase in the metal over what it was in separate atoms leads to a repulsion.

Metallic bonding occurs when the attraction between the positive metal ions

Table 10.3.

CRYSTAL TYPES.

Type		Bond	Example	Properties
Ionic		Electrostatic attraction	Sodium chloride NaCl $E_{\text{cohesive}} = 3.28$ eV/atom	Hard; high melting points; may be soluble in polar liquids such as water
Covalent		Shared electrons	Diamond C $E_{\text{cohesive}} = 7.4$ eV/atom	Very hard; high melting points; insoluble in nearly all solvents
Metallic		Electron gas	Sodium Na $E_{\text{cohesive}} = 1.1$ eV/atom	Ductile; metallic luster; high electrical and thermal conductivity
Molecular		Van der Waals forces	Methane CH_4 $E_{\text{cohesive}} = 0.1$ eV/atom	Soft; low melting and boiling points; soluble in covalent liquids

and the electron gas exceeds the mutual repulsion of the electrons in that gas; that is, when the reduction in electron potential energy exceeds in magnitude the concomitant increase in electron kinetic energy. The greater the number of valence electrons per atom, the higher the average kinetic energy will be in a metallic crystal, but without a commensurate drop in the potential energy. For this reason the metallic elements are nearly all found in the first three groups of the periodic table. Some elements are right on the border line and may form both metallic and covalent crystals. Tin is a notable example. Above 13.2°C the metal "white tin" exists whose atoms each have six nearest neighbors. Below 13.2°C the covalent solid "gray tin" exists whose structure is the same as that of diamond. Gray tin and white tin are quite different substances; they have the respective densities of 5.8 and 7.3 g/cm³, for instance, and gray tin is a semiconductor whereas white tin has the typically high electric conductivity of a metal.

10.6 THE BAND THEORY OF SOLIDS

The atoms in almost every crystalline solid, whether a metal or not, are so close together that their valence electrons constitute a single system of electrons common to the entire crystal. The exclusion principle is obeyed by such an electron system because the energy states of the outer electron shells of the atoms are all altered somewhat by their mutual interactions. In place of each precisely defined characteristic energy level of an individual atom, the entire crystal possesses an energy band composed of myriad separate levels very close together. Since there are as many of these separate levels as there are atoms in the crystal, the band cannot be distinguished from a continuous spread of permitted energies. The presence of energy bands, the gaps that may occur between them, and the extent to which they are filled by electrons not only determine the electrical behavior of a solid but also have important bearing on other of its properties.

There are two ways to consider the origin of energy bands. The simplest is to look into what happens to the energy levels of isolated atoms as they are brought closer and closer together to form a solid. We shall introduce the subject in this way, and then examine some of the consequences of the notion of energy bands. Later in the chapter we shall analyze energy bands in terms of the restrictions imposed by the periodicity of a crystal lattice on the motion of electrons, a more powerful approach that provides the basis of much of the modern theory of solids.

Figure 10-12 shows the energy levels in sodium plotted versus internuclear distance. The 3s level is the first occupied level in the sodium atom to broaden into a band; the 2p level does not begin to spread out until a quite small

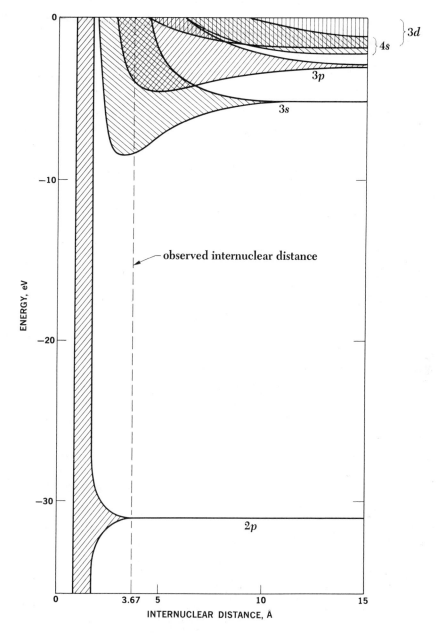

FIGURE 10-12 The energy levels of sodium atoms become bands as their internuclear distance decreases. The observed internuclear distance in solid sodium is 3.67 Å.

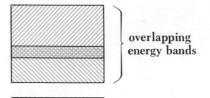

overlapping
energy bands

FIGURE 10-13 The energy bands in a solid may overlap.

internuclear separation. This behavior reflects the order in which the electron subshells of sodium atoms interact as the atoms are brought together. The average energies in the $3p$ and $3s$ bands drop at first, implying attractive forces. The actual internuclear distance in solid sodium is indicated, and it corresponds, as it should, to a situation of minimum average energy.

The energy bands in a solid correspond to the energy levels in an atom, and an electron in a solid can possess only those energies that fall within these energy bands. The various energy bands in a solid may overlap, as in Fig. 10-13, in which case its electrons have a continuous distribution of permitted energies. In other solids the bands may *not* overlap (Fig. 10-14), and the intervals between them represent energies which their electrons can not possess. Such intervals are called *forbidden bands*. The electrical behavior of a crystalline solid is determined both by its energy-band structure and by how these bands are normally filled by electrons.

Figure 10-15 is a simplified diagram of the energy levels of a sodium atom and the energy bands of solid sodium. A sodium atom has a single $3s$ electron in its outer shell. This means that the $3s$ band in a sodium crystal is only half occupied, since each level in the band, like each level in the atom, is able to contain *two* electrons. When an electric field is set up across a piece of solid

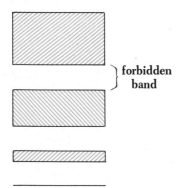

forbidden
band

FIGURE 10-14 A forbidden band separates nonoverlapping energy bands.

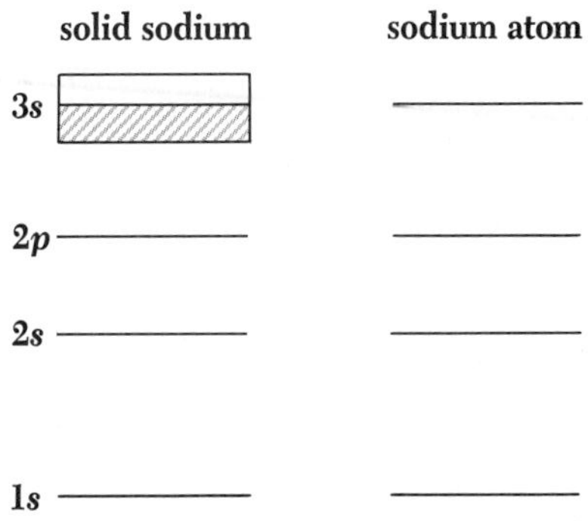

3s

2p

2s

1s

sodium, electrons easily acquire additional energy while remaining in their original energy band. The additional energy is in the form of kinetic energy, and the moving electrons constitute an electric current. Sodium is therefore a good conductor of electricity, as are other crystalline solids with energy bands that are only partially filled.

Figure 10-16 is a simplified diagram of the energy bands of diamond. There is an energy band completely filled with electrons separated by a gap of 6 eV from an empty band above it. This means that at least 6 eV of additional energy must be provided to an electron in a diamond crystal if it is to have any kinetic energy, since it cannot have an energy lying in the forbidden band. An energy increment of this magnitude cannot readily be given to an electron in a crystal by an electric field. An electron moving through a crystal undergoes a collision with an imperfection in the crystal lattice an average of every $\sim 10^{-8}$ m, and it loses much of the energy it gained from any electric field in the collision. An electric-field intensity of 6×10^8 V/m is necessary if an electron is to gain 6 eV in a path length of 10^{-8} m, well over 10^{10} times greater than the electric field intensity needed to cause a current to flow in sodium. Diamond is therefore a very poor conductor of electricity and is accordingly classed as an insulator.

Silicon has a crystal structure resembling that of diamond, and, as in diamond, a gap separates the top of a filled energy band from a vacant higher band. The forbidden band in silicon, however, is only 1.1 eV wide. At low temperatures

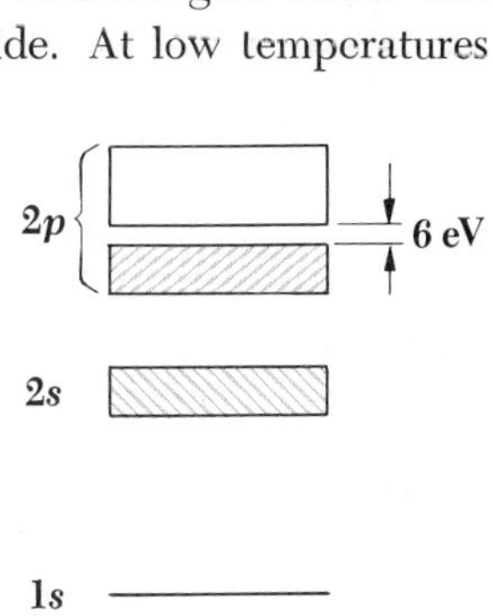

2p   6 eV

FIGURE 10-16   Energy bands in diamond (not to scale).

2s

1s

silicon is little better than diamond as a conductor, but at room temperature a small proportion of its electrons have sufficient kinetic energy of thermal origin to jump the forbidden band and enter the energy band above it. These electrons are sufficient to permit a limited amount of current to flow when an electric field is applied. Thus silicon has an electrical resistivity intermediate between those of conductors and those of insulators, and it is termed a *semiconductor*.

The resistivity of semiconductors can be altered considerably by small amounts of impurity. Let us incorporate a few arsenic atoms in a silicon crystal. Arsenic atoms have five electrons in their outermost shells, while silicon atoms have four. (These shells have the configurations $4s^2 4p^3$ and $3s^2 3p^2$ respectively.) When an arsenic atom replaces a silicon atom in a silicon crystal, four of its electrons are incorporated in covalent bonds with its nearest neighbors. The fifth electron requires little energy to be detached and move about in the crystal (see Probs. 23 and 24). As shown in Fig. 10-17, the presence of arsenic as an impurity provides energy levels just below the band which electrons must occupy for conduction to take place. Such levels are termed *donor levels*, and the substance is called an *n-type* semiconductor because electric current in it is carried by negative charges.

If we alternatively incorporate gallium atoms in a silicon crystal, a different effect occurs. Gallium atoms have only three electrons in their outer shells, whose configuration is $4s^2 4p$, and their presence leaves vacancies called *holes* in the electron structure of the crystal. An electron needs relatively little energy to enter a hole, but as it does so, it leaves a new hole in its former location. When an electric field is applied across a silicon crystal containing a trace of gallium, electrons move toward the anode by successively filling holes. The flow of current here is conveniently described with reference to the holes, whose behavior is like that of positive charges since they move toward the negative electrode. A substance of this kind is called a *p-type* semiconductor. (Certain metals, such as zinc, conduct current primarily by the motion of holes.) In the energy-band diagram of Fig. 10-18 we see that the presence of gallium provides energy levels, termed *acceptor levels*, just above the highest filled band. Any electrons that occupy these levels leave behind them, in the formerly filled band, vacancies which permit electric current to flow.

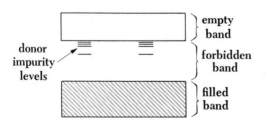

donor impurity levels

empty band

forbidden band

filled band

FIGURE 10-17 A trace of arsenic in a silicon crystal provides donor levels in the normally forbidden band, producing an *n*-type semiconductor.

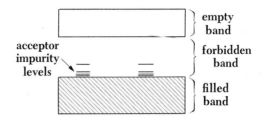

FIGURE 10-18 A trace of gallium in a silicon crystal provides acceptor levels in the normally forbidden band, producing a *p*-type semiconductor.

empty band

forbidden band

filled band

acceptor impurity levels

*10.7 THE FERMI ENERGY

We shall now look more closely into the properties of the free electrons in a metal. Electrons are Fermi particles since they obey the exclusion principle, and hence the electron gas in a metal has a Fermi-Dirac distribution of energies (Sec. 9.9). The Fermi-Dirac distribution law for the number of electrons n_i with the energy ε_i is

10.6 $$n_i = \frac{g_i}{e^{\alpha} e^{\varepsilon_i / kT} + 1}$$

It is more convenient to consider a continuous distribution of electron energies than the discrete distribution of Eq. 10.6, so that the distribution law becomes

10.7 $$n(\varepsilon)\, d\varepsilon = \frac{g(\varepsilon)\, d\varepsilon}{e^{\alpha} e^{\varepsilon / kT} + 1}$$

To find $g(\varepsilon)\, d\varepsilon$, the number of quantum states available to electrons with energies between ε and $\varepsilon + d\varepsilon$, we use the same reasoning as for the photon gas involved in black-body radiation. The correspondence is exact because there are two possible spin states, $m_s = +\frac{1}{2}$ and $m_s = -\frac{1}{2}$, for electrons, thus doubling the number of available phase-space cells just as the existence of two possible directions of polarization for otherwise identical photons doubles the number of cells for a photon gas. In terms of momentum we found in Sec. 9.8 that

$$g(p)\, dp = \frac{8\pi V p^2\, dp}{h^3}$$

For nonrelativistic electrons

$$p^2\, dp = (2m^3 \varepsilon)^{1/2}\, d\varepsilon$$

with the result that

10.8 $$g(\varepsilon)\, d\varepsilon = \frac{8\sqrt{2}\pi V m^{3/2}}{h^3}\, \varepsilon^{1/2}\, d\varepsilon$$

The next step is to evaluate the parameter α. In order to do this, we consider the condition of the electron gas at low temperatures. As observed in Sec. 9.10, the occupation index when T is small is 1 from $\varepsilon = 0$ until near the Fermi energy ε_F, where it drops rapidly to 0. This situation reflects the effect of the exclusion principle: no states can contain more than one electron, and so the minimum energy configuration of an electron gas is one in which the lowest states are filled and the remaining ones are empty. If we set

10.9 $$\alpha = -\frac{\varepsilon_F}{kT}$$

the occupation index becomes

10.10 $$f(\varepsilon) = \frac{n(\varepsilon)}{g(\varepsilon)} = \frac{1}{e^{(\varepsilon - \varepsilon_F)/kT} + 1}$$

The formula is in accord with the exclusion principle. At $T = 0$ K,

$$f(\varepsilon) = 1 \text{ when } \varepsilon < \varepsilon_F$$
$$= 0 \text{ when } \varepsilon > \varepsilon_F$$

As the temperature increases, the occupation index changes from 1 to 0 more and more gradually, as in Fig. 10-19. At all temperatures

$$f(\varepsilon) = \tfrac{1}{2} \text{ when } \varepsilon = \varepsilon_F$$

If a particular metal sample contains N free electrons, we can calculate its Fermi energy ε_F by filling up its energy states with these electrons in order of increasing energy starting from $\varepsilon = 0$. The highest energy state to be filled will then have the energy $\varepsilon = \varepsilon_F$ by definition. The number of electrons that can have the same energy ε is equal to the number of states $g(\varepsilon)$ that have this energy, since each state is limited to one electron. Hence

10.11 $$\int_0^{\varepsilon_F} g(\varepsilon)\, d\varepsilon = N$$

Substituting Eq. 10.8 for $g(\varepsilon)\, d\varepsilon$ yields

$$N = \frac{8\sqrt{2}\pi V m^{3/2}}{h^3} \int_0^{\varepsilon_F} \varepsilon^{1/2}/d\varepsilon$$

$$= \frac{16\sqrt{2}\pi V m^{3/2}}{3h^3} \varepsilon_F^{3/2}$$

and

10.12 $$\varepsilon_F = \frac{h^2}{2m}\left(\frac{3N}{8\pi V}\right)^{2/3}$$ **Fermi energy**

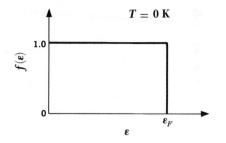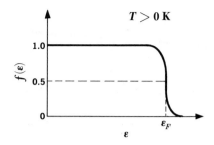

FIGURE 10-19 The occupation index for a Fermi-Durac distribution at absolute zero and at a higher temperature.

The quantity N/V is the density of free electrons; hence ε_F is independent of the dimensions of whatever metal sample is being considered.

Let us use Eq. 10.12 to calculate the Fermi energy in copper. The electron configuration of the ground state of the copper atom is $1s^2 2s^2 2p^6 3s^2 3p^6 3d^{10} 4s$; that is, each atom has a single $4s$ electron outside closed inner shells. It is therefore reasonable to assume that each copper atom contributes one free electron to the electron gas. The electron density $\eta = N/V$ is accordingly equal to the number of copper atoms per unit volume, which is given by

$$\frac{\text{Atoms}}{\text{Volume}} = \frac{(\text{atoms/kmol}) \times (\text{mass/volume})}{\text{mass/kmol}}$$

10.13
$$= \frac{N_0 \rho}{w}$$

Here

$$N_0 = \text{Avogadro's number} = 6.02 \times 10^{26} \text{ atoms/kmol}$$
$$\rho = \text{density of copper} = 8.94 \times 10^3 \text{ kg/m}^3$$
$$w = \text{atomic mass of copper} = 63.5 \text{ kg/kmol}$$

so that

$$\eta = \frac{6.02 \times 10^{26} \text{ atoms/kmol} \times 8.94 \times 10^3 \text{ kg/m}^3}{63.5 \text{ kg/kmol}}$$

$$= 8.5 \times 10^{28} \text{ atoms/m}^3$$
$$= 8.5 \times 10^{28} \text{ electrons/m}^3$$

The corresponding Fermi energy is, from Eq. 10.12,

$$\varepsilon_F = \frac{(6.63 \times 10^{-34} \text{ J-s})^2}{2 \times 9.11 \times 10^{-31} \text{ kg/electron}} \left(\frac{3 \times 8.5 \times 10^{28} \text{ electrons/m}^3}{8\pi} \right)^{2/3}$$

$$= 1.13 \times 10^{-18} \text{ J}$$
$$= 7.04 \text{ eV}$$

At absolute zero, $T = 0$ K, there would be electrons with energies of up to 7.04 eV in copper. By contrast, *all* the molecules in an ideal gas at absolute zero would have zero energy. Because of its decidedly nonclassical behavior, the electron gas in a metal is said to be *degenerate*. Table 10.4 gives the Fermi energies of several common metals.

*10.8 ELECTRON-ENERGY DISTRIBUTION

We may now substitute for α and $g(\varepsilon)\,d\varepsilon$ in Eq. 10.7 to obtain a formula for the number of electrons in an electron gas having energies between some value ε and $\varepsilon + d\varepsilon$. This formula is

10.14
$$n(\varepsilon)\,d\varepsilon = \frac{(8\sqrt{2}\pi\ Vm^{3/2}/h^3)\varepsilon^{1/2}\,d\varepsilon}{e^{(\varepsilon-\varepsilon_F)/kT} + 1}$$

If we express the numerator of Eq. 10.14 in terms of the Fermi energy ε_F, we obtain

10.15
$$n(\varepsilon)\,d\varepsilon = \frac{(3N/2)\varepsilon_F^{-3/2}\varepsilon^{1/2}\,d\varepsilon}{e^{(\varepsilon-\varepsilon_F)/kT} + 1}$$

Equation 10.15 is plotted in Fig. 10-20 for the temperatures $T = 0$, 300, and 1200 K.

It is interesting to determine the average electron energy at absolute zero. To do this, we first obtain the total energy U_0 at 0 K, which is

$$U_0 = \int_0^{\varepsilon_F} \varepsilon n(\varepsilon)\,d\varepsilon$$

Table 10.4.

SOME FERMI ENERGIES.

Metal		Fermi energy, eV
Lithium	Li	4.72
Sodium	Na	3.12
Aluminum	Al	11.8
Potassium	K	2.14
Cesium	Cs	1.53
Copper	Cu	7.04
Zinc	Zn	11.0
Silver	Ag	5.51
Gold	Au	5.54

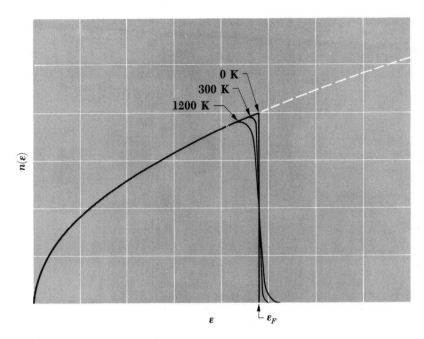

FIGURE 10-20 Distribution of electron energies in a metal at various temperatures.

Since at $T = 0$ K all the electrons have energies less than or equal to the Fermi energy ε_F, we may let

$$e^{(\varepsilon-\varepsilon_F)/kT} = 0$$

and

$$U_0 = \frac{3N}{2}\varepsilon_F^{-3/2} \int_0^{\varepsilon_F} \varepsilon^{3/2}\,d\varepsilon$$

10.16
$$= \frac{3}{5}N\varepsilon_F$$

The average electron energy $\bar{\varepsilon}_0$ is this total energy divided by the number of electrons present N, which yields

10.17
$$\bar{\varepsilon}_0 = \frac{3}{5}\varepsilon_F$$

Since Fermi energies for metals are usually several electron volts, the average electron energy in them at 0 K will also be of this order of magnitude. The temperature of an ideal gas whose molecules have an average kinetic energy of 1 eV is 11,600 K; this means that, if free electrons behaved classically, a sample of copper would have to be at a temperature of about 50,000 K for its electrons to have the same average energy they actually have at 0 K.

THE SOLID STATE   343

The considerable amount of kinetic energy possessed by the valence electrons in the electron gas of a metal represents a repulsive influence, as noted in Sec. 10.5. The act of assembling a group of metal atoms into a solid requires that additional energy be given to the valence electrons in order to elevate them to the higher energy states required by the exclusion principle. The atoms in a metallic solid, however, are closer together because of their bonds than they would be otherwise. As a result the valence electrons are, on the average, closer to an atomic nucleus in a metallic solid than they are in an isolated metal atom. These electrons accordingly have lower potential energies in the former case than in the latter, sufficiently lower to lead to a net cohesive force even when the added electron kinetic energy is taken into account.

*10.9 BRILLOUIN ZONES

We now turn to a more detailed examination of how allowed and forbidden bands in a solid originate. The fundamental idea is that an electron in a crystal moves in a region of periodically varying potential (Fig. 10-21) rather than one of constant potential, and as a result diffraction effects occur that limit the electron to certain ranges of momenta that correspond to the allowed energy bands. In this way of thinking the interactions among the atoms influence valence-electron behavior indirectly through the lattice of the crystal these interactions bring about, instead of directly as in the approach described in Sec. 10.6. An intuitive approach will be used here, rather than a formal treatment based on Schrödinger's equation.

The de Broglie wavelength of a free electron of momentum p is

10.18 $$\lambda = \frac{h}{p}$$ **Free electron**

FIGURE 10-21   The potential energy of an electron in a periodic array of positive ions.

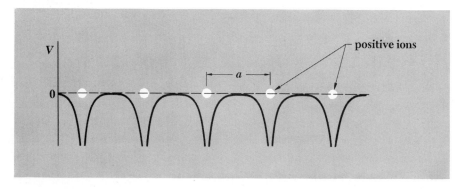

Unbound low-energy electrons can travel freely through a crystal since their wavelengths are long relative to the lattice spacing a. More energetic electrons, such as those with the Fermi energy in a metal, have wavelengths comparable with a, and such electrons are diffracted in precisely the same way as X rays (Sec. 2.4) or electrons in a beam (Sec. 3.5) directed at the crystal from the outside. (When λ is near a, $2a$, $3a$, . . . in value, Eq. 10.18 no longer holds, as discussed later.) An electron of wavelength λ undergoes Bragg "reflection" from one of the atomic planes in a crystal when it approaches the plane at the angle θ, where from Eq. 2.8

10.19 $\qquad n\lambda = 2a \sin \theta \qquad n = 1, 2, 3, \ldots$

It is customary to treat the situation of electron waves in a crystal by replacing λ by the wave number k introduced in Sec. 3.3, where

10.20 $\qquad k = \dfrac{2\pi}{\lambda}$ $\qquad\qquad\qquad\qquad\qquad$ **Wave number**

The wave number is equal to the number of radians per meter in the wave train it describes. Since the wave train moves in the same direction as the particle, we can describe the wave train by means of a vector **k**. Bragg's formula in terms of k is

10.21 $\qquad k = \dfrac{n\pi}{a \sin \theta}$

Figure 10-22 shows Bragg reflection in a two-dimensional square lattice. Evidently we can express the Bragg condition by saying that reflection from the vertical rows of ions occurs when the component of **k** in the x direction, k_x, is equal to $n\pi/a$. Similarly, reflection from the horizontal rows occurs when $k_y = n\pi/a$.

Let us consider first those electrons whose wave numbers are sufficiently small for them to avoid diffraction. If k is less than π/a, the electron can move freely through the lattice in any direction. When $k = \pi/a$, they are prevented from moving in the x or y directions by diffraction. The more k exceeds π/a, the more limited the possible directions of motion, until when $k = \pi/a \sin 45° = \sqrt{2}\pi/a$ the electrons are diffracted even when they move diagonally through the lattice.

The region in k-space that low-k electrons can occupy without being diffracted is called the *first Brillouin zone* and is shown in Fig. 10-23. The second Brillouin zone is also shown; it contains electrons with $k > \pi/a$ that do not fit into the first zone yet which have sufficiently small propagation constants to avoid diffraction by the diagonal sets of atomic planes in Fig. 10-22. The second zone contains electrons with k values from π/a to $2\pi/a$ for electrons moving in the

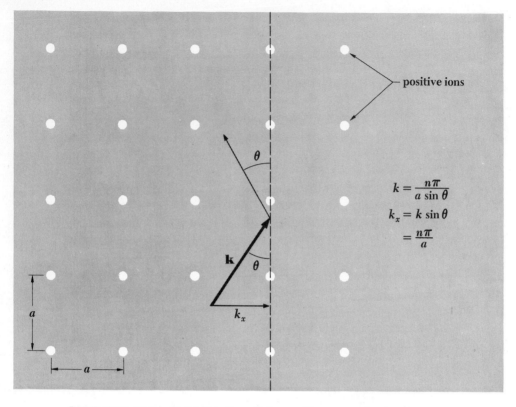

$$k = \frac{n\pi}{a \sin \theta}$$

$$k_x = k \sin \theta$$

$$= \frac{n\pi}{a}$$

FIGURE 10-22 Bragg reflection from the vertical rows of ions occurs when $k_x = n\pi/a$.

$\pm x$ and $\pm y$ directions, with the possible range of k values narrowing as the diagonal directions are approached. Further Brillouin zones can be constructed in the same manner. The extension of this analysis to actual three-dimensional structures leads to the Brillouin zones shown in Fig. 10-24.

*10.10 ORIGIN OF FORBIDDEN BANDS

The significance of the Brillouin zones becomes apparent when we examine the energies of the electrons in each zone. The energy of a free electron is related to its momentum p by

$$E = \frac{p^2}{2m}$$

and hence to its wave number k by

10.22
$$E = \frac{\hbar^2 k^2}{2m}$$

In the case of an electron in a crystal for which $k \ll \pi/a$, there is practically no interaction with the lattice, and Eq. 10.22 is valid. Since the energy of such an electron depends upon k^2, the contour lines of constant energy in a two-dimensional k-space are simply circles of constant k, as in Fig. 10-25. With increasing k the constant-energy contour lines become progressively closer together and also more and more distorted. The reason for the first effect is merely that E varies with k^2. The reason for the second is almost equally straightforward. The closer an electron is to the boundary of a Brillouin zone in k-space, the closer it is to being diffracted by the actual crystal lattice. But

FIGURE 10-23 The first and second Brillouin zones of a two-dimensional square lattice.

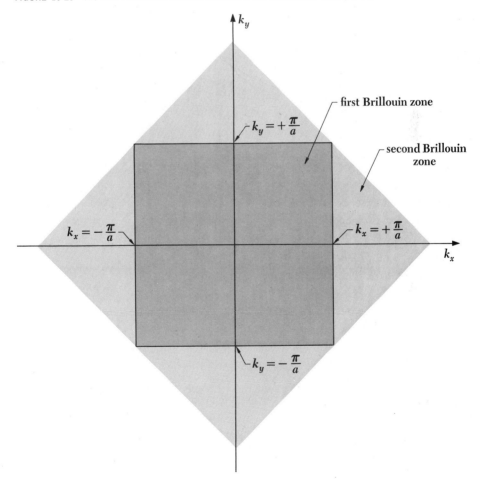

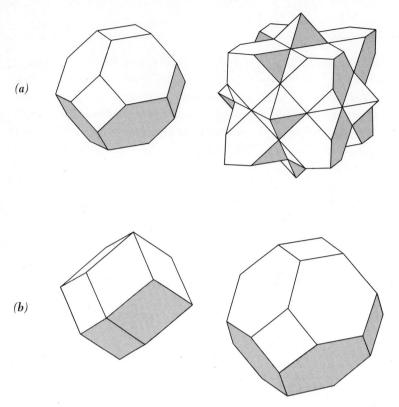

(a)

(b)

FIGURE 10-24 First and second Brillouin zones in (a) face-centered cubic structure and (b) body-centered cubic structure (see Figs. 10-4 and 10-5).

in particle terms the diffraction occurs by virtue of the interaction of the electron with the periodic array of positive ions that occupy the lattice points, and the stronger the interaction, the more the electron's energy is affected.

Figure 10-26 shows how E varies with k in the x direction. As k approaches π/a, E increases more slowly than $\hbar^2 k^2/2m$, the free-particle figure. At $k = \pi/a$, E has two values, the lower belonging to the first Brillouin zone and the higher to the second zone. There is a definite gap between the possible energies in the first and second Brillouin zones which corresponds to the forbidden band spoken of earlier. The same pattern continues as successively higher Brillouin zones are reached.

The energy discontinuity at the boundary of a Brillouin zone follows from the fact that the limiting values of k correspond to standing waves rather than traveling waves. For clarity we shall consider electrons moving in the x direction; the extension of the argument to any other direction is straightforward. When

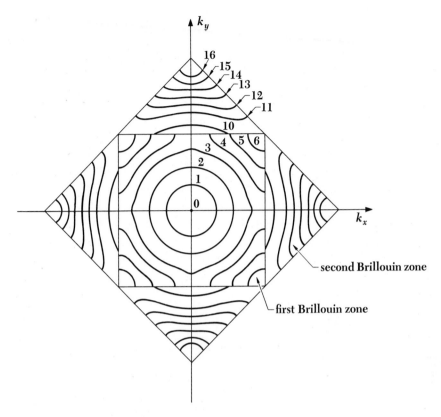

FIGURE 10-25 Energy contours in electron volts in the first and second Brillouin zones of a hypothetical square lattice.

FIGURE 10-26 Electron energy E versus wave number k in the k_x direction. The dashed line shows how E varies with k for a free electron.

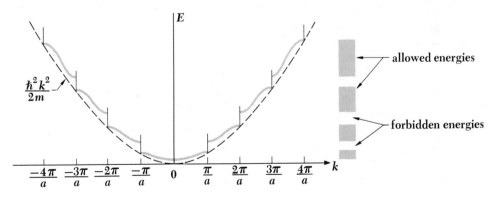

$k = \pm\pi/a$, as we have seen, the waves are Bragg-reflected back and forth, and so the only solutions of Schrödinger's equation consist of standing waves whose wavelength is equal to the periodicity of the lattice. There are two possibilities for these standing waves for $n = 1$, namely

10.23 $$\psi_1 = A \sin \frac{\pi x}{a}$$

10.24 $$\psi_2 = A \cos \frac{\pi x}{a}$$

The probability densities $|\psi_1|^2$ and $|\psi_2|^2$ are plotted in Fig. 10-27. Evidently $|\psi_1|^2$ has its minima at the lattice points occupied by the positive ions, while $|\psi_2|^2$ has its maxima at the lattice points. Since the charge density corresponding to an electron wave function ψ is $e|\psi|^2$, the charge density in the case of ψ_1 is concentrated *between* the positive ions while in the case of ψ_2 it is concentrated

FIGURE 10-27 Distributions of the probability densities $|\psi_1|^2$ and $|\psi_2|^2$.

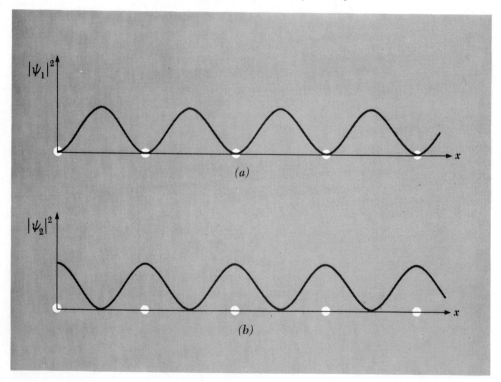

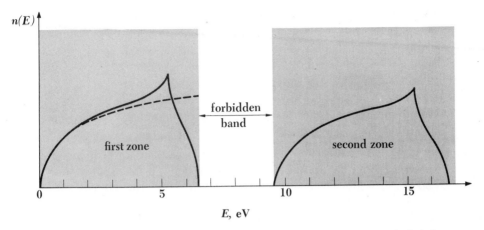

FIGURE 10-28 The distributions of electron energies in the Brillouin zones of Fig. 10-25. The dashed line is the distribution predicted by the free-electron theory.

at the positive ions. The potential energy of an electron in a lattice of positive ions is greatest midway between each pair of ions and least at the ions themselves, so the electron energies E_1 and E_2 associated with the standing waves ψ_1 and ψ_2 are different. No other solutions are possible when $k = \pm\pi/a$, and accordingly no electron can have an energy between E_1 and E_2.

Figure 10-28 shows the distribution of electron energies that corresponds to the Brillouin zones pictured in Fig. 10-25. At low energies (in this hypothetical situation for $E < \sim 2$ eV) the curve is almost exactly the same as that of Fig. 10-20 based on the free-electron theory, which is not surprising since at low energies k is small and the electrons in a periodic lattice then *do* behave like free electrons. With increasing energy, however, the number of available energy states goes beyond that of the free-electron theory owing to the distortion of the energy contours by the lattice: there are more different k values for each energy. Then, when $k = \pm\pi/a$, the energy contours reach the boundaries of the first zone, and energies higher than about 4 eV (in this particular model) are forbidden for electrons in the k_x and k_y directions although permitted in other directions. As the energy goes farther and farther beyond 4 eV, the available energy states become restricted more and more to the corners of the zone, and $n(E)$ falls. Finally, at approximately $6\frac{1}{2}$ eV, there are no more states and $n(E) = 0$. The lowest possible energy in the second zone is somewhat less than 10 eV, and another curve similar in shape to the first begins. Here the gap between the possible energies in the two zones is about 3 eV, and so the forbidden band is about 3 eV wide.

Although there must be an energy gap between successive Brillouin zones in any given direction, the various gaps may overlap permitted energies in other directions so that there is no forbidden band in the crystal as a whole. Figure 10-29 contains graphs of E versus k for three directions in a crystal that has a forbidden band and in a crystal whose allowed bands overlap sufficiently to avoid having a forbidden band.

The electrical behavior of a solid depends upon the degree of occupancy of its energy bands as well as upon the nature of the band structure, as noted earlier. There are two available energy states (one for each spin) in each band for each structural unit in the crystal. (By "structural unit" in this context is meant an atom in a metal or covalent elemental solid such as diamond, a molecule in a molecular solid, and an ion pair in an ionic solid.) A solid will be an insulator if two conditions are met: (1) It must have an even number of valence electrons per structural unit, and (2) the band that contains the highest-energy electrons must be separated from the allowed band above it by an energy gap large compared with kT. The reason for condition (1) is that it ensures that the highest-energy band be completely filled, and the reason for (2) is that none of the electrons be able to cross the gap to reach unfilled states. Thus diamond, with four valence electrons per atom, solid hydrogen, with two valence electrons per H_2 molecule, and NaCl, with eight valence electrons per Na^+—Cl^- ion pair, all have wide forbidden bands in addition and are insulators. Figure 10-30a shows the energy contours of a hypothetical insulator.

A conductor is characterized by its violation of either (or both) of the above conditions. Thus the alkali metals, with an odd number of valence electrons per structural unit (namely one per atom), are conductors, as are such divalent metals as magnesium and zinc which have overlapping energy bands. Figure 10-30b and c respectively show the energy contours of these two types of metal. When the forbidden band in an insulator is narrow or the amount of overlap in a metal is small, the electrical conductivity falls in the semiconductor region, and it is not really correct to speak of the substance as either a metal or a nonmetal. The boundary between filled and empty electron energy states in three-dimensional k-space is called the *Fermi surface*.

Experiments indicate that the conductivity of the divalent metals beryllium, zinc, and cadmium is largely due to positive charge carriers, not to electrons. This unexpected finding is readily accounted for on the basis of the band picture by assuming that the overlap of the Fermi surface into the highest band is small, leaving vacant states—which are holes—in the band below it. The holes in the lower band carry the bulk of the current while the electrons in the upper band play a minor role.

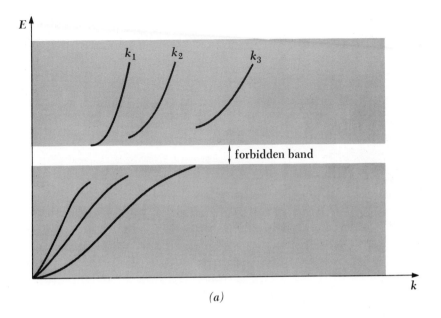

(a)

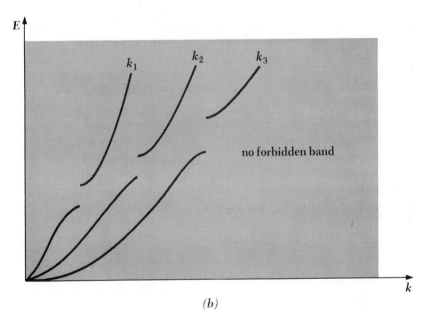

(b)

FIGURE 10-29 E versus k curves for three directions in two crystals. In (a) there is a forbidden band, while in (b) the allowed energy bands overlap and there is no forbidden band.

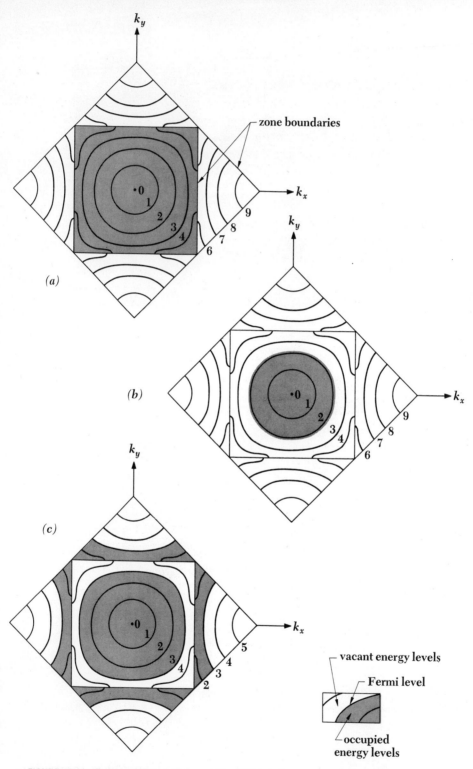

FIGURE 10-30 Electron energy contours and Fermi levels in three types of solid: (a) insulator; (b) mono-valent metal; (c) divalent metal. Energies are in electron volts.

10.11 EFFECTIVE MASS

An electron in a crystal interacts with the crystal lattice, and because of this interaction its response to external forces is not, in general, the same as that of a free electron. There is nothing unusual about this phenomenon—no particle subject to constraints behaves like a free particle. What *is* unusual is that the deviations of an electron in a crystal from free-electron behavior under the influence of external forces can all be incorporated into the simple statement that the *effective mass* of such an electron is not the same as its actual mass.

The most important results of the free-electron theory of metals discussed in Sections 10.7 and 10.8 can be incorporated in the more realistic band theory merely by replacing the electron mass m by the average effective mass $m^°$ at the Fermi surface. Thus the Fermi energy in a metal is given by

10.25 $$u_F = \frac{h^2}{2m^°} \left(\frac{3N}{8\pi V} \right)^{2/3}$$ **Fermi energy**

where N/V is the density of valence electrons. Table 10.5 is a list of effective mass ratios $m^°/m$ in metals.

Table 10.5.

EFFECTIVE MASS RATIOS $m^°/m$ AT THE FERMI SURFACE IN SOME METALS.

Metal		$m^°/m$
Lithium	Li	1.2
Beryllium	Be	1.6
Sodium	Na	1.2
Aluminum	Al	0.97
Cobalt	Co	14
Nickel	Ni	28
Copper	Cu	1.01
Zinc	Zn	0.85
Silver	Ag	0.99
Platinum	Pt	13

Problems

1. What is the effect on the cohesive energy of ionic and covalent crystals of (a) van der Waals forces and (b) zero-point oscillations of the ions and atoms about their equilibrium positions?

2. The van der Waals attraction between two He atoms leads to a binding energy of about 6×10^{-5} eV at an equilibrium separation of about 3 Å. Use the uncertainty principle to show that, at ordinary pressures (<25 atm), solid He cannot exist.

3. The *Joule-Thomson effect* refers to the drop in temperature a gas undergoes when it passes slowly from a full container to an empty one through a porous plug. Since the expansion is into a rigid container, no mechanical work is done. Explain the Joule-Thomson effect in terms of the van der Waals attraction between molecules.

4. The ion spacings and melting points of the sodium halides are as follows:

	NaF	NaCl	NaBr	NaI
Ion spacing, Å	2.3	2.8	2.9	3.2
Melting point, °C	988	801	740	660

Explain the regular variation in these quantities with halogen atomic number.

5. Use the notion of energy bands to explain the following optical properties of solids: (*a*) All metals are opaque to light of all wavelengths. (*b*) Semiconductors are transparent to infrared light although opaque to visible light. (*c*) Most insulators are transparent to visible light.

6. The energy gap in silicon is 1.1 eV and in diamond it is 6 eV. Discuss the transparency of these substances to visible light.

7. A small proportion of indium is incorporated in a germanium crystal. Is the crystal an *n*-type or a *p*-type semiconductor?

8. A small proportion of antimony is incorporated in a germanium crystal. Is the crystal an *n*-type or a *p*-type semiconductor?

9. What is the connection between the fact that the free electrons in a metal obey Fermi statistics and the fact that the photoelectric effect is virtually temperature-independent?

10. (*a*) How much energy is required to form a K^+ and I^- ion pair from a pair of these atoms? (*b*) What must the separation be between a K^+ and an I^- ion if their total energy is to be zero?

11. (*a*) How much energy is required to form a Li^+ and Br^- ion pair from a pair of these atoms? (*b*) What must the separation be between a Li^+ and a Br^- ion if their total energy is to be zero?

12. Show that the first five terms in the series for the Madelung constant of

NaCl are

$$\alpha = 6 - \frac{12}{\sqrt{2}} + \frac{8}{\sqrt{3}} - \frac{6}{2} + \frac{24}{\sqrt{5}} - \cdots$$

13. (a) The ionization energy of potassium is 4.34 eV and the electron affinity of chlorine is 3.61 eV. The Madelung constant for the KCl structure is 1.748 and the distance between ions of opposite sign is 3.14 Å. On the basis of these data only, compute the cohesive energy of KCl. (b) The observed cohesive energy of KCl is 6.42 eV per ion pair. On the assumption that the difference between this figure and that obtained in (a) is due to the exclusion-principle repulsion, find the exponent n in the formula Br^{-n} for the potential energy arising from this source.

14. Repeat Prob. 13 for LiCl, in which the Madelung constant is 1.748, the ion spacing is 2.57 Å, and the observed cohesive energy is 6.8 eV per ion pair. The ionization energy of Li is 5.4 eV.

15. The potential energy $V(x)$ of a pair of atoms in a solid that are displaced by x from their equilibrium separation at 0 K may be written $V(x) = ax^2 - bx^3 - cx^4$, where the anharmonic terms $-bx^3$ and $-cx^4$ represent, respectively, the asymmetry introduced by the repulsive forces between the atoms and the leveling off of the attractive forces at large displacements. At a temperature T the likelihood that a displacement x will occur relative to the likelihood of no displacement is $e^{-V/kT}$, so that the average displacement x at this temperature is

$$\bar{x} = \frac{\displaystyle\int_{-\infty}^{\infty} xe^{-V/kT}\,dx}{\displaystyle\int_{-\infty}^{\infty} e^{-V/kT}\,dx}$$

Show that, for small displacements, $\bar{x} \approx 3bkT/4a^2$. (This is the reason that the change in length of a solid when its temperature changes is proportional to ΔT.)

° 16. The Fermi energy in copper is 7.04 eV. (a) Approximately what percentage of the free electrons in copper are in excited states at room temperature? (b) At the melting point of copper, 1083°C?

° 17. The Fermi energy in silver is 5.51 eV. (a) What is the average energy of the free electrons in silver at 0 K? (b) What temperature is necessary for the average molecular energy in an ideal gas to have this value? (c) What is the speed of an electron with this energy?

° 18. The density of aluminum is 2.70 g/cm³ and its atomic weight is 26.97. The electronic structure of aluminum is given in Table 7.2 (note: the energy difference between $3s$ and $3p$ electrons is very small), and the effective mass of an electron in aluminum is $0.97\,m_e$. Calculate the Fermi energy in aluminum.

°19. The density of zinc is $7.13\ \text{g/cm}^3$ and its atomic weight is 65.4. The electronic structure of zinc is given in Table 7.2, and the effective mass of an electron in zinc is $0.85\,m_e$. Calculate the Fermi energy in zinc.

20. Explain why the free electrons in a metal make only a minor contribution to its specific heat.

°21. Find the ratio between the kinetic energies of an electron in a two-dimensional square lattice which has $k_x = k_y = \pi/a$ and an electron which has $k_x = \pi/a$, $k_y = 0$.

°22. Draw the third Brillouin zone of the two-dimensional square lattice whose first two Brillouin zones are shown in Fig. 10-23.

23. Phosphorus is present in a germanium sample. Assume that one of its five valence electrons revolves in a Bohr orbit around each P^+ ion in the germanium lattice. (a) If the effective mass of the electron is $0.17\,m_e$ and the dielectric constant of germanium is 16, find the radius of the first Bohr orbit of the electron. (b) The energy gap between the valence and conduction bands in germanium is 0.65 eV. How does the ionization energy of the above electron compare with this energy and with kT at room temperature?

24. Repeat Prob. 23 for a silicon sample that contains arsenic. The effective mass of an electron in silicon is about $0.31\,m_e$, the dielectric constant of silicon is 12, and the energy gap in silicon is 1.1 eV.

°25. The calculation of the Fermi energy in copper made in Sec. 10.7 did not take into account the difference between m_e and m_e°, and yet the u_F value obtained was approximately correct. Why?

26. The effective mass m° of a current carrier in a semiconductor can be directly determined by means of a *cyclotron resonance* experiment in which the carriers (whether electrons or holes) move in spiral orbits about the direction of an externally applied magnetic field **B**. An alternating electric field is applied perpendicular to **B**, and resonant absorption of energy from this field occurs when its frequency ν is equal to the frequency of revolution ν_c of the carrier. (a) Derive an equation for ν_c in terms of m°, e, and B. (b) In a certain experiment, $B = 0.1$ T and maximum absorption is found to occur at $\nu = 1.4 \times 10^{10}$ Hz. Find m°. (c) Find the maximum orbital radius of a charge carrier in this experiment whose speed is 3×10^4 m/s.

THE NUCLEUS 4

THE ATOMIC NUCLEUS 11

Thus far we have regarded the atomic nucleus solely as a point mass that possesses positive charge. The behavior of atomic electrons is responsible for the chief properties (except mass) of atoms, molecules, and solids, not the behavior of atomic nuclei. But the nucleus itself is far from insignificance in the grand scheme of things. For instance, the elements exist because of the ability of nuclei to possess multiple electric charges, and explaining this ability is the central problem of nuclear physics. Furthermore, the energy that powers the continuing evolution of the universe apparently can all be traced to nuclear reactions and transformations. And, of course, the mundane applications of nuclear energy are familiar enough.

11.1 ATOMIC MASSES

The nucleus of an atom contains nearly all its mass, and a good deal of information on nuclear properties can be inferred from a knowledge of atomic masses. An instrument used to measure atomic masses is called a *mass spectrometer;* modern spectrometers and techniques are capable of precisions of better than 1 part in 10^6.

Atomic masses refer to the masses of neutral atoms, not of stripped nuclei. Thus the masses of the orbital electrons and the mass equivalent of their binding energies are incorporated in the figures. Atomic masses are conventionally expressed in *mass units* (u) such that the mass of the most abundant type of carbon atom is, by definition, exactly 12.000 . . . u. The value of a mass unit is, to five significant figures,

$$1\,u = 1.6604 \times 10^{-27}\,kg$$

and its energy equivalent is 931.48 MeV.

Not long after the development of methods for determining atomic masses early in this century, it was discovered that not all of the atoms of a particular element have the same mass. The different varieties of the same element are called its *isotopes*. Another widely used term, *nuclide*, refers to a particular species of nucleus; thus each isotope of an element is a nuclide.

The atomic masses listed in Table 7.1 refer to the *average* atomic mass of each element, which is the quantity of primary interest to chemists. Table 11.1 contains the atomic masses and relative abundances of the five stable isotopes of zinc. The individual masses range from 63.92914 u to 69.92535 u, and the relative abundances range from 0.62 percent to 48.89 percent. The average mass is 65.38 u, which is accordingly the atomic mass of zinc. Twenty elements are composed of only a single nuclide each; beryllium, fluorine, sodium, and aluminum are examples.

Even hydrogen is found to have isotopes, though the two heavier ones make up only about 0.015 percent of natural hydrogen. Their atomic masses are 1.007825, 2.014102, and 3.01605 u; the heavier isotopes are known as *deuterium* and *tritium* respectively. (Tritium nuclei, called *tritons*, are unstable and decay radioactively into an isotope of helium.) The nucleus of the lightest isotope is the *proton*, whose mass of

$$m_p = 1.0072766 \text{ u}$$
$$= 1.6725 \times 10^{-27} \text{ kg}$$

is, within experimental error, equal to the mass of the entire atom minus the mass of the electron it also contains. The proton, like the electron, is an elementary particle rather than a composite of other particles. (The notion of elementary particle is considered in some detail in Chap. 13.)

An interesting regularity is apparent in listings of nuclide masses: the values are always very close to being integral multiples of the mass of the hydrogen atom, 1.007825 u. For example, the deuterium atom is approximately twice as massive as the hydrogen atom, and the tritium atom approximately three times as massive. The masses of the zinc isotopes listed in Table 11.1 further illustrate this pattern, being quite near to 64, 66, 67, 68, and 70 times the hydrogen-atom

Table 11.1.

PROPERTIES OF THE STABLE ISOTOPES OF ZINC, Z = 30.

Mass number of isotope	Atomic mass, U	Relative abundance, %
64	63.92914	48.89
66	65.92605	27.81
67	66.92715	4.11
68	67.92486	18.56
70	69.92535	0.62

mass. It is therefore tempting to regard all nuclei as consisting of protons—hydrogen nuclei—somehow bound together. However, a closer look rules out this notion, since a nuclide mass is invariably greater than the mass of a number of hydrogen atoms equal to its atomic number Z—and the nuclear charge of an atom is $+Ze$. The atomic number of zinc is 30, but its isotopes all have masses more than double that of 30 hydrogen atoms.

Another possibility comes to mind. Perhaps electrons may be present in nuclei which neutralize the positive charge of some of the protons. Thus the helium nucleus, whose mass is four times that of the proton though its charge is only $+2e$, would be regarded as being composed of four protons and two electrons. This explanation is buttressed by the fact that certain radioactive nuclei spontaneously emit electrons, a phenomenon called *beta decay*, whose occurence is easy to account for if electrons are present in nuclei.

Despite the superficial attraction of the hypothesis of nuclear electrons, however, there are a number of strong arguments against it:

1. *Nuclear size.* Nuclei are only $\sim 10^{-14}$ m across. To confine an electron to so small a region requires, by the uncertainty principle, an uncertainty in its momentum of $\Delta p \geqslant 1.1 \times 10^{-20}$ kg-m/s, as was calculated in Sec. 3.7. The electron momentum must be at least as large as the minimum value of Δp. The electron kinetic energy that corresponds to a momentum of 1.1×10^{-20} kg-m/s is 21 MeV. (This figure may also be obtained by calculating the lowest energy level of an electron in a box of nuclear dimensions; since $T \gg m_0 c^2$, the latter calculation must be made relativistically.) However, the electrons emitted during beta decay have energies of only 2 or 3 MeV—an order of magnitude smaller than the energies they must have had within the nucleus if they were to have existed there.

We might remark that the uncertainty principle yields a very different result when applied to protons within a nucleus. For a proton with a momentum of 1.1×10^{-20} kg-m/s, $T \ll m_0 c^2$, and its kinetic energy can be calculated classically. We have

$$T = \frac{p^2}{2m}$$
$$= \frac{(1.1 \times 10^{-20} \text{ kg-m/s})^2}{2 \times 1.67 \times 10^{-27} \text{ kg}}$$
$$= 3.6 \times 10^{-14} \text{ J}$$
$$= 0.23 \text{ MeV}$$

The presence of protons with such kinetic energies in a nucleus is entirely plausible.

2. *Nuclear spin.* Protons and electrons are Fermi particles with spins of $\frac{1}{2}$, that is, angular momenta of $\frac{1}{2}\hbar$. Thus nuclei with an even number of protons plus electrons should have integral spins, while those with an odd number of protons plus electrons should have half-integral spins. This prediction is not obeyed. The fact that the deuteron, which is the nucleus of an isotope of hydrogen, has an atomic number of 1 and a mass number of 2, would be interpreted as implying the presence of two protons and one electron. Depending upon the orientations of the particles, the nuclear spin of ^2_1H should therefore be $\frac{1}{2}$ or $\frac{3}{2}$. However, the observed spin of the deuteron is 1, something that cannot be reconciled with the hypothesis of nuclear electrons.

3. *Magnetic moment.* The proton has a magnetic moment only about 1.5×10^{-3} that of the electron, so that nuclear magnetic moments ought to be of the same order of magnitude as that of the electron if electrons are present in nuclei. However, the observed magnetic moments of nuclei are comparable with that of the proton, not with that of the electron, a discrepancy that cannot be understood if electrons are nuclear constituents.

4. *Electron-nuclear interaction.* It is observed that the forces that act between nuclear particles lead to binding energies of the order of 8 MeV per particle. It is therefore hard to see why, if electrons can interact strongly enough with protons to form nuclei, the orbital electrons in an atom interact only electrostatically with its nucleus. That is, how can half the electrons in an atom escape the strong binding of the other half? Furthermore, when fast electrons are scattered by nuclei, they behave as though acted upon solely by electrostatic forces, while the nuclear scattering of fast protons reveals departures from electrostatic influences that can be ascribed only to a specifically nuclear force.

The difficulties of the nuclear electron hypothesis were known for some time before the correct explanation for nuclear masses came to light, but there seemed to be no serious alternative. The problem of the mysterious ingredient besides the proton in atomic nuclei was not solved until 1932.

11.2 THE NEUTRON

The composition of atomic nuclei was finally understood in 1932. Two years earlier the German physicists W. Bothe and H. Becker had bombarded beryllium with alpha particles from a sample of polonium and found that radiation was emitted which was able to penetrate matter readily. Bothe and Becker ascertained that the radiation did not consist of charged particles and assumed, quite naturally, that it consisted of gamma rays. (Gamma rays are electromagnetic waves of extremely short wavelength.) The ability of the radiation to pass through as much as several centimeters of lead without being absorbed suggested gamma rays of unprecedentedly short wavelength. Other physicists became interested in this radiation, and a number of experiments were performed to determine its properties in detail. In one such experiment Irène Curie and F. Joliot observed that when the radiation struck a slab of paraffin, a hydrogen-rich substance, protons were knocked out. At first glance this is not very surprising: X rays can give energy to electrons in Compton collisions, and there is no reason why shorter-wavelength gamma rays cannot give energy to protons in similar processes.

Curie and Joliot found proton recoil energies of up to about 5.3 MeV. From Eq. 2.15 for the Compton effect the minimum gamma-ray photon energy $E = h\nu$ needed to transfer the kinetic energy T to a proton can be calculated. The result is a minimum initial gamma-ray photon energy of 53 MeV. This seemed peculiar because no nuclear radiation known at the time had more than a small fraction of this considerable energy. The peculiarity became even more striking when it was calculated that the presumed reaction of an alpha particle and a beryllium nucleus to yield a carbon nucleus would result in a mass decrease of 0.01144 u, which is equivalent to only 10.7 MeV—one-fifth the energy needed by a gamma-ray photon if it is to knock 5.3 MeV protons out of paraffin.

In 1932 James Chadwick, an associate of Rutherford, proposed an alternative hypothesis for the now-mysterious radiation emitted by beryllium when bombarded by alpha particles. He assumed that the radiation consisted of neutral *particles* whose mass is approximately the same as that of the proton. The electrical neutrality of these particles, which were called *neutrons*, accounted for their ability to penetrate matter readily. Their mass accounted nicely for the observed proton recoil energies: a moving particle colliding head-on with one at rest whose mass is the same can transfer all of its kinetic energy to the latter. A maximum proton energy of 5.3 MeV thus implies a neutron energy of 5.3 MeV, not the 53 MeV required by a gamma ray to cause the same effect. Other experiments had shown that such light nuclei as those of helium, carbon, and nitrogen could also be knocked out of appropriate absorbers by the beryllium radiation, and the measurements made of the energies of these nuclei fit in well with the neutron hypothesis. In fact, Chadwick arrived at the neutron-mass figure of $m_n \approx m_p$ from an analysis of observed proton and nitrogen nuclei recoil energies; no other mass gave as good agreement with the experimental data.

Before we consider the role of the neutron in nuclear structure, we should note that it is not a stable particle outside nuclei. The free neutron decays radioactively into a proton, an electron, and an antineutrino; the half life of the free neutron is 10.8 min.

Immediately after its discovery the neutron was recognized as the missing ingredient in atomic nuclei. Its mass of

$$m_n = 1.0086654 \, \text{u}$$
$$= 1.6748 \times 10^{-27} \, \text{kg}$$

which is slightly more than that of the proton, its electrical neutrality, and its spin of $\frac{1}{2}$ all fit in perfectly with the observed properties of nuclei when it is assumed that nuclei are composed solely of neutrons and protons.

The following terms and symbols are widely used to describe a nucleus:

Z = atomic number = number of protons
N = neutron number = number of neutrons
$A = Z + N$ = mass number = total number of neutrons and protons

The term *nucleon* refers to both protons and neutrons, so that the mass number A is the number of nucleons in a particular nucleus. Nuclides are identified according to the scheme

$$_Z^A\text{X}$$

where X is the chemical symbol of the species. Thus the arsenic isotope of mass number 75 is denoted by

$$_{33}^{75}\text{As}$$

since the atomic number of arsenic is 33. Similarly a nucleus of ordinary hydrogen, which is a proton, is denoted by

$$_1^1\text{H}$$

Here the atomic and mass numbers are the same because no neutrons are present.

The fact that nuclei are composed of neutrons as well as protons immediately explains the existence of isotopes: the isotopes of an element all contain the same numbers of protons but have different numbers of neutrons. Since its nuclear charge is what is ultimately responsible for the characteristic properties of an atom, the isotopes of an element all have identical chemical behavior and differ conspicuously only in mass.

11.3 STABLE NUCLEI

Not all combinations of neutrons and protons form stable nuclei. In general, light nuclei ($A < 20$) contain approximately equal numbers of neutrons and protons, while in heavier nuclei the proportion of neutrons becomes progressively greater. This is evident from Fig. 11-1, which is a plot of N versus Z for stable nuclei. The tendency for N to equal Z follows from the existence of nuclear energy levels, whose origin and properties we shall examine shortly. Nucleons,

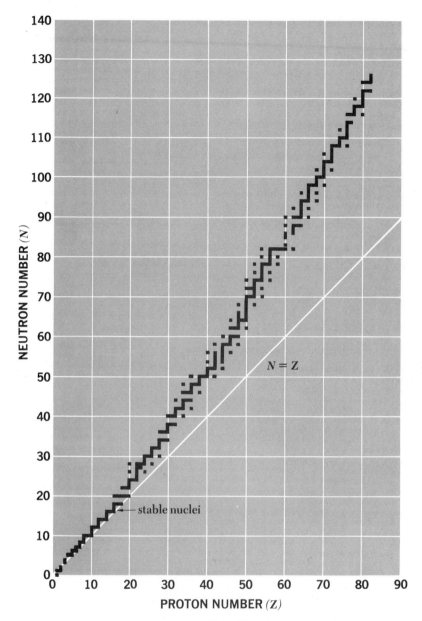

FIGURE 11-1 Neutron-proton diagram for stable nuclides. There are no stable nuclides with $Z = 43$ or 61, with $N = 19$, 35, 39, 45, 61, 89, 115, or 126, or with $A = Z + N = 5$ or 8. All nuclides with $Z > 83$, $N > 126$, and $A > 209$ are unstable.

which have spins of $\frac{1}{2}$, obey the exclusion principle. As a result, each nuclear energy level can contain two neutrons of opposite spins and two protons of opposite spins. Energy levels in nuclei are filled in sequence, just as energy levels in atoms are, to achieve configurations of minimum energy and therefore maximum stability. A nucleus with, say, three neutrons and one proton outside filled inner levels will have more energy than one with two neutrons and two protons in the same situation, since in the former case one of the neutrons must go into a higher level while in the latter case all four nucleons fit into the lowest available level. Figure 11-2 shows how this notion accounts for the absence of a stable $^{12}_{5}B$ isotope while permitting $^{12}_{6}C$ to exist.

The preceding argument is only part of the story. Protons are positively charged and repel one another electrostatically. This repulsion becomes so great in nuclei with more than 10 protons or so that an excess of neutrons, which produce only attractive forces, is required for stability; thus the curve of Fig. 11-1 departs more and more from the $N = Z$ line as Z increases. Even in light nuclei N may exceed Z, but is never smaller; $^{11}_{5}B$ is stable, for instance, but not $^{11}_{6}C$.

Nuclear forces are limited in range, and as a result nucleons interact strongly only with their nearest neighbors. This effect is referred to as the *saturation* of nuclear forces. Because the coulomb repulsion of the protons is appreciable throughout the entire nucleus, there is a limit to the ability of neutrons to prevent the disruption of a large nucleus. This limit is represented by the bismuth isotope $^{209}_{83}Bi$, which is the heaviest stable nuclide. All nuclei with $Z > 83$ and $A > 209$ spontaneously transform themselves into lighter ones through the emission of one or more alpha particles, which are $^{4}_{2}He$ nuclei. Since an alpha particle

FIGURE 11-2 Simplified energy-level diagrams of stable boron and carbon isotopes. The exclusion principle limits the occupancy of each level to two neutrons of opposite spin and two protons of opposite spin.

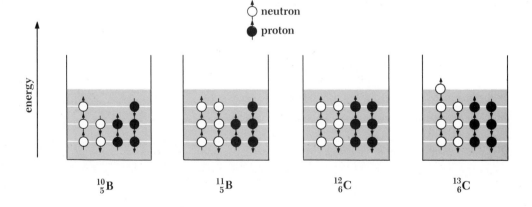

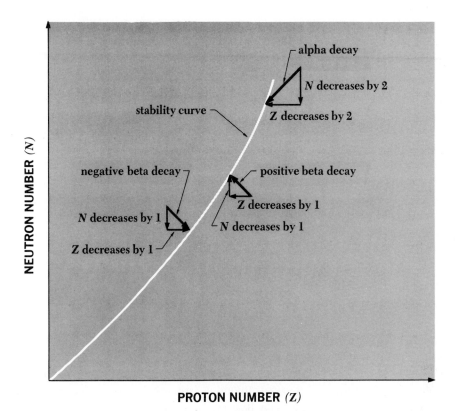

FIGURE 11-3 Alpha and beta decays permit an unstable nucleus to reach a stable configuration.

consists of two protons and two neutrons, an alpha decay reduces the Z and the N of the original nucleus by two each. If the resulting daughter nucleus has either too small or too large a neutron/proton ratio for stability, it may beta decay to a more appropriate configuration. In negative beta decay, a neutron is transformed into a proton and an electron:

$$n \rightarrow p + e^-$$

The electron leaves the nucleus and is observed as a "beta particle." In positive beta decay, a proton becomes a neutron and a positron is emitted:

$$p \rightarrow n + e^+$$

Thus negative beta decay decreases the proportion of neutrons and positive beta decay increases it. Figure 11-3 shows how alpha and beta decays enable stability to be achieved. Radioactivity is considered in more detail in Chap. 12.

11.4 NUCLEAR SIZES AND SHAPES

The Rutherford scattering experiment provided the first evidence that nuclei are of finite size. In that experiment, as we saw in Chap. 4, an incident alpha particle is deflected by a target nucleus in a manner consistent with Coulomb's law provided the distance between them exceeds about 10^{-14} m. For smaller separations the predictions of Coulomb's law are not obeyed because the nucleus no longer appears as a point charge to the alpha particle.

Since Rutherford's time a variety of experiments have been performed to determine nuclear dimensions, with particle scattering still a favored technique. Fast electrons and neutrons are ideal for this purpose, since an electron interacts with a nucleus only through electrical forces while a neutron interacts only through specifically nuclear forces. Thus electron scattering provides information on the distribution of charge in a nucleus and neutron scattering provides information on the distribution of nuclear matter. In both cases the de Broglie wavelength of the particle must be smaller than the radius of the nucleus under study (see Prob. 3 of Chap. 3).

Actual experiments on the sizes of nuclei have employed electrons of several hundred MeV to over 1 GeV (1 GeV = 1,000 MeV = 10^9 eV) and neutrons of 20 MeV and up. In every case it is found that the volume of a nucleus is directly proportional to the number of nucleons it contains, which is its mass number A. If a nuclear radius is R, the corresponding volume is $\frac{4}{3}\pi R^3$ and so R^3 is proportional to A. This relationship is usually expressed in inverse form as

11.1 $\qquad R = R_0 A^{1/3}$ **Nuclear radii**

The value of R_0 is

$$R_0 \approx 1.2 \times 10^{-15} \text{ m}$$

The indefiniteness in R_0 is a consequence, not just of experimental error, but of the characters of the various experiments: electrons and neutrons interact differently with nuclei. The value of R_0 is slightly smaller when it is deduced from electron scattering, which implies that nuclear matter and nuclear charge are not identically distributed throughout a nucleus.

As we saw in the earlier part of this book, the angstrom unit (1 Å = 10^{-10} m) is a convenient unit of length for expressing distances in the atomic realm. For example, the radius of the hydrogen atom is 0.53 Å, the C and O atoms in a CO molecule are 1.13 Å apart, and the Na$^+$ and Cl$^-$ ions in crystalline NaCl are 2.81 Å apart. Nuclei are so small that the *fermi* (fm), only 10^{-5} the size of the angstrom, is an appropriate unit of length:

$$1 \text{ fermi} = 1 \text{ fm} = 10^{-15} \text{ m}$$

Hence we can write

11.2 $\qquad R \approx 1.2 \, A^{1/3} \text{ fm}$

for nuclear radii. From this formula we find that the radius of the $^{12}_{6}C$ nucleus is

$$R \approx 1.2 \times (12)^{1/3} \text{ fm} \approx 2.7 \text{ fm}$$

Similarly, the radius of the $^{107}_{47}Ag$ nucleus is 5.7 fm and that of the $^{238}_{92}U$ nucleus is 7.4 fm.

Now that we know nuclear sizes as well as masses, we can compute the density of nuclear matter. In the case of $^{12}_{6}C$, whose atomic mass is 12.0 u, we have for the nuclear density (the masses and binding energies of the six electrons may be neglected here)

$$\begin{aligned}
\rho &= \frac{m}{\frac{4}{3}\pi R^3} \\
&= \frac{12.0 \text{ u} \times 1.66 \times 10^{-27} \text{ kg/u}}{\frac{4}{3}\pi \times (2.7 \times 10^{-15} \text{ m})^3} \\
&= 2 \times 10^{17} \text{ kg/m}^3
\end{aligned}$$

This figure—equivalent to 3 billion tons per cubic inch!—is essentially the same for all nuclei. Certain stars, known as "white dwarfs," are composed of atoms whose electron shells have collapsed owing to enormous pressure, and the densities of such stars approach that of pure nuclear matter.

We have been assuming that nuclei are spherical. How can nuclear shapes be determined? If the distribution of charge in a nucleus is not spherically symmetric, the nucleus will have an electric quadrupole moment. A nuclear quadrupole moment will interact with the orbital electrons of the atom, and the consequent shifts in atomic energy levels will lead to hyperfine splitting of the spectral lines. Of course, this source of hyperfine structure must be distinguished from that due to the magnetic moment of the nucleus, but when this is done it is found that deviations from sphericity actually do occur in nuclei whose spin quantum numbers are 1 or more. Such nuclei may be prolate or oblate spheroids, but the difference between major and minor axes never exceeds ~20 percent and is usually much less. For almost all purposes it is sufficient to regard nuclei as being spherical; nevertheless the departures from sphericity, small as they are, furnish valuable information on nuclear structure.

11.5 BINDING ENERGY

A stable atom invariably has a smaller mass than the sum of the masses of its constituent particles. The deuterium atom ^2_1H, for example, has a mass of 2.014102 u, while the mass of a hydrogen atom (^1_1H) plus that of a neutron is

$$m_{hydrogen} + m_n = 1.007825\,u + 1.008665\,u$$
$$= 2.016490\,u$$

which is 0.002388 u greater. Since a deuterium nucleus—called a *deuteron*—is composed of a proton and a neutron, and both ^1_1H and ^2_1H have single orbital electrons, it is evident that the *mass defect* of 0.002388 u is related to the binding of a proton and a neutron to form a deuteron. A mass of 0.002388 u is equal to

$$0.002388\,u \times 931\,MeV/u = 2.23\,MeV$$

When a deuteron is formed from a free proton and neutron, then, 2.23 MeV of energy is liberated. Conversely, 2.23 MeV must be supplied from an external source to break a deuteron up into a proton and a neutron. This inference is supported by experiments on the photodisintegration of the deuteron, which show that a gamma-ray photon must have an energy of at least 2.23 MeV to disrupt a deuteron (Fig. 11-4).

The energy equivalent of the mass defect in a nucleus is called its *binding energy* and is a measure of the stability of the nucleus. Binding energies arise from the action of the forces that hold nucleons together to form nuclei, just as ionization energies of atoms, which must be provided to remove electrons from them, arise from the action of electrostatic forces. Binding energies range from 2.23 MeV for the deuteron, which is the smallest compound nucleus, up to 1,640 MeV for $^{209}_{83}Bi$, the heaviest stable nucleus.

FIGURE 11-4 The binding energy of the deuteron is 2.23 MeV, which is confirmed by experiments that show that a gamma-ray photon with a minimum energy of 2.23 MeV can split a deuteron into a free neutron and a free proton.

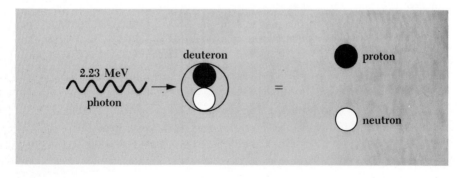

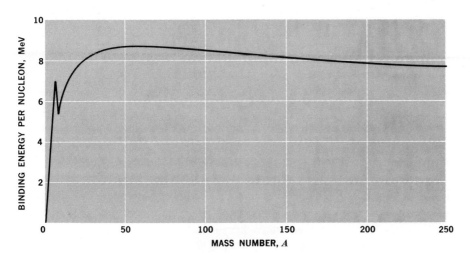

FIGURE 11-5 Binding energy per nucleon as a function of mass number. The peak at $A = 4$ corresponds to the $^4_2$He nucleus.

The binding energy per nucleon, arrived at by dividing the total binding energy of a nucleus by the number of nucleons it contains, is a most interesting quantity. The binding energy per nucleon is plotted as a function of mass number A in Fig. 11-5. The curve rises steeply at first and then more gradually until it reaches a maximum of 8.79 MeV at $A = 56$, corresponding to the iron nucleus $^{56}_{26}$Fe, and then drops slowly to about 7.6 MeV at the highest mass numbers. Evidently nuclei of intermediate mass are the most stable, since the greatest amount of energy must be supplied to liberate each of their nucleons. This fact suggests that energy will be evolved if heavy nuclei can somehow be split into lighter ones or if light nuclei can somehow be joined to form heavier ones. The former process is known as *nuclear fission* and the latter as *nuclear fusion,* and both indeed occur under proper circumstances and do evolve energy as predicted.

Nuclear binding energies are strikingly large. To appreciate their magnitude, it is helpful to convert the figures from MeV/nucleon to more familiar units, say kcal/kg. Since 1 eV $= 1.60 \times 10^{-19}$ J and 1 J $= 2.39 \times 10^{-4}$ kcal, we find that 1 MeV $= 3.83 \times 10^{-17}$ kcal. One mass unit is equal to 1.66×10^{-27} kg, and each nucleon in a nucleus has a mass of very nearly 1 u. Hence

$$1 \frac{\text{MeV}}{\text{nucleon}} = \frac{3.83 \times 10^{-17} \text{ kcal}}{1.66 \times 10^{-27} \text{ kg}} = 2.31 \times 10^{10} \frac{\text{kcal}}{\text{kg}}$$

A binding energy of 8 MeV/nucleon, a typical value, is therefore equivalent to 1.85×10^{11} kcal/kg. By contrast, the heat of vaporization of water is a mere 540 kcal/kg, and even the heat of combustion of gasoline, 1.13×10^4 kcal/kg, is 10 million times smaller.

The unique short-range forces that bind nucleons so securely into nuclei constitute by far the strongest class of forces known. Unfortunately nuclear forces are nowhere near as well understood as electrical forces, and in consequence the theory of nuclear structure is still primitive as compared with the theory of atomic structure. However, even without a satisfactory understanding of nuclear forces, considerable progress has been made in recent years in interpreting the properties and behavior of nuclei in terms of detailed models, and we shall examine some of the concepts embodied in these models in this chapter. Before looking into any of these theories, though, it is instructive to see what can be revealed by even a very general approach. The simplest nucleus containing more than one nucleon is the *deuteron*, which consists of a proton and a neutron. The deuteron binding energy is 2.23 MeV, a figure that can be obtained either from the discrepancy in mass between $m_{deuteron}$ and $m_p + m_n$ or from photo-disintegration experiments which show that only gamma rays with $h\nu \geqslant 2.23$ MeV can disrupt deuterons into their constituent nucleons. In Chap. 6 we analyzed another two-body system, the hydrogen atom, with the help of quantum mechanics, but in that case the precise nature of the force between the proton and the electron was known. If a force law is known for an interaction, the corresponding potential energy function V can be found and substituted into Schrödinger's equation. Our understanding of nuclear forces is less complete than our understanding of coulomb forces, however, and so it is not possible to discuss the deuteron in as much quantitative detail as the hydrogen atom.

The actual potential energy V of the deuteron, that is, the potential energy of either nucleon with respect to the other, depends upon the distance r between the centers of the neutron and proton more or less as shown by the solid line in Fig. 11-6. (The repulsive "core" perhaps 0.4×10^{-15} m in radius expresses the inability of nucleons to mesh together more than a certain amount.) We shall approximate this $V(r)$ by the "square well" shown as a dashed line in the figure. This approximation means that we consider the nuclear force between neutron and proton to be zero when they are more than r_0 apart, and to have a constant magnitude, leading to the constant potential energy $-V_0$, when they are closer together than r_0. Thus the parameters V_0 and r_0 are representative of the strength and range, respectively, of the interaction holding the deuteron together, and the square-well potential itself is representative of the short-range character of the interaction.

A square-well potential means that V is a function of r alone, and therefore, as in the case of other central-force potentials, it is easiest to examine the problem in a spherical polar-coordinate system (see Fig. 6-1). In spherical polar coordinates Schrödinger's equation for a particle of mass m is, with $\hbar = h/2\pi$,

$$\frac{1}{r^2}\frac{\partial}{\partial r}\left(r^2\frac{\partial\psi}{\partial r}\right) + \frac{1}{r^2\sin\theta}\frac{\partial}{\partial\theta}\left(\sin\theta\,\frac{\partial\psi}{\partial\theta}\right)$$

$$+ \frac{1}{r^2\sin^2\theta}\frac{\partial^2\psi}{\partial\phi^2} + \frac{2m}{\hbar^2}(E - V)\psi = 0$$

Let us choose the particle in question to be the neutron, so that we imagine it moving in the force field of the proton. (The opposite choice would, of course, yield identical results.) We note from Fig. 11-6 that E, the total energy of the neutron, is negative and is the same as the binding energy of the deuteron.

In analyzing the hydrogen atom, where one particle is much heavier than the other, it is still necessary to consider the effects of nuclear motion, and we did this in Chaps. 4 and 6 by replacing the electron mass m_e by its reduced mass m'. In this way the problem of a proton and an electron moving about a common center of mass is replaced by the problem of a single particle of mass m' moving about a fixed point. A similar procedure is even more appropriate here, since neutron and proton masses are almost the same. According to Eq. 4.27, the reduced mass of a neutron-proton system is

11.4 $$m' = \frac{m_n m_p}{m_n + m_p}$$

and so we replace the m of Eq. 11.3 with m' as given above.

FIGURE 11-6 The actual potential energy of either proton or neutron in a deuteron and the square-well approximation to this potential energy as functions of the distance between proton and neutron.

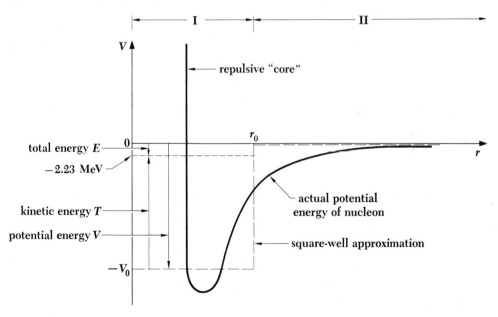

We now assume that the solution of Eq. 11.3 can be written as the product of radial and angular functions,

11.5 $\psi(r, \theta, \phi) = R(r)\Theta(\theta)\Phi(\phi)$

As before, the function $R(r)$ describes how the wave function ψ varies along a radius vector from the nucleus, with θ and ϕ constant; the function $\Theta(\theta)$ describes how ψ varies with zenith angle θ along a meridian on a sphere centered at $r = 0$, with r and ϕ constant; and the function $\Phi(\phi)$ describes how ψ varies with azimuth angle ϕ along a parallel on a sphere centered at $r = 0$, with r and θ constant.

Although angular motion can occur in a square-well potential, our interest for the moment is in radial motion, that is, in oscillations of the neutron and proton about their center of mass. If there is no angular motion, Θ and Φ are both constant and their derivatives are zero. With $\partial\psi/\partial\theta = \partial^2\psi/\partial\phi^2 = 0$, Eq. 11.3 becomes

11.6 $$\frac{1}{r^2}\frac{d}{dr}\left(r^2\frac{dR}{dr}\right) + \frac{2m'}{\hbar^2}(E - V)R = 0$$

A further simplification can be made by letting

11.7 $u(r) = rR(r)$

In terms of the new function u the wave equation becomes

11.8 $$\frac{d^2u}{dr^2} + \frac{2m'}{\hbar^2}(E - V)u = 0$$

Because V has two different values, $V = -V_0$ inside the well and $V = 0$ outside it, there are two different solutions to Eq. 11.8, u_{I} for $r \leqslant r_0$ and u_{II} for $r \geqslant r_0$. Inside the well the wave equation is

$$\frac{d^2u_{\mathrm{I}}}{dr^2} + \frac{2m'}{\hbar^2}(E + V_0)u_{\mathrm{I}} = 0$$

or, if we let

11.9 $$a^2 = \frac{2m'}{\hbar^2}(E + V_0)$$

it becomes simply

11.10 $$\frac{d^2u_{\mathrm{I}}}{dr^2} + a^2u_{\mathrm{I}} = 0$$

(We note from Fig. 11-6 that, since $|V_0| > |E|$, the quantities $E + V_0$ and hence a^2 are positive.) Equation 11.10 is the same as the wave equation for a particle in a box of Chap. 5, and again the solution is

11.11 $\qquad u_\text{I} = A \cos ar + B \sin ar$

We recall that the radial wave function R is given by $R = u/r$, which means that the cosine solution must be discarded if R is not to be infinite at $r = 0$. Hence $A = 0$ and u_I inside the well is just

11.12 $\qquad u_\text{I} = B \sin ar$

Outside the well $V = 0$ and so

11.13 $\qquad \dfrac{d^2 u_\text{II}}{dr^2} + \dfrac{2m'}{\hbar^2} E u_\text{II} = 0$

The total energy E of the neutron is a negative quantity since it is bound to the proton. Therefore

11.14 $\qquad b^2 = \dfrac{2m'}{\hbar^2}(-E)$

is a positive quantity, and we can write

11.15 $\qquad \dfrac{d^2 u_\text{II}}{dr^2} - b^2 u_\text{II} = 0$

The solution of Eq. 11.15 is

11.16 $\qquad u_\text{II} = Ce^{-br} + De^{br}$

Because it must be true that $u \to 0$ as $r \to \infty$, we conclude that $D = 0$. Hence outside the well

11.17 $\qquad u_\text{II} = Ce^{-br}$

*11.7 GROUND STATE OF THE DEUTERON

We now have expressions for u (and hence for ψ) both inside and outside the well, and it remains to match these expressions and their first derivatives at the well boundary since it is necessary that both u and du/dr be continuous everywhere in the region for which they are defined. At $r = r_0$, then,

$$u_\text{I} = u_\text{II}$$
11.18 $\qquad B \sin ar_0 = Ce^{-br_0}$

and

$$\frac{du_\text{I}}{dr} = \frac{du_\text{II}}{dr}$$
11.19 $\qquad aB \cos ar_0 = -bCe^{-br_0}$

By dividing Eq. 11.18 by Eq. 11.19 we eliminate the coefficients B and C and obtain the transcendental equation

11.20 $$\tan ar_0 = -\frac{a}{b}$$

Equation 11.20 cannot be solved analytically, but it can be solved either graphically or numerically to any desired degree of accuracy. We note that

11.21 $$\frac{a}{b} = \frac{\sqrt{2m'(E + V_0)}/\hbar}{\sqrt{2m'(-E)}/\hbar} = \sqrt{\frac{E + V_0}{-E}}$$

where $-E$ is the binding energy of the deuteron and V_0 is the depth of the potential well. Since $|V_0| > |E|$, in order to obtain a first crude approximation we might assume that a/b is so large that

$$\tan ar_0 \approx \infty$$

Since $\tan \theta$ becomes infinite at $\theta = \pi/2, \pi, 3\pi/2, \ldots$, in this approximation the ground state of the deuteron, for which $n = 1$, corresponds to

$$ar_0 \approx \frac{\pi}{2}$$

(In fact, this is the only bound state of the deuteron.) Hence

$$\frac{\sqrt{2m'(E + V_0)}}{\hbar}r_0 \approx \frac{\sqrt{2m'V_0}}{\hbar}r_0 \approx \frac{\pi}{2}$$

since we are assuming that E is negligible compared with V_0, and

11.22 $$V_0 \approx \frac{\pi^2\hbar^2}{8m'r_0^2}$$

The above approximation is equivalent to assuming that the function u_{I} inside the well is at its maximum (corresponding to $ar = 90°$) at the boundary of the well. Actually, u_{I} must be somewhat past its maximum there in order to join smoothly with the function u_{II} outside the well, as shown in Fig. 11-7; a more detailed calculation shows that $ar \approx 116°$ at $r = r_0$. The difference between the two results is due to our neglect of the binding energy $-E$ relative to V_0 in obtaining Eq. 11.22, and when this neglect is remedied, the better approximation

11.23 $$V_0 \approx \frac{\pi^2\hbar^2}{8m'r_0^2} + \frac{2\hbar}{r_0}\sqrt{\frac{-E}{2m'}}$$

is obtained.

Equation 11.23 is a relationship among r_0, the radius of the potential well

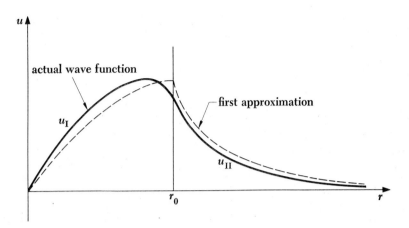

FIGURE 11-7   The wave function $u(r)$ of either proton or neutron in a deuteron.

and therefore representative of the range of nuclear forces, V_0, the depth of the well and therefore representative of the interaction potential energy between two nucleons due to nuclear forces, and $-E$, the binding energy of the deuteron. What we now ask is whether this relationship corresponds to reality in the sense that a reasonable choice of r_0 leads to a reasonable value for V_0. (There is nothing in our simple model that can point to a unique value for either r_0 or V_0.) Such a choice for r_0 might be 2 fm. Substituting for the known quantities in Eq. 11.23 and expressing energies in MeV yields

$$V_0 \approx \frac{1.0 \times 10^{-28} \text{ MeV-m}^2}{r_0{}^2} + \frac{1.9 \times 10^{-14} \text{ MeV-m}}{r_0}$$

and so, for $r_0 = 2$ fm $= 2 \times 10^{-15}$ m,

$$V_0 \approx 35 \text{ MeV}$$

This is a plausible figure for V_0, from which we conclude that the basic features of our model—nucleons that maintain their identities in the nucleus instead of fusing together, strong nuclear forces with a short range and relatively minor angular dependence—are valid.

11.8   TRIPLET AND SINGLET STATES

Equation 11.23 contains all the information about the neutron-proton force that can be obtained from the observation that the deuteron is a stable system with a binding energy of $-E$. To go further we require additional experimental data. Among the most significant such data is that concerning angular momentum, which was ignored in the preceding analysis of the deuteron when it was assumed that the neutron-proton potential is a function of r alone. The latter is not correct; in general, angular momentum *does* play a significant role in nuclear structure, although certain aspects of this structure are virtually independent of it. In the case of the deuteron, for example, the proton and neutron interact in such a manner that binding occurs only when their spins are parallel to produce a *triplet state*, not when their spins are antiparallel to produce a *singlet state*. Evidently the neutron-proton force depends on the relative orientation of the spins and is weaker when the spins are antiparallel.

The difference between the triplet and singlet potentials, together with the Pauli exclusion principle, makes it possible to see why diprotons and dineutrons do not exist despite the observed stability of the deuteron and the charge-independence of nuclear forces. The exclusion principle prevents a diproton or a dineutron from occurring in a triplet state, since with parallel spins both nucleons in each system would be in identical quantum states. No such restriction applies to the deuteron, since the neutron and proton of which it consists are distinguishable particles even with parallel spins. However, while a diproton or dineutron could occur in principle in a singlet state, the singlet nuclear force is not sufficiently strong to produce binding—and, indeed, diprotons and dineutrons have never been found.

11.9 THE LIQUID-DROP MODEL

While the attractive forces that nucleons exert upon one another are very strong, their range is so small that each particle in a nucleus interacts solely with its nearest neighbors. This situation is the same as that of atoms in a solid, which ideally vibrate about fixed positions in a crystal lattice, or that of molecules in a liquid, which ideally are free to move about while maintaining a fixed inter-molecular distance. The analogy with a solid cannot be pursued because a calculation shows that the vibrations of the nucleons about their average positions would be too great for the nucleus to be stable. The analogy with a liquid, on the other hand, turns out to be extremely useful in understanding certain aspects of nuclear behavior.

Let us see how the picture of a nucleus as a drop of liquid accounts for the observed variation of binding energy per nucleon with mass number. We shall start by assuming that the energy associated with each nucleon-nucleon bond has some value U; this energy is really negative, since attractive forces are involved, but is usually written as positive because binding energy is considered a positive quantity for convenience. Because each bond energy U is shared by two nucleons, each has a binding energy of $\frac{1}{2}U$. When an assembly of spheres of the same size is packed together into the smallest volume, as we suppose is the case of nucleons within a nucleus, each interior sphere has 12 other spheres in contact with it (Fig. 11-8). Hence each interior nucleon in a nucleus has a binding energy of $12 \times \frac{1}{2}U$ or $6U$. If all A nucleons in a nucleus were in its interior, the total binding energy of the nucleus would be

11.24 $\qquad E_v = 6AU$

Equation 11.24 is often written simply as

11.25 $\qquad E_v = a_1 A$

The energy E_v is called the *volume energy* of a nucleus and is directly proportional to A.

Actually, of course, some nucleons are on the surface of every nucleus and therefore have fewer than 12 neighbors. The number of such nucleons depends upon the surface area of the nucleus in question. A nucleus of radius R has an area of

$$4\pi R^2 = 4\pi R_0^2 A^{2/3}$$

Hence the number of nucleons with fewer than the maximum number of bonds is proportional to $A^{2/3}$, reducing the total binding energy by

11.26 $\qquad E_s = -a_2 A^{2/3}$

The negative energy E_s is called the *surface energy* of a nucleus; it is most significant for the lighter nuclei since a greater fraction of their nucleons are on the surface. Because natural systems always tend to evolve toward configurations of minimum potential energy, nuclei tend toward configurations of maxi-

FIGURE 11-8 In a tightly packed assembly of identical spheres, each interior sphere is in contact with twelve others.

mum binding energy. (We recall that binding energy is the mass-energy difference between a nucleus and the same numbers of free neutrons and protons.) Hence a nucleus should exhibit the same surface-tension effects as a liquid drop, and in the absence of external forces it should be spherical since a sphere has the least surface area for a given volume.

The electrostatic repulsion between each pair of protons in a nucleus also contributes toward decreasing its binding energy. The *coulomb energy* E_c of a nucleus is the work that must be done to bring together Z protons from infinity into a volume equal to that of the nucleus. Hence E_c is proportional to $Z(Z - 1)/2$, the number of proton pairs in a nucleus containing Z protons, and inversely proportional to the nuclear radius $R = R_0 A^{1/3}$:

11.27
$$E_c = -a_3 \frac{Z(Z - 1)}{A^{1/3}}$$

The coulomb energy is negative because it arises from a force that opposes nuclear stability.

The total *binding energy* E_b of a nucleus is the sum of its volume, surface, and coulomb energies:

$$E_b = E_v + E_s + E_c$$

11.28
$$= a_1 A - a_2 A^{2/3} - a_3 \frac{Z(Z - 1)}{A^{1/3}}$$

The binding energy *per nucleon* is therefore

11.29
$$\frac{E_b}{A} = a_1 - \frac{a_2}{A^{1/3}} - a_3 \frac{Z(Z - 1)}{A^{4/3}}$$

Each of the terms of Eq. 11.29 is plotted in Fig. 11-9 versus A, together with their sum, E_b/A. The latter is reasonably close to the empirical curve of E_b/A shown in Fig. 11-5. Hence the analogy of a nucleus with a liquid drop has some validity at least, and we may be encouraged to see what further aspects of nuclear behavior it can illuminate. We shall do this in Chap. 12 in connection with nuclear reactions.

Before leaving the subject of nuclear binding energy, it should be noted that effects other than those we have here considered also are involved. For instance, nuclei with equal numbers of protons and neutrons are especially stable, as are nuclei with even numbers of protons and neutrons. Thus such nuclei as ^4_2He, $^{12}_6\text{C}$, and $^{16}_8\text{O}$ appear as peaks on the empirical binding energy per nucleon curve. These peaks imply that the energy states of neutrons and protons in a nucleus are almost identical and that each state can be occupied by two particles of opposite spin, as discussed in Sec. 11.3.

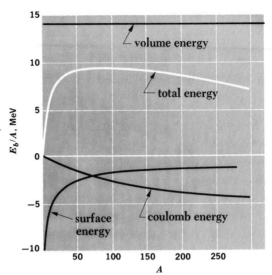

FIGURE 11-9 The binding energy per nucleon is the sum of the volume, surface, and coulomb energies.

11.10 THE SHELL MODEL

The basic assumption of the liquid-drop model is that the constituents of a nucleus interact only with their nearest neighbors, like the molecules of a liquid. There is a good deal of empirical support for this assumption. There is also, however, extensive experimental evidence for the contrary hypothesis that the nucleons in a nucleus interact primarily with a general force field rather than directly with one another. This latter situation resembles that of electrons in an atom, where only certain quantum states are permitted and no more than two electrons, which are Fermi particles, can occupy each state. Nucleons are also Fermi particles, and several nuclear properties vary periodically with Z and N in a manner reminiscent of the periodic variation of atomic properties with Z.

The electrons in an atom may be thought of as occupying positions in "shells" designated by the various principal quantum numbers, and the degree of occupancy of the outermost shell is what determines certain important aspects of an atom's behavior. For instance, atoms with 2, 10, 18, 36, 54, and 86 electrons have all their electron shells completely filled. Such electron structures are stable, thereby accounting for the chemical inertness of the rare gases. The same kind of situation is observed with respect to nuclei; nuclei having 2, 8, 20, 28, 50, 82, and 126 neutrons or protons are more abundant than other nuclei of similar mass numbers, suggesting that their structures are more stable. Since complex nuclei arose from reactions among lighter ones, the evolution of heavier and heavier nuclei became retarded when each relatively inert nucleus was formed; this accounts for their abundance.

Other evidence also points up the significance of the numbers 2, 8, 20, 28, 50, 82, and 126, which have become known as *magic numbers,* in nuclear structure. An example is the observed pattern of nuclear electric quadrupole moments, which are measures of the departures of nuclear charge distributions from sphericity. A spherical nucleus has no quadrupole moment, while one shaped like a football has a positive moment and one shaped like a pumpkin has a negative moment. Nuclei of magic N and Z are found to have zero quadrupole moments and hence are spherical, while other nuclei are distorted in shape.

The *shell model* of the nucleus is an attempt to account for the existence of magic numbers and certain other nuclear properties in terms of interactions between an individual nucleon and a force field produced by all the other nucleons. A potential energy function is used that corresponds to a square well about 50 MeV deep with rounded corners, so that there is a more realistic gradual change from $V = V_0$ to $V = 0$ than the sudden change of the pure square-well potential we used in treating the deuteron. Schrödinger's equation for a particle in a potential well of this kind is then solved, and it is found that stationary states of the system occur characterized by quantum numbers n, l, and m_l whose significance is the same as in the analogous case of stationary states of atomic electrons. Neutrons and protons occupy separate sets of states in a nucleus since the latter interact electrically as well as through the specifically nuclear charge.

In order to obtain a series of energy levels that leads to the observed magic numbers, it is merely necessary to assume a spin-orbit interaction whose magnitude is such that the consequent splitting of energy levels into sublevels is large for large l, that is, for large orbital angular momenta. It is assumed that LS coupling holds only for the very lightest nuclei, in which the l values are necessarily small in their normal configurations. In this scheme, as we saw in Chap. 7, the intrinsic spin angular momenta S_i of the particles concerned (the neutrons form one group and the protons another) are coupled together into a total spin momentum S, and the orbital angular momenta L_i are separately coupled together into a total orbital momentum L; S and L are then coupled to form a total angular momentum J of magnitude $\sqrt{J(J + 1)}\,\hbar$. After a transition region in which an intermediate coupling scheme holds, the heavier nuclei exhibit jj coupling. In this case the S_i and L_i of each particle are first coupled to form a J_i for that particle of magnitude $\sqrt{j(j + 1)}\,\hbar$, and the various J_i then couple together to form the total angular momentum J. The jj coupling scheme holds for the great majority of nuclei.

When an appropriate strength is assumed for the spin-orbit interaction, the energy levels of either class of nucleon fall into the sequence shown in Fig. 11-10. The levels are designated by a prefix equal to the total quantum number n, a letter that indicates l for each particle in that level according to the usual pattern

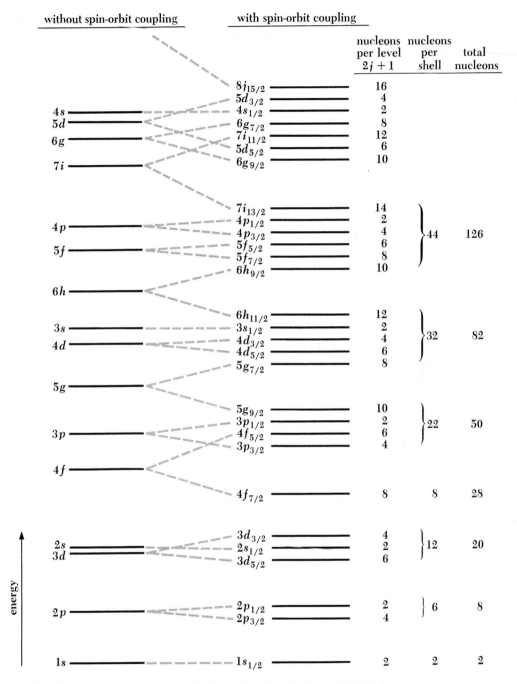

FIGURE 11-10 Sequence of nucleon energy levels according to the shell model (not to scale).

($s, p, d, f, g, \ldots$ correspond respectively to $l = 0, 1, 2, 3, 4, \ldots$), and a subscript equal to j. The spin-orbit interaction splits each state of given j into $2j + 1$ substates, since there are $2j + 1$ allowed orientations of $\mathbf{J}_i$. Large energy gaps appear in the spacing of the levels at intervals that are consistent with the notion of separate shells. The number of available nuclear states in each nuclear shell is, in ascending order of energy, 2, 6, 12, 8, 22, 32, and 44; hence shells are filled when there are 2, 8, 20, 28, 50, 82, or 126 neutrons or protons in a nucleus.

The shell model is able to account for several nuclear phenomena in addition to magic numbers. Since each energy sublevel can contain two particles (spin up and spin down), only filled sublevels are present when there are even numbers of neutrons and protons in a nucleus ("even-even" nucleus). At the other extreme, a nucleus with odd numbers of neutrons and protons ("odd-odd" nucleus) contains unfilled sublevels for both kinds of particle. The stability we expect to be conferred by filled sublevels is borne out by the fact that 160 stable even-even nuclides are known, as against only four stable odd-odd nuclides: ^2_1H, ^6_3Li, $^{10}_5\text{B}$, and $^{14}_7\text{N}$.

Further evidence in favor of the shell model is its ability to predict total nuclear angular momenta. In even-even nuclei, all the protons and neutrons should pair off so as to cancel out one another's spin and orbital angular momenta. Thus even-even nuclei ought to have zero nuclear angular momenta, as observed. In even-odd (even Z, odd N) and odd-even (odd Z, even N) nuclei, the half-integral spin of the single "extra" nucleon should be combined with the integral angular momentum of the rest of the nucleus for a half-integral total angular momentum, and odd-odd nuclei each have an extra neutron and an extra proton whose half-integral spins should yield integral total angular momenta. Both of these predictions are experimentally confirmed.

If the nucleons in a nucleus are so close together and interact so strongly that the nucleus can be considered as analogous to a liquid drop, how can these same nucleons be regarded as moving independently of each other in a common force field as required by the shell model? It would seem that the points of view are mutually exclusive, since a nucleon moving about in a liquid-drop nucleus must surely undergo frequent collisions with other nucleons.

A closer look shows that there is no contradiction. In the ground state of a nucleus, the neutrons and protons fill the energy levels available to them in order of increasing energy in such a way as to obey the exclusion principle (see Fig. 11-2). In a collision, energy is transferred from one nucleon to another, leaving the former in a state of reduced energy and the latter in one of increased energy. But all the available levels of lower energy are already filled, so such an energy transfer can take place only if the exclusion principle is violated. Of course, it is possible for two nucleons of the same kind to merely exchange their respective energies, but such a collision is hardly significant since the system remains in exactly the same state it was in initially. In essence, then, the exclusion principle prevents nucleon-nucleon collisions even in a tightly packed nucleus, and thereby justifies the independent-particle approach to nuclear structure.

Both the liquid-drop and shell models of the nucleus are, in their very different ways, able to account for much that is known of nuclear behavior. Recently attempts have been made to devise theories that combine the best features of each of these models in a consistent scheme, and some success has been achieved in the endeavor. The resulting model includes the possibility of a nucleus vibrating and rotating as a whole. The situation is complicated by the non-spherical shape of all but even-even nuclei and the centrifugal distortion experienced by a rotating nucleus; the detailed theory is consistent with the spacing of excited nuclear levels inferred from the gamma-ray spectra of nuclei and in other ways.

Problems

1. A beam of singly charged ions of $^6_3$Li with energies of 400 eV enters a uniform magnetic field of flux density 0.08 T. The ions move perpendicular to the field direction. Find the radius of their path in the magnetic field. (The $^6_3$Li atomic mass is 6.01513 u.)

2. A beam of singly charged boron ions with energies of 1,000 eV enters a uniform magnetic field of flux density 0.2 T. The ions move perpendicular to the field direction. Find the radii of the path of the $^{10}_5$B (10.013 u) and $^{11}_5$B (11.009 u) isotopes in the magnetic field.

3. Ordinary boron is a mixture of the $^{10}_5$B and $^{11}_5$B isotopes and has a composite atomic weight of 10.82 u. What percentage of each isotope is present in ordinary boron?

4. Show that the nuclear density of $^1_1$H is 10^{14} times greater than its atomic density. (Assume the atom to have the radius of the first Bohr orbit.)

5. The binding energy of $^{35}_{17}$Cl is 298 MeV. Find its mass in u.

6. The mass of $^{20}_{10}$Ne is 19.9924 u. Find its binding energy in MeV.

7. Find the average binding energy per nucleon in $^{16}_8$O (the mass of the neutral $^{16}_8$O atom is 15.9949 u).

8. How much energy is required to remove one proton from $^{16}_8$O? (The mass of the neutral $^{15}_7$N atom is 15.0001 u; that of the neutral $^{15}_8$O atom is 15.0030 u.)

9. How much energy is required to remove one neutron from $^{16}_8$O?

10. Compare the minimum energy a gamma-ray photon must possess if it is to disintegrate an alpha particle into a triton and a proton with that it must possess if it is to disintegrate an alpha particle into a $^3_2$He nucleus and a neutron. (The atomic masses of $^3_1$H and $^3_2$He are respectively 3.01605 u and 3.01603 u.)

11. Show that the electrostatic potential energy of two protons 1.7×10^{-15} m apart is of the correct order of magnitude to account for the difference in binding energy between $^3_1$H and $^3_2$He. How does this result bear upon the problem of the charge-independence of nuclear forces?

12. Protons, neutrons, and electrons all have spins of $\frac{1}{2}$. Why do $^4_2$He atoms obey Bose-Einstein statistics while $^3_2$He atoms obey Fermi-Dirac statistics?

13. Show that the results of Sec. 11.7—a potential well for the deuteron that is about 35 MeV deep and 2 fm in radius—are consistent with the uncertainty principle.

14. Calculate the approximate value of a_3 in Eq. 11.27 using whatever assumptions seem appropriate.

15. According to the *Fermi gas model* of the nucleus, its protons and neutrons exist in a box of nuclear dimensions and fill the lowest available quantum states to the extent permitted by the exclusion principle. Since both neutrons and protons have spins of $\frac{1}{2}$, they are Fermi particles and must obey Fermi-Dirac statistics. (a) Starting from Eq. 10.12, derive an equation for the Fermi energy in a nucleus under the assumption that equal numbers of neutrons and protons are present. (b) Find the Fermi energy for such a nucleus in which $R_0 = 1.2$ fm.

NUCLEAR TRANSFORMATIONS 12

Despite the strength of the forces that hold their constituent nucleons together, atomic nuclei are not immutable. Many nuclei are unstable and spontaneously alter their compositions through radioactive decay. And all nuclei can be transformed by reactions with nucleons or other nuclei that collide with them. In fact, complex nuclei came into being in the first place through successive nuclear reactions, probably in stellar interiors. The principal aspects of radioactivity and nuclear reactions are discussed in this chapter.

12.1 RADIOACTIVE DECAY

Perhaps no single phenomenon has played so significant a role in the development of both atomic and nuclear physics as radioactivity. A nucleus undergoing radioactive decay spontaneously emits a $^4_2$He nucleus (alpha particle), an electron (beta particle), or a photon (gamma ray), thereby ridding itself of nuclear excitation energy or achieving a configuration that is or will lead to one of greater stability.

The *activity* of a sample of any radioactive material is the rate at which the nuclei of its constituent atoms decay. If N is the number of nuclei present at a certain time in the sample, its activity R is given by

12.1
$$R = -\frac{dN}{dt}$$

The minus sign is inserted to make R a positive quantity, since dN/dt is, of course, intrinsically negative. While the natural units for activity are disintegrations per second, it is customary to express R in terms of the *curie* (Ci) and its submultiples, the *millicurie* (mCi) and *microcurie* (μCi). By definition,

$$1 \text{ Ci} = 3.70 \times 10^{10} \text{ disintegrations/s}$$
$$1 \text{ mc} = 10^{-3} \text{ Ci} = 3.70 \times 10^{7} \text{ disintegrations/s}$$
$$1 \text{ } \mu\text{c} = 10^{-6} \text{ Ci} = 3.70 \times 10^{4} \text{ disintegrations/s}$$

Experimental measurements on the activities of radioactive samples indicate that, in every case, they fall off exponentially with time. Figure 12-1 is a graph of R versus t for a typical radioisotope. We note that in every 5-h period, regardless of when the period starts, the activity drops to half of what it was at the start of the period. Accordingly the *half life* $T_{1/2}$ of the isotope is 5 h. Every radioisotope has a characteristic half life; some have half lives of a millionth of a second, others have half lives that range up to billions of years. When the observations plotted in Fig. 12-1 began, the activity of the sample was R_0. Five h later it decreased to $0.5R_0$. After another 5 h, R again decreased by a factor of 2 to $0.25R_0$. That is, the activity of the sample was only 0.25 its initial value after an interval of $2T_{1/2}$. With the lapse of another half life of 5 h, corresponding to a total interval of $3T$, R became $\frac{1}{2}(0.25R_0)$, or $0.125R_0$.

The behavior illustrated in Fig. 12-1 indicates that we can express our empirical information about the time variation of activity in the form

12.2 $R = R_0 e^{-\lambda t}$

where λ, called the *decay constant,* has a different value for each radioisotope. The connection between decay constant λ and half life $T_{1/2}$ is easy to establish. After a half life has elapsed, that is, when $t = T_{1/2}$, the activity R drops to $\frac{1}{2}R_0$ by definition. Hence

$$R = R_0 e^{-\lambda t}$$
$$\tfrac{1}{2}R_0 = R_0 e^{-\lambda T_{1/2}}$$
$$e^{\lambda T_{1/2}} = 2$$

Taking natural logarithms of both sides of this equation,

$$\lambda T_{1/2} = \ln 2$$

12.3 $$T_{1/2} = \frac{\ln 2}{\lambda} = \frac{0.693}{\lambda}$$ **Half life**

The decay constant of the radioisotope whose half life is 5 h is therefore

$$\lambda = \frac{0.693}{T_{1/2}}$$
$$= \frac{0.693}{5 \text{ h} \times 3,600 \text{ s/h}}$$
$$= 3.85 \times 10^{-5} \text{ s}^{-1}$$

The fact that radioactive decay follows the exponential law of Eq. 12.2 is strong evidence that this phenomenon is statistical in nature: every nucleus in a sample of radioactive material has a certain probability of decaying, but there is no way of knowing in advance *which* nuclei will actually decay in a particular time span.

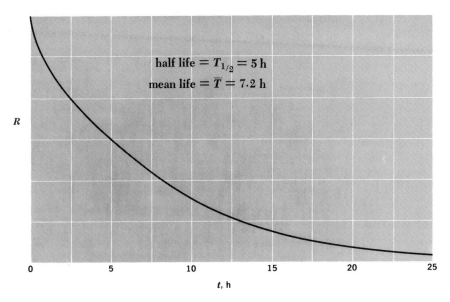

$$\text{half life} = T_{1/2} = 5\,\text{h}$$
$$\text{mean life} = \overline{T} = 7.2\,\text{h}$$

FIGURE 12-1 The activity of a radioisotope decreases exponentially with time.

If the sample is large enough—that is, if many nuclei are present—the actual fraction of it that decays in a certain time span will be very close to the probability for any individual nucleus to decay. The statement that a certain radioisotope has a half life of 5 h, then, signifies that every nucleus of this isotope has a 50 percent chance of decaying in any 5-h period. This does *not* mean a 100 percent probability of decaying in 10 h; a nucleus does not have a memory, and its decay probability per unit time is constant until it actually does decay. A half life of 5 h implies a 75 percent probability of decay in 10 h, which increases to 87.5 percent in 15 h, to 93.75 percent in 20 h, and so on, because in every 5-h interval the probability is 50 percent.

The empirical activity law of Eq. 12.2 follows directly from the assumption of a constant probability λ per unit time for the decay of each nucleus of a given isotope. Since λ is the probability per unit time, $\lambda\,dt$ is the probability that any nucleus will undergo decay in a time interval dt. If a sample contains N undecayed nuclei, the number dN that decay in a time dt is the product of the number of nuclei N and the probability $\lambda\,dt$ that each will decay in dt. That is,

12.4 $$dN = -N\lambda\,dt$$

where the minus sign is required because N decreases with increasing t. Equation 12.4 can be rewritten

$$\frac{dN}{N} = -\lambda\,dt$$

and each side can now be integrated:

$$\int_{N_0}^{N} \frac{dN}{N} = -\lambda \int_{0}^{t} dt$$

$$\ln N - \ln N_0 = -\lambda t$$

12.5
$$N = N_0 e^{-\lambda t}$$

Equation 12.5 is a formula that gives the number N of undecayed nuclei at the time t in terms of the decay probability per unit time λ of the isotope involved and the number N_0 of undecayed nuclei at $t = 0$.

Since the activity of a radioactive sample is defined as

$$R = -\frac{dN}{dt}$$

we see that, from Eq. 12.5,

$$R = \lambda N_0 e^{-\lambda t}$$

This agrees with the empirical activity law if

$$R_0 = \lambda N_0$$

or, in general, if

12.6
$$R = \lambda N$$

Evidently the decay constant λ of a radioisotope is the same as the probability per unit time for the decay of a nucleus of that isotope.

Equation 12.6 permits us to calculate the activity of a radioisotope sample if we know its mass, atomic mass, and decay constant. As an example, let us determine the activity of a 1-gm sample of $^{90}_{38}\text{Sr}$, whose half life against beta decay is 28 yr. The decay constant of $^{90}_{38}\text{Sr}$ is

$$\lambda = \frac{0.693}{T_{1/2}}$$

$$= \frac{0.693}{28 \text{ yr} \times 3.16 \times 10^7 \text{ s/yr}}$$

$$= 7.83 \times 10^{-10} \text{ s}^{-1}$$

A kmol of an isotope has a mass very nearly equal to the mass number of that isotope expressed in kilograms. Hence 1 gm of $^{90}_{38}\text{Sr}$ contains

$$\frac{10^{-3} \text{ kg}}{90 \text{ kg/kmol}} = 1.11 \times 10^{-5} \text{ kmol}$$

One kmol of any isotope contains Avogadro's number of atoms, and so 1 gm of $^{90}_{38}$Sr contains

$$1.11 \times 10^{-5} \text{ kmol} \times 6.025 \times 10^{26} \text{ atoms/kmol}$$
$$= 6.69 \times 10^{21} \text{ atoms}$$

Thus the activity of the sample is

$$R = \lambda N$$
$$= 7.83 \times 10^{-10} \times 6.69 \times 10^{21} \text{ s}^{-1}$$
$$= 5.23 \times 10^{12} \text{ s}^{-1}$$
$$= 141 \text{ Ci}$$

It is worth keeping in mind that the half life of a radioisotope is not the same as its *mean lifetime* $\overline{T}$. The mean lifetime of an isotope is the reciprocal of its decay probability per unit time:

12.7
$$\overline{T} = \frac{1}{\lambda}$$

Hence

12.8
$$\overline{T} = \frac{1}{\lambda} = \frac{T_{1/2}}{0.693} = 1.44T_{1/2} \qquad\qquad \textbf{Mean lifetime}$$

$\overline{T}$ is nearly half again more than $T_{1/2}$. The mean lifetime of an isotope whose half life is 5 h is 7.2 h.

12.2   RADIOACTIVE SERIES

Most of the radioactive elements found in nature are members of four *radioactive series*, with each series consisting of a succession of daughter products all ultimately derived from a single parent nuclide. The reason that there are exactly four such series follows from the fact that alpha decay reduces the mass number of a nucleus by 4. Thus the nuclides whose mass numbers are all given by

12.9
$$A = 4n$$

where n is an integer, can decay into one another in descending order of mass number. Radioactive nuclides whose mass numbers obey Eq. 12.9 are said to be members of the $4n$ series. The members of the $4n + 1$ series have mass numbers specified by

12.10
$$A = 4n + 1$$

Table 12.1.

FOUR RADIOACTIVE SERIES.

Mass numbers	Series	Parent	Half life, yr	Stable end product
$4n$	Thorium	$^{232}_{90}\text{Th}$	1.39×10^{10}	$^{208}_{82}\text{Pb}$
$4n + 1$	Neptunium	$^{237}_{93}\text{Np}$	2.25×10^{6}	$^{209}_{83}\text{Bi}$
$4n + 2$	Uranium	$^{238}_{92}\text{U}$	4.51×10^{9}	$^{206}_{82}\text{Pb}$
$4n + 3$	Actinium	$^{235}_{92}\text{U}$	7.07×10^{8}	$^{207}_{82}\text{Pb}$

and members of the $4n + 2$ and $4n + 3$ series have mass numbers specified respectively by

12.11 $A = 4n + 2$

12.12 $A = 4n + 3$

The members of each of these series, too, can decay into one another in descending order of mass number.

Table 12.1 is a list of the names of four important radioactive series, their parent nuclides and the half lives of these parents, and the stable daughters which are end products of the series. The half life of neptunium is so short compared with the estimated age ($\sim 10^{10}$ yr) of the universe that the members of this series are not found in nature today. They have, however, been produced in the laboratory by the neutron bombardment of other heavy nuclei; a brief discussion is given in Sec. 12.12. The sequences of alpha and beta decays that lead

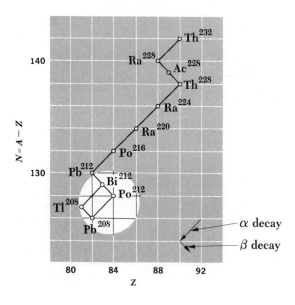

FIGURE 12-2 The thorium decay series ($A = 4n$). The decay of $^{212}_{83}\text{Bi}$ may proceed either by alpha emission and then beta emission or in the reverse order.

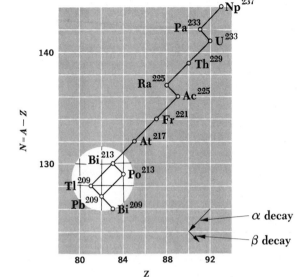

FIGURE 12-3 The neptunium decay series ($A = 4n + 1$). The decay of $^{213}_{83}$Bi may proceed either by alpha emission and then beta emission or in the reverse order.

from parent to stable end product in each series are shown in Figs. 12-2 to 12-5. Some nuclides may decay either by beta or alpha emission, so that the decay chain *branches* at them. Thus $^{212}_{83}$Bi, a member of the thorium series, has a 66.3 percent chance of beta decaying into $^{212}_{84}$Po and a 33.7 percent chance of alpha

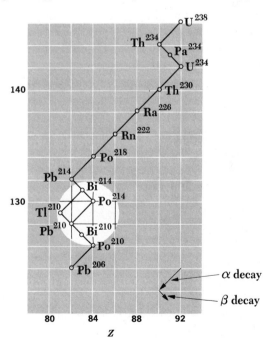

FIGURE 12-4 The uranium decay series ($A = 4n + 2$). The decay of $^{214}_{83}$Bi may proceed either by alpha emission and then beta emission or in the reverse order.

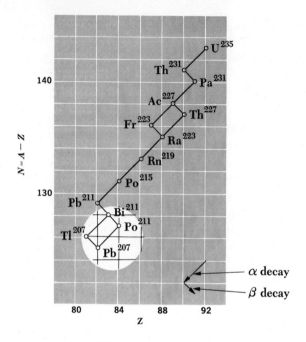

FIGURE 12-5 The actinium decay series ($A = 4n + 3$). The decays of $^{227}_{89}$Ac and $^{211}_{83}$Bi may proceed either by alpha emission and then beta emission or in the reverse order.

decaying into $^{208}_{81}$Tl. The beta decay is followed by a subsequent alpha decay and the alpha decay is followed by a subsequent beta decay, so that both branches lead to $^{208}_{82}$Pb.

Several alpha-radioactive nuclides whose atomic numbers are less than 82 are found in nature, though they are not very abundant.

12.3 ALPHA DECAY

Because the attractive forces between nucleons are of short range, the total binding energy in a nucleus is approximately proportional to its mass number A, the number of nucleons it contains. The repulsive electrostatic forces between protons, however, are of unlimited range, and the total disruptive energy in a nucleus is approximately proportional to Z^2. Nuclei which contain 210 or more nucleons are so large that the short-range nuclear forces that hold them together are barely able to counterbalance the mutual repulsion of their protons. Alpha decay occurs in such nuclei as a means of increasing their stability by reducing their size.

Why are alpha particles almost invariably emitted rather than, say, individual protons or $^3_2$He nuclei? The answer follows from the high binding energy of the alpha particle. To escape from a nucleus, a particle must have kinetic energy, and the alpha-particle mass is sufficiently smaller than that of its constituent nucleons for such energy to be available. To illustrate this point, we can compute, from the known masses of each particle and the parent and daughter nuclei, the kinetic energy Q released when various particles are emitted by a heavy nucleus. This is given by

$$Q = (m_i - m_f - m_\alpha)c^2$$

where m_i is the mass of the initial nucleus, m_f the mass of the final nucleus, and m_α the alpha-particle mass. We find that *only* the emission of an alpha particle is energetically possible; other decay modes would require energy to be supplied from outside the nucleus. Thus alpha decay in $^{232}_{92}U$ is accompanied by the release of 5.4 MeV, while 6.1 MeV would somehow have to be furnished if a proton is to be emitted and 9.6 MeV if a ^3_2He nucleus is to be emitted. The observed disintegration energies in alpha decay agree with the corresponding predicted values based upon the nuclear masses involved.

The kinetic energy T_α of the emitted alpha particle is never quite equal to the disintegration energy Q because, since momentum must be conserved, the nucleus recoils with a small amount of kinetic energy when the alpha particle emerges. It is easy to show that, as a consequence of momentum and energy conservation, T_α is related to Q and the mass number A of the original nucleus by

$$T_\alpha \approx \frac{A - 4}{A} Q$$

The mass numbers of nearly all alpha emitters exceed 210, and so most of the disintegration energy appears as the kinetic energy of the alpha particle. In the decay of $^{222}_{86}Rn$, $Q = 5.587$ MeV while $T_\alpha = 5.486$ MeV.

While a heavy nucleus can, in principle, spontaneously reduce its bulk by alpha decay, there remains the problem of *how* an alpha particle can actually escape from the nucleus. Figure 12-6 is a plot of the potential energy V of an alpha particle as a function of its distance r from the center of a certain heavy nucleus. The height of the potential barrier is about 25 MeV, which is equal to the work that must be done against the repulsive electrostatic force to bring an alpha particle from infinity to a position adjacent to the nucleus but just outside the range of its attractive forces. We may therefore regard an alpha particle in such a nucleus as being inside a box whose walls require an energy of 25 MeV to be surmounted. However, decay alpha particles have energies that range from 4 to 9 MeV, depending upon the particular nuclide involved—16 to 21 MeV short of the energy needed for escape.

Although alpha decay is inexplicable on the basis of classical arguments, quantum mechanics provides a straightforward explanation. In fact, the theory of alpha decay developed independently in 1928 by Gamow and by Gurney and

Condon was greeted as an especially striking confirmation of quantum mechanics. In the following two sections we shall show how even a simplified treatment of the problem of the escape of an alpha particle from a nucleus gives results in agreement with experiment.

The basic notions of this theory are:

1. An alpha particle may exist as an entity within a heavy nucleus;
2. Such a particle is in constant motion and is contained in the nucleus by the surrounding potential barrier;
3. There is a small—but definite—likelihood that the particle may pass through the barrier (despite its height) each time a collision with it occurs.

Thus the decay probability per unit time λ can be expressed as

$$\lambda = \nu P$$

where ν is the number of times per second an alpha particle within a nucleus strikes the potential barrier around it and P is the probability that the particle will be transmitted through the barrier. If we suppose that at any moment only one alpha particle exists as such in a nucleus and that it moves back and forth along a nuclear diameter,

$$\nu = \frac{v}{2R}$$

FIGURE 12-6 The potential energy of an alpha particle as a function of its distance from the center of a nucleus.

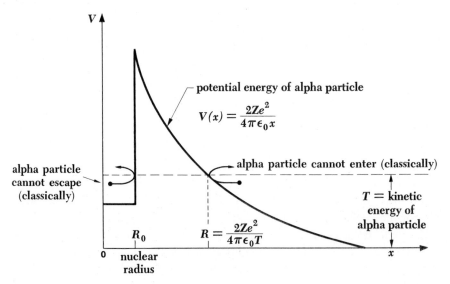

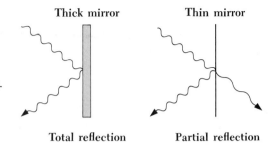

FIGURE 12-7 An incident wave penetrates the surface of even a perfect reflecting surface for a short distance and may pass through it if the surface is sufficiently thin.

Thick mirror

Thin mirror

Total reflection

Partial reflection

where v is the alpha-particle velocity when it eventually leaves the nucleus and R is the nuclear radius. Typical values for v and R might be 2×10^7 m/s and 10^{-14} m respectively, so that

$$\nu \approx 10^{21} \text{ s}^{-1}$$

The alpha particle knocks at its confining wall 10^{21} times per second and yet may have to wait an average of as much as 10^{10} yr to escape from some nuclei! Since $V > E$, classical physics predicts a transmission probability P of zero. In quantum mechanics a moving alpha particle is regarded as a wave, and the result is a small but definite value for P. The optical analog of this effect is well known: a light wave undergoing reflection from even a perfect mirror nevertheless penetrates it with an exponentially decreasing amplitude before reversing direction (Fig. 12-7).

*12.4 BARRIER PENETRATION

Let us consider the case of a beam of particles of kinetic energy T incident from the left on a potential barrier of height V and width L, as in Fig. 12-8. On both sides of the barrier $V = 0$, which means that no forces act upon the particles there. In these regions Schrödinger's equation for the particles is

12.13
$$\frac{\partial^2 \psi_{\text{I}}}{\partial x^2} + \frac{2m}{\hbar^2} E\psi_{\text{I}} = 0$$

and

12.14
$$\frac{\partial^2 \psi_{\text{III}}}{\partial x^2} + \frac{2m}{\hbar^2} E\psi_{\text{III}} = 0$$

Let us assume that

12.15
$$\psi_{\text{I}} = Ae^{iax} + Be^{-iax}$$

12.16
$$\psi_{\text{III}} = Ee^{iax} + Fe^{-ia \cdot x}$$

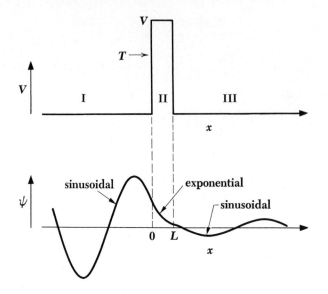

FIGURE 12-8 A beam of particles can "leak" through a finite barrier.

are solutions to Eqs. 12.13 and 12.14 respectively. The various terms in these solutions are not hard to interpret. As shown schematically in Fig. 12-8, Ae^{iax} is a wave of amplitude A incident from the left on the barrier. That is,

12.17 $\qquad \psi_{I+} = Ae^{iax}$

This wave corresponds to the incident beam of particles in the sense that $|\psi_{I+}|^2$ is their probability density. If v is the group velocity of the wave, which equals the particle velocity,

12.18 $\qquad |\psi_{I+}|^2 v$

is the flux of particles that arrive at the barrier. At $x = 0$ the incident wave strikes the barrier and is partially reflected, with

12.19 $\qquad \psi_{I-} = Be^{-iax}$

representing the reflected wave (Fig. 12-9). Hence

12.20 $\qquad \psi_I = \psi_{I+} + \psi_{I-}$

On the far side of the barrier $(x > L)$ there can be only a wave

$$\psi_{III+} = Ee^{iax}$$

traveling in the $+x$ direction, since, by hypothesis, there is nothing in region III that could reflect the wave. Hence

$$F = 0$$

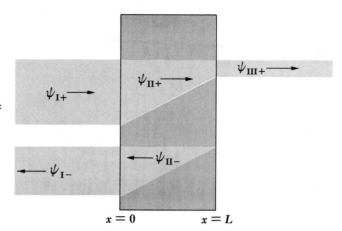

FIGURE 12-9 Schematic representation of barrier penetration.

$x = 0$ $x = L$

and

$$\psi_{III} = \psi_{III+}$$
12.21 $$= Ee^{iax}$$

By substituting ψ_I and ψ_{III} back into their respective differential equations, we find that

12.22 $$a = \sqrt{\frac{2mT}{\hbar^2}}$$

It is evident that the transmission probability P for a particle to pass through the barrier is the ratio

12.23 $$P = \frac{|\psi_{III}|^2}{|\psi_I|^2} = \frac{EE^\circ}{AA^\circ}$$

between its probability density in region III and its probability density in region I. Classically $P = 0$ because the particle cannot exist inside the barrier; let us see what the quantum-mechanical result is.

In region II Schrödinger's equation for the particles is

12.24 $$\frac{\partial^2 \psi_{II}}{\partial x^2} + \frac{2m}{\hbar^2}(T - V)\psi_{II} = 0$$

Its solution is

12.25 $$\psi_{II} = Ce^{ibx} + De^{-ibx}$$
where

12.26 $$b = \sqrt{\frac{2m(T - V)}{\hbar^2}}$$

Since $V > T$, b is imaginary and we may define a new wave number b' by

$$b' = -ib$$

12.27
$$= \sqrt{\frac{2m(V - T)}{\hbar^2}}$$

Hence

12.28
$$\psi_{II} = Ce^{-b'x} + De^{b'x}$$

The term

12.29
$$\psi_{II+} = Ce^{-b'x}$$

is an exponentially decreasing wave function that corresponds to a nonoscillatory disturbance moving to the right through the barrier. Within the barrier part of the disturbance is reflected, and

12.30
$$\psi_{II-} = De^{b'x}$$

is an exponentially decreasing wave function that corresponds to the reflected disturbance moving to the left.

Even though ψ_{II} does not oscillate, and therefore does not represent a moving particle of positive kinetic energy, the probability density $|\psi_{II}|^2$ is not zero. There is a finite probability of finding a particle within the barrier. A particle at the far end of the barrier that is not reflected there will emerge into region III with the same kinetic energy T it originally had, and its wave function will be ψ_{III} as it continues moving unimpeded in the $+x$ direction. In the limit of an infinitely thick barrier, $\psi_{III} = 0$, which implies that all the incident particles are reflected. The reflection process takes place *within* the barrier, however, not at its left-hand wall, and a barrier of finite width therefore permits a fraction P of the initial beam to pass through it.

In order to calculate P, we must apply certain boundary conditions to ψ_I, ψ_{II}, and ψ_{III}. Figure 12-8 is a schematic representation of the wave functions in regions I, II, and III which may help in visualizing the boundary conditions. As discussed earlier, both ψ and its derivative $\partial\psi/\partial x$ must be continuous everywhere. With reference to Fig. 12-8, these conditions mean that, at each wall of the barrier, the wave functions inside and outside must not only have the same value but also the same slope, so that they match up perfectly. Hence at the left-hand wall of the barrier

12.31a
$$\psi_I = \psi_{II}$$

12.31b
$$\frac{\partial\psi_I}{\partial x} = \frac{\partial\psi_{II}}{\partial x}$$
$$\Big\} \; x = 0$$

and at the right-hand wall

12.32a $\qquad \psi_{II} = \psi_{III}$

12.32b $\qquad \dfrac{\partial \psi_{II}}{\partial x} = \dfrac{\partial \psi_{III}}{\partial x}$ $\left. \right\} x = L$

Substituting ψ_I, ψ_{II}, and ψ_{III} from Eqs. 12.15, 12.28, and 12.21 into the above equations yields

12.33 $\qquad\qquad A + B = C + D$

12.34 $\qquad\qquad iaA - iaB = -b'C + b'D$

12.35 $\qquad\qquad Ce^{-b'L} + De^{b'L} = Ee^{iaL}$

12.36 $\qquad -aCe^{-b'L} + aDe^{b'L} = iaEe^{iaL}$

Equations 12.33 to 12.36 may be readily solved to yield

12.37 $\qquad \left(\dfrac{A}{E}\right) = \left[\dfrac{1}{2} + \dfrac{i}{4}\left(\dfrac{b'}{a} - \dfrac{a}{b'}\right)\right] e^{(ia+b')L} + \left[\dfrac{1}{2} - \dfrac{i}{4}\left(\dfrac{b'}{a} - \dfrac{a}{b'}\right)\right] e^{(ia-b')L}$

The complex conjugate of A/E, which we require to compute the transmission probability P, is found by replacing i by $-i$ wherever it occurs in A/E:

12.38 $\qquad \left(\dfrac{A}{E}\right)^{*} = \left[\dfrac{1}{2} - \dfrac{i}{4}\left(\dfrac{b'}{a} - \dfrac{a}{b'}\right)\right] e^{(-ia+b')L}$

$$+ \left[\dfrac{1}{2} + \dfrac{i}{4}\left(\dfrac{b'}{a} - \dfrac{a}{b'}\right)\right] e^{(-ia-b')L}$$

Let us assume that the potential barrier is high relative to the kinetic energy of an incident particle; this means that $b' > a$ and

12.39 $\qquad \left(\dfrac{b'}{a} - \dfrac{a}{b'}\right) \approx \dfrac{b'}{a}$

Let us also assume that the barrier is wide enough for ψ_{II} to be severely attenuated between $x = 0$ and $x = L$; this means that $b'L \gg 1$ and

12.40 $\qquad e^{b'L} \gg e^{-b'L}$

Hence Eqs. 12.37 and 12.38 may be approximated by

12.41 $\qquad \left(\dfrac{A}{E}\right) = \left(\dfrac{1}{2} + \dfrac{ib'}{4a}\right) e^{(ia+b')L}$

and

12.42 $\qquad \left(\dfrac{A}{E}\right)^{*} = \left(\dfrac{1}{2} - \dfrac{ib'}{4a}\right) e^{(-ia+b')L}$

Multiplying (A/E) and $(A/E)^\circ$ yields

$$\left(\frac{A}{E}\right)\left(\frac{A}{E}\right)^\circ = \left(\frac{1}{4} + \frac{b'^2}{16a^2}\right)e^{2b'L}$$

and so the transmission probability P is

$$P = \frac{EE^\circ}{AA^\circ} = \left[\left(\frac{A}{E}\right)\left(\frac{A}{E}\right)^\circ\right]^{-1}$$

12.43
$$= \left[\frac{16}{4 + (b'/a)^2}\right]e^{-2b'L}$$

Since from the definitions of a (Eq. 12.22) and b' (Eq. 12.27)

$$\left(\frac{b'}{a}\right)^2 = \frac{V}{T} - 1$$

the variation in the coefficient of the exponential of Eq. 12.43 with T and V is negligible compared with the variation in the exponential itself. The coefficient, furthermore, is never far from unity, and so

12.44 $P \approx e^{-2b'L}$

is a good approximation for the transmission probability. We shall find it convenient to write Eq. 12.44 as

12.45 $\ln P = -2b'L$

*12.5 THEORY OF ALPHA DECAY

Equation 12.45 is derived for a rectangular potential barrier, while an alpha particle inside a nucleus is faced with a barrier of varying height, as in Fig. 12-6. We must therefore replace $\ln P = -2b'L$ by

12.46 $\ln P = -2 \int_0^L b'(x)\, dx = -2 \int_{R_0}^{R} b'(x)\, dx$

where R_0 is the radius of the nucleus and R the distance from its center at which $V = T$. Beyond R the kinetic energy of the alpha particle is positive, and it is able to move freely (Fig. 12-10). Now

$$V(x) = \frac{2Ze^2}{4\pi\epsilon_0 x}$$

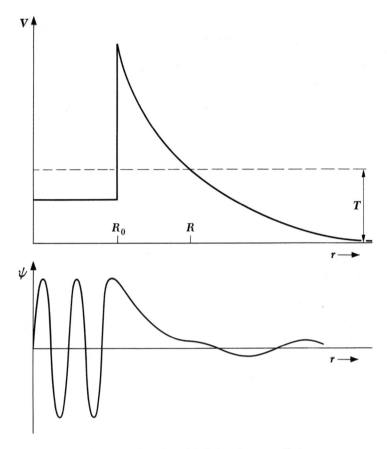

FIGURE 12-10 Alpha decay from the point of view of wave mechanics.

is the electrostatic potential energy of an alpha particle at a distance x from the center of a nucleus of charge Ze (that is, Ze is the nuclear charge *minus* the alpha-particle charge of $2e$). We therefore have

$$b' = \sqrt{\frac{2m(V - T)}{\hbar^2}}$$

$$= \left(\frac{2m}{\hbar^2}\right)^{1/2} \left(\frac{2Ze^2}{4\pi\varepsilon_0 x} - T\right)^{1/2}$$

and, since $T = V$ when $x = R$,

$$b' = \left(\frac{2mT}{\hbar^2}\right)^{1/2} \left(\frac{R}{x} - 1\right)^{1/2}$$

Hence

$$\ln P = -2 \int_{R_0}^{R} b'(x) \, dx$$

$$= -2 \left(\frac{2mT}{\hbar^2}\right)^{1/2} \int_{R_0}^{R} \left(\frac{R}{x} - 1\right)^{1/2} dx$$

12.47
$$= -2 \left(\frac{2mT}{\hbar^2}\right)^{1/2} R \left[\cos^{-1}\left(\frac{R_0}{R}\right)^{1/2} - \left(\frac{R_0}{R}\right)^{1/2} \left(1 - \frac{R_0}{R}\right)^{1/2}\right]$$

Because the potential barrier is relatively wide, $R \gg R_0$, and

$$\cos^{-1}\left(\frac{R_0}{R}\right)^{1/2} \approx \frac{\pi}{2} - \left(\frac{R_0}{R}\right)^{1/2}$$

$$\left(1 - \frac{R_0}{R}\right)^{1/2} \approx 1$$

with the result that

$$\ln P = -2 \left(\frac{2mT}{\hbar^2}\right)^{1/2} R \left[\frac{\pi}{2} - 2\left(\frac{R_0}{R}\right)^{1/2}\right]$$

Replacing R by

$$R = \frac{2Ze^2}{4\pi\varepsilon_0 T}$$

we obtain

12.48
$$\ln P = \frac{4e}{\hbar}\left(\frac{m}{\pi\varepsilon_0}\right)^{1/2} Z^{1/2} R_0^{1/2} - \frac{e^2}{\hbar\varepsilon_0}\left(\frac{m}{2}\right)^{1/2} Z T^{-1/2}$$

The result of evaluating the various constants in Eq. 12.48 is

$$\ln P = 2.97 Z^{1/2} R_0^{1/2} - 3.95 Z T^{-1/2}$$

where T (the alpha-particle kinetic energy) is expressed in MeV, R_0 (the nuclear radius) is expressed in fm($1 \text{ fm} = 10^{-15}$ m), and Z is the atomic number of the nucleus minus the alpha particle. The decay constant λ, given by

$$\lambda = \nu P$$

$$= \frac{v}{2R} P$$

may therefore be written

12.49
$$\ln \lambda = \ln\left(\frac{v}{2R_0}\right) + 2.97 Z^{1/2} R_0^{1/2} - 3.95 Z T^{-1/2} \qquad \textbf{Alpha decay}$$

To express Eq. 12.49 in terms of common logarithms, we note that

$$\ln A = \frac{\log_{10} A}{\log_{10} e} = \frac{\log_{10} A}{0.4343}$$

and so

$$\log_{10} \lambda = \log_{10}\left(\frac{v}{2R_0}\right) + 0.4343(2.97Z^{1/2} R_0^{1/2} - 3.95ZT^{-1/2})$$

$$= \log_{10}\left(\frac{v}{2R_0}\right) + 1.29Z^{1/2} R_0^{1/2} - 1.72ZT^{-1/2}$$

Figure 12-11 is a plot of $\log_{10} \lambda$ versus $ZT^{-1/2}$ for a number of alpha-radio-active nuclides. The straight line fitted to the experimental data has the -1.72 slope predicted throughout the entire range of decay constants. We can use the position of the line to determine R_0, the nuclear radius. The result is just about what is obtained from nuclear scattering experiments like that of Rutherford, namely, ~ 10 fm in such very heavy nuclei. This approach constitutes an independent means for determining nuclear sizes.

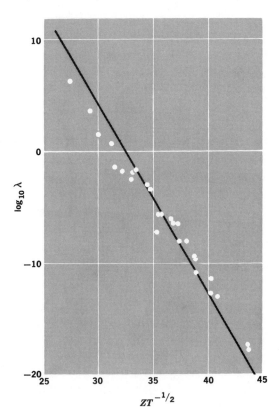

FIGURE 12-11 Experimental verification of the theory of alpha decay.

The quantum-mechanical analysis of alpha-particle emission, which is in complete accord with the observed data, is significant on two grounds. First, it makes understandable the enormous variation in half life with disintegration energy. The slowest decay is that of $^{232}_{90}$Th, whose half life is 1.3×10^{10} years, and the fastest decay is that of $^{212}_{84}$Po, whose half life is 3.0×10^{-7} sec. While its half life is 10^{24} greater, the disintegration energy of $^{232}_{90}$Th (4.05 MeV) is only about half that of $^{212}_{84}$Po (8.95 MeV)—behavior predicted by Eq. 12.49.

The second significant feature of the theory of alpha decay is its explanation of this phenomenon in terms of the penetration of a potential barrier by a particle which does not have enough energy to surmount the barrier. In classical physics such penetration cannot occur: a baseball thrown against the Great Wall of China has, classically, a zero probability of getting through. In quantum mechanics the probability is not much more than zero, but it is *not* identically equal to zero.

12.6 BETA DECAY

Beta decay, like alpha decay, is a means whereby a nucleus can alter its Z/N ratio to achieve greater stability. Beta decay, however, presents a rather different problem to the physicist who seeks to understand natural phenomena. The most obvious difficulty is that in beta decay a nucleus emits an electron, while, as we have seen in the previous chapter, there are strong arguments against the presence of electrons in nuclei. Since beta decay is essentially the spontaneous conversion of a nuclear neutron into a proton and electron, this difficulty is disposed of if we simply assume that the electron leaves the nucleus immediately after its creation. A more serious problem is that observations of beta decay reveal that three conservation principles, those of energy, momentum, and angular momentum, are apparently being violated.

The electron energies observed in the beta decay of a particular nuclide are found to vary *continuously* from 0 to a maximum value T_{max} characteristic of the nuclide. Figure 12-12 shows the energy spectrum of the electrons emitted in the beta decay of $^{210}_{83}$Bi; here $T_{max} = 1.17$ MeV. In every case the maximum energy

$$E_{max} = m_0 c^2 + T_{max}$$

carried off by the decay electron is equal to the energy equivalent of the mass difference between the parent and daughter nuclei. Only seldom, however, is an emitted electron found with an energy of T_{max}.

It was suspected at one time that the "missing" energy was lost during collisions between the emitted electron and the atomic electrons surrounding the nucleus.

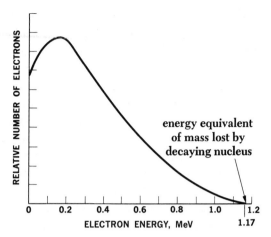

FIGURE 12-12 Energy spectrum of electrons from the beta decay of $^{212}_{83}\text{Bi}$.

energy equivalent of mass lost by decaying nucleus

RELATIVE NUMBER OF ELECTRONS

0   0.2   0.4   0.6   0.8   1.0   1.2
1.17

ELECTRON ENERGY, MeV

An experiment first performed in 1927 showed that this hypothesis is not correct. In the experiment a sample of a beta-radioactive nuclide is placed in a calorimeter, and the heat evolved after a given number of decays is measured. The evolved heat divided by the number of decays gives the average energy per decay. In the case of $^{210}_{83}\text{Bi}$ the average evolved energy was found to be 0.35 MeV, which is very close to the 0.39-MeV average of the spectrum in Fig. 12-12 but far away indeed from the T_{max} value of 1.17 MeV. The conclusion is that the observed continuous spectra represent the actual energy distributions of the electrons emitted by beta-radioactive nuclei.

Linear and angular momenta are also found not to be conserved in beta decay. In the beta decay of certain nuclides the directions of the emitted electrons and of the recoiling nuclei can be observed; they are almost never exactly opposite as required for momentum conservation. The nonconservation of angular momentum follows from the known spins of $\frac{1}{2}$ of the electron, proton, and neutron. Beta decay involves the conversion of a nuclear neutron into a proton:

$$n \rightarrow p + e^-$$

Since the spin of each of the particles involved is $\frac{1}{2}$, this reaction cannot take place if spin (and hence angular momentum) is to be conserved.

In 1930 Pauli proposed that if an uncharged particle of small or zero mass and spin $\frac{1}{2}$ is emitted in beta decay together with the electron, the energy, momentum, and angular-momentum discrepancies discussed above would be removed. It was supposed that this particle, later christened the *neutrino*, carries off an energy equal to the difference between T_{max} and the actual electron kinetic energy (the recoil nucleus carries away negligible kinetic energy) and, in so doing, has a momentum exactly balancing those of the electron and the recoiling daughter nucleus. Subsequently it was found that there are *two* kinds of neutrino involved in beta decay, the neutrino itself (symbol ν) and the antineutrino (symbol $\bar{\nu}$). We shall discuss the distinction between them in Chap. 13. In ordinary beta decay it is an antineutrino that is emitted:

12.50 $n \rightarrow p + e^- + \bar{\nu}$ **Beta decay**

The neutrino hypothesis has turned out to be completely successful. The neutrino mass was not expected to be more than a small fraction of the electron mass because T_{max} is observed to be equal (within experimental error) to the value calculated from the parent-daughter mass difference; the neutrino mass is now believed to be zero. The reason neutrinos were not experimentally detected until recently is that their interaction with matter is extremely feeble. Lacking charge and mass, and not electromagnetic in nature as is the photon, the neutrino can pass unimpeded through vast amounts of matter. A neutrino would have to pass through over 100 *light-years* of solid iron on the average before interacting! The only interaction with matter a neutrino can experience is through a process called inverse beta decay, which we shall consider shortly.

Positive electrons, usually called *positrons*, were discovered in 1932 and two years later were found to be spontaneously emitted by certain nuclei. The properties of the positron are identical with those of the electron except that it carries a charge of $+e$ instead of $-e$. Positron emission corresponds to the conversion of a nuclear proton into a neutron, a positron, and a neutrino:

12.51 $p \rightarrow n + e^+ + \nu$ **Positron emission**

While a neutron outside a nucleus can undergo negative beta decay into a proton because its mass is greater than that of the proton, the lighter proton cannot be transformed into a neutron except within a nucleus. Positron emission leads to a daughter nucleus of lower atomic number Z while leaving the mass number A unchanged.

Closely connected with positron emission is the phenomenon of *electron capture*. In electron capture a nucleus absorbs one of its inner orbital electrons, with the result that a nuclear proton becomes a neutron and a neutrino is emitted. Thus the fundamental reaction in electron capture is

12.52 $p + e^- \rightarrow n + \nu$ **Electron capture**

Electron capture is competitive with positron emission since both processes lead to the same nuclear transformation. Electron capture occurs more often than positron emission in heavy elements because the electron orbits in such elements have smaller radii; the closer proximity of the electrons promotes their interaction with the nucleus. Since almost the only unstable nuclei found in nature are of high Z, positron emission was not discovered until several decades after electron emission had been established.

12.7 INVERSE BETA DECAY

The beta decay of a proton within a nucleus follows the scheme

12.53 $$p \rightarrow n + e^+ + \nu$$

Because the absorption of an electron by a nucleus is equivalent to its emission of a positron, the electron capture reaction

12.54 $$p + e^- \rightarrow n + \nu$$

is essentially the same as the beta decay of Eq. 12.53. Similarly, the absorption of an antineutrino is equivalent to the emission of a neutrino, so that the reaction

12.55 $$p + \bar{\nu} \rightarrow n + e^+$$

also involves the same physical process as that of Eq. 12.53. This latter reaction, called *inverse beta decay*, is interesting because it provides a method for establishing the actual existence of neutrinos.

Starting in 1953, a series of experiments were begun by F. Reines, C. L. Cowan, and others to detect the immense flux of neutrinos from the beta decays that occur in a nuclear reactor. A tank of water containing a cadmium compound in solution supplied the protons which were to interact with the incident neutrinos. Surrounding the tank were gamma-ray detectors. Immediately after a proton absorbed a neutrino to yield a positron and a neutron, the positron encountered an electron and both were annihilated. The gamma-ray detectors responded to the resulting pair of 0.51-MeV photons. Meanwhile the newly formed neutron migrated through the solution until, after a few microseconds, it was captured by a cadmium nucleus. The new, heavier cadmium nucleus then released about 8 MeV of excitation energy divided among three or four photons, which were picked up by the detectors several microseconds after those from the positron-electron annihilation. In principle, then, the arrival of the above sequence of photons at the detector is a sure sign that the reaction of Eq. 12.55 has occurred. To avoid any uncertainty, the experiment was performed with the reactor alternately on and off, and the expected variation in the frequency of neutrino-capture events was observed. Thus the existence of the neutrino may be regarded as experimentally established.

Inverse beta decay is the sole known means whereby neutrinos and antineutrinos interact with matter:

$$p + \bar{\nu} \rightarrow n + e^+$$
$$n + \nu \rightarrow p + e^-$$

The probability for these reactions is almost vanishingly small; this is why neutrinos are able to traverse freely such vast amounts of matter. Once liberated, neutrinos travel freely through space and matter indefinitely, constituting a kind of independent universe within the universe of other particles.

12.8 GAMMA DECAY

Nuclei can exist in states of definite energies, just as atoms can. An excited nucleus is denoted by an asterisk after its usual symbol; thus $^{87}_{38}\text{Sr}^*$ refers to $^{87}_{38}\text{Sr}$ in an excited state. Excited nuclei return to their ground states by emitting photons whose energies correspond to the energy differences between the various initial and final states in the transitions involved. The photons emitted by nuclei range in energy up to several MeV, and are traditionally called *gamma rays*.

A simple example of the relationship between energy levels and decay schemes is shown in Fig. 12-13, which pictures the beta decay of $^{27}_{12}\text{Mg}$ to $^{27}_{13}\text{Al}$. The half life of the decay is 9.5 min, and it may take place to either of the two excited states of $^{27}_{13}\text{Al}$. The resulting $^{27}_{13}\text{Al}^*$ nucleus then undergoes one or two gamma decays to reach the ground state.

As an alternative to gamma decay, an excited nucleus in some cases may return to its ground state by giving up its excitation energy to one of the orbital electrons around it. While we can think of this process, which is known as *internal conversion*, as a kind of photoelectric effect in which a nuclear photon is absorbed by an atomic electron, it is in better accord with experiment to regard internal conversion as representing a direct transfer of excitation energy from a nucleus to an electron. The emitted electron has a kinetic energy equal to the lost nuclear excitation energy minus the binding energy of the electron in the atom.

Most excited nuclei have very short half lives against gamma decay, but a few remain excited for as long as several hours. A long-lived excited nucleus is called an *isomer* of the same nucleus in its ground state. The excited nucleus $^{87}_{38}\text{Sr}^*$ has a half life of 2.8 h and is accordingly an isomer of $^{87}_{38}\text{Sr}$.

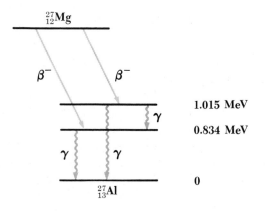

1.015 MeV

0.834 MeV

0

FIGURE 12-13 Successive beta and gamma emission in the decay of $^{27}_{12}\text{Mg}$ to $^{27}_{13}\text{Al}$.

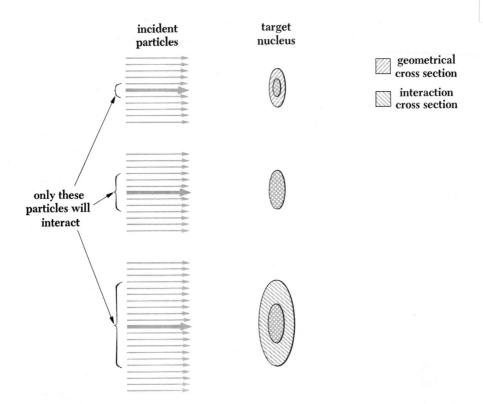

incident particles

target nucleus

◩ geometrical cross section

◳ interaction cross section

only these particles will interact

FIGURE 12-14 The concept of cross section. The interaction cross section may be smaller than, equal to, or larger than the geometrical cross section.

12.9 CROSS SECTION

Nuclear reactions, like chemical reactions, provide both information and a means of utilizing this information in a practical way. Most of what we know about atomic nuclei has come from experiments in which energetic bombarding particles collide with stationary target nuclei. A very convenient way to express the probability that a bombarding particle will interact in a certain way with a target particle employs the idea of *cross section* that was introduced in Chap. 4 in connection with the Rutherford scattering experiment. What we do is visualize each target particle as presenting a certain area, called its cross section, to the incident particles, as in Fig. 12-14. Any incident particle that is directed at this area interacts with the target particle. Hence the greater the cross section, the greater the likelihood of an interaction. The interaction cross section of a target particle varies with the nature of the process involved and with the energy of the incident particle; it may be greater or less than the geometrical cross section of the particle.

Suppose we have a slab of some material whose area is A and whose thickness is dx (Fig. 12-15). If the material contains n atoms per unit volume, there are a total of $nA\,dx$ nuclei in the slab, since its volume is $A\,dx$. Each nucleus has a cross section of σ for some particular interaction, so that the aggregate cross section of all the nuclei in the slab is $nA\sigma\,dx$. If there are N incident particles in a bombarding beam, the number dN that interact with nuclei in the slab is therefore specified by

$$\frac{\text{Interacting particles}}{\text{incident particles}} = \frac{\text{aggregate cross section}}{\text{target area}}$$

$$\frac{dN}{N} = \frac{nA\sigma\,dx}{A}$$

12.56 $$= n\sigma\,dx \qquad\qquad \textbf{Cross section}$$

Equation 12.56 is valid only for a slab of infinitesimal thickness. To find the proportion of incident particles that interact with nuclei in a slab of finite thickness, we must integrate dN/N. If we assume that each incident particle is capable of only a single interaction, dN particles may be thought of as being removed from the beam in passing through the first dx of the slab. Hence we must introduce a minus sign in Eq. 12.56, which becomes

$$-\frac{dN}{N} = n\sigma\,dx$$

Denoting the initial number of incident particles by N_0, we have

$$\int_{N_0}^{N} \frac{dN}{N} = -n\sigma \int_0^x dx$$

$$\ln N - \ln N_0 = -n\sigma x$$

12.57 $$N = N_0 e^{-n\sigma x}$$

The number of surviving particles N decreases exponentially with increasing slab thickness x.

While cross sections, which are areas, should be expressed in m², it is convenient and customary to express them in *barns* (b), where $1\text{ b} = 10^{-28}\text{ m}^2$. The barn is of the order of magnitude of the geometrical cross section of a nucleus. The cross sections for most nuclear reactions depend upon the energy of the incident particle. Figure 12-16 shows how the neutron-absorption cross section of $^{113}_{48}\text{Cd}$ varies with neutron energy; the narrow peak at 0.176 eV is associated with a specific energy level in the resulting $^{114}_{48}\text{Cd}$ nucleus.

The *mean free path* l of a particle in a material is the average distance it can travel in the material before interacting with a target nucleus. The probability f that an incident particle will undergo an interaction in a slab Δx thick

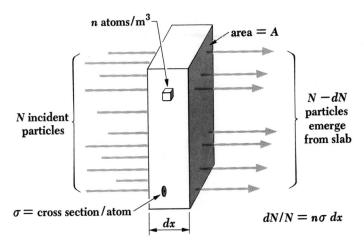

FIGURE 12-15   The relationship between cross section and beam intensity.

FIGURE 12-16   The cross section for neutron absorption $^{113}_{48}Cd$ varies with neutron energy.

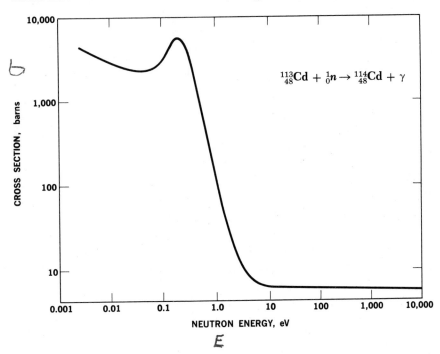

of the material is

12.58 $$f = n\sigma \, \Delta x$$

The number of times H it must traverse the slab before interacting is therefore

12.59 $$H = \frac{1}{n\sigma \, \Delta x}$$

on the average. Accordingly the average distance the particle travels before interacting is

$$H \, \Delta x = \frac{1}{n\sigma}$$

which is, by definition, the mean free path. Hence

12.60 $$l = \frac{1}{n\sigma}$$ **Mean free path**

The cross section for the interaction of a neutrino with matter has been found to be approximately 10^{-47} m^2. Let us use Eq. 12.60 to find the mean free path of neutrinos in solid iron. The atomic mass of iron is 55.9, so that the mass of an iron atom is, on the average,

$$m_{Fe} = 55.9 \text{ u/atom} \times 1.66 \times 10^{-27} \text{ kg/u}$$
$$= 9.3 \times 10^{-26} \text{ kg/atom}$$

Since the density of iron is 7.8×10^3 kg/m^3, the number of atoms per m^3 in iron is

$$n = \frac{7.8 \times 10^3 \text{ kg/m}^3}{9.3 \times 10^{-26} \text{ kg/atom}}$$
$$= 8.4 \times 10^{28} \text{ atoms/m}^3$$

The mean free path for neutrinos in iron is therefore

$$l = \frac{1}{n\sigma} = \frac{1}{8.4 \times 10^{28} \text{ atoms/m}^3 \times 10^{-47} \text{ m}^2}$$
$$= 1.2 \times 10^{18} \text{ m}$$

A light-year (the distance light travels in free space in a year) is equal to 9.46×10^{15} m, and so the mean free path turns out to be

$$l = \frac{1.2 \times 10^{18} \text{ m}}{9.46 \times 10^{15} \text{ m/light-year}} = 130 \text{ light-years}$$

in solid iron! An immense flux of neutrinos is produced in the sun and other stars in the course of the nuclear reactions that occur within them, and this flux moves practically unimpeded through the universe. There are already far more neutrinos than atoms in the universe, and their number continues to increase. The energy these neutrinos carry is—apparently—lost forever in the sense of being unavailable for conversion into other forms.

12.10 THE COMPOUND NUCLEUS

Many nuclear reactions actually involve two separate stages. In the first, an incident particle strikes a target nucleus and the two combine to form a new nucleus, called a *compound nucleus*, whose atomic and mass numbers are respectively the sum of the atomic numbers of the original particles and the sum of their mass numbers. The compound nucleus has no "memory" of how it was formed, since its nucleons are mixed together regardless of origin and the energy brought into it by the incident particle is shared among all of them. A given compound nucleus may therefore be formed in a variety of ways. To illustrate this, Table 12.2 shows six reactions whose product is the compound nucleus $^{14}_{7}N^{*}$. (The asterisk signifies an excited state; compound nuclei are invariably excited by amounts equal to at least the binding energies of the incident particles in them.) While $^{13}_{7}N$ and $^{11}_{6}C$ are beta-radioactive with such short half lives as to preclude the detailed study of their reactions to form $^{14}_{7}N^{*}$, there is no doubt that these reactions can occur.

Compound nuclei have lifetimes of the order of 10^{-16} s or so, which, while so short as to prevent actually observing such nuclei directly, are nevertheless long relative to the 10^{-21} s or so required for a nuclear particle with an energy of several MeV to pass through a nucleus. A given compound nucleus may decay

Table 12.2.

NUCLEAR REACTIONS WHOSE PRODUCT IS THE COMPOUND NUCLEUS $^{14}_{7}N^{*}$. The excitation energies given are calculated from the masses of the particles involved; the kinetic energy of an incident particle will add to the excitation energy of its reaction by an amount depending upon the dynamics of the reaction.

$$^{13}_{7}N + {}^{1}_{0}n \rightarrow {}^{14}_{7}N^{*} \text{ (10.5 MeV)}$$
$$^{13}_{6}C + {}^{1}_{1}H \rightarrow {}^{14}_{7}N^{*} \text{ (7.5 MeV)}$$
$$^{12}_{6}C + {}^{2}_{1}H \rightarrow {}^{14}_{7}N^{*} \text{ (10.3 MeV)}$$
$$^{11}_{6}C + {}^{3}_{1}H \rightarrow {}^{14}_{7}N^{*} \text{ (22.7 MeV)}$$
$$^{11}_{5}B + {}^{3}_{2}He \rightarrow {}^{14}_{7}N^{*} \text{ (20.7 MeV)}$$
$$^{10}_{5}B + {}^{4}_{2}He \rightarrow {}^{14}_{7}N^{*} \text{ (11.6 MeV)}$$

in one or more different ways, depending upon its excitation energy. Thus $^{14}_{7}\text{N}^\circ$ with an excitation energy of, say, 12 MeV can decay via the reactions

$$^{14}_{7}\text{N}^\circ \rightarrow {}^{13}_{7}\text{N} + {}^{1}_{0}n$$
$$^{14}_{7}\text{N}^\circ \rightarrow {}^{13}_{6}\text{C} + {}^{1}_{1}\text{H}$$
$$^{14}_{7}\text{N}^\circ \rightarrow {}^{12}_{6}\text{C} + {}^{2}_{1}\text{H}$$
$$^{14}_{7}\text{N}^\circ \rightarrow {}^{10}_{5}\text{B} + {}^{4}_{2}\text{He}$$

or simply emit one or more gamma rays whose energies total 12 MeV, but it *cannot* decay by the emission of a triton ($^{3}_{1}\text{H}$) or a helium-3 ($^{3}_{2}\text{He}$) particle since it does not have enough energy to liberate them. Usually a particular decay mode is favored by a compound nucleus in a specific excited state.

The formation and decay of a compound nucleus has an interesting interpretation on the basis of the liquid-drop nuclear model described in Chap. 11. In terms of this model, an excited nucleus is analogous to a drop of hot liquid, with the binding energy of the emitted particles corresponding to the heat of vaporization of the liquid molecules. Such a drop of liquid will eventually evaporate one or more molecules, thereby cooling down. The evaporation process occurs when statistical fluctuations in the energy distribution within the drop cause a particular molecule to have enough energy for escape. Similarly, a compound nucleus persists in its excited state until a particular nucleon or group of nucleons momentarily happens to have a sufficiently large fraction of the excitation energy to leave the nucleus. The time interval between the formation and decay of a compound nucleus fits in nicely with this picture.

The analysis of the reaction that occurs when a moving nucleon or nucleus strikes another one at rest is greatly simplified by the use of a coordinate system moving with the center of mass of the colliding particles. To an observer located at the center of mass, the particles have equal and opposite momenta (Fig. 12-17). Hence if a particle of mass m_1 and speed v is incident upon a stationary particle of mass m_2 as viewed by an observer in the laboratory, the speed V of the center of mass is defined by the condition

$$m_1(v - V) = m_2 V$$
$$V = \left(\frac{m_1}{m_1 + m_2}\right)v$$

In most nuclear reactions, $v \ll c$, and so a nonrelativistic treatment is satisfactory.

In the laboratory system, the total kinetic energy is that of the incident particle only:

$$T_{\text{lab}} = \frac{1}{2}m_1 v^2$$

In the center-of-mass system, both particles are moving and contribute to the total kinetic energy:

$$T_{cm} = \frac{1}{2}m_1(v - V)^2 + \frac{1}{2}m_2V^2$$

$$= \frac{1}{2}m_1v^2 - \frac{1}{2}(m_1 + m_2)V^2$$

$$= T_{lab} - \frac{1}{2}(m_1 + m_2)V^2$$

12.61
$$= \left(\frac{m_2}{m_1 + m_2}\right)T_{lab}$$

The total kinetic energy of the particles relative to the center of mass is their total kinetic energy in the laboratory system minus the kinetic energy $\frac{1}{2}(m_1 + m_2)V^2$ of the moving center of mass. Thus we can regard T_{cm} as the kinetic energy of the relative motion of the particles. When the particles collide, the maximum amount of kinetic energy that can be converted to excitation energy of the resulting compound nucleus while still conserving momentum is T_{cm}, which is always less than T_{lab}.

Information about the excited states of nuclei can be gained from nuclear reactions as well as from radioactive decay. The presence of an excited state may be detected by a peak in the cross section versus energy curve of a particular reaction, as in the neutron-capture reaction of Fig. 12-16. Such a peak is called

FIGURE 12-17 Laboratory and center-of-mass coordinate systems.

(*a*) **Motion in the laboratory coordinate system before collision.**

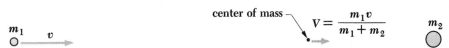

(*b*) **Motion in the center-of-mass coordinate system before collision.**

(*c*) **A completely inelastic collision as seen in laboratory and center-of-mass coordinate systems.**

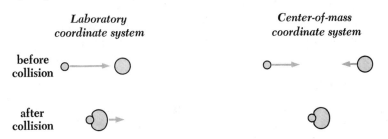

a *resonance* by analogy with ordinary acoustic or ac circuit resonances: a compound nucleus is more likely to be formed when the excitation energy provided exactly matches one of its energy levels than if the excitation energy has some other value.

The reaction of Fig. 12-16 has a resonance at 0.176 eV whose width (at half-maximum) is $\Gamma = 0.115$ eV. The uncertainty principle in the form $\Delta E \, \Delta t \geqslant \hbar$ enables us to relate the level width Γ of an excited state with the mean lifetime τ of the state. The width Γ evidently corresponds to the uncertainty ΔE in the energy of the state, and the mean lifetime τ corresponds to the uncertainty Δt in the time when the state will decay, in the present example by the emission of a gamma ray. The mean lifetime of an excited state is defined in general as

12.62 $$\tau = \frac{\hbar}{\Gamma}$$ **Mean lifetime of excited state**

In the case of the above reaction, the level width of 0.115 eV implies a mean lifetime for the compound nucleus of

$$\tau = \frac{1.054 \times 10^{-34} \text{ J-s}}{0.115 \text{ eV} \times 1.60 \times 10^{-19} \text{ J/eV}}$$
$$= 5.73 \times 10^{-15} \text{ s}$$

12.11 NUCLEAR FISSION

Another type of nuclear-reaction phenomenon that can be analyzed with the help of the liquid-drop model is *fission*, in which a heavy nucleus ($A > \sim 230$) splits into two lighter ones. When a liquid drop is suitably excited, it may oscillate in a variety of ways. A simple one is shown in Fig. 12-18: the drop successively becomes a prolate spheroid, a sphere, an oblate spheroid, a sphere, a prolate spheroid again, and so on. The restoring force of its surface tension always returns the drop to spherical shape, but the inertia of the moving liquid molecules causes the drop to overshoot sphericity and go to the opposite extreme of distortion.

While nuclei may be regarded as exhibiting surface tension, and so can vibrate like a liquid drop when in an excited state, they also are subject to disruptive forces due to the mutual electrostatic repulsion of their protons. When a nucleus is distorted from a spherical shape, the short-range restoring force of surface tension must cope with the latter long-range repulsive force as well as with the inertia of the nuclear matter. If the degree of distortion is small, the surface

time ———————→

FIGURE 12-18   The oscillations of a liquid drop.

tension is adequate to do both, and the nucleus vibrates back and forth until it eventually loses its excitation energy by gamma decay. If the degree of distortion is sufficiently great, however, the surface tension is not adequate to bring back together the now widely separated groups of protons, and the nucleus splits into two parts. This picture of fission is illustrated in Fig. 12-19.

The new nuclei that result from fission are called *fission fragments*. Usually fission fragments are of unequal size (Fig. 12-20), and, because heavy nuclei have a greater neutron/proton ratio than lighter ones, they contain an excess of neutrons. To reduce this excess, two or three neutrons are emitted by the fragments as soon as they are formed, and subsequent beta decays bring their neutron/proton ratios to stable values.

A heavy nucleus undergoes fission when it acquires enough excitation energy (5 MeV or so) to oscillate violently. Certain nuclei, notably $^{235}_{92}U$, are sufficiently excited by the mere absorption of an additional neutron to split in two. Other nuclei, notably $^{238}_{92}U$ (which composes 99.3 percent of natural uranium, with $^{235}_{92}U$ composing the remainder), require more excitation energy for fission than the binding energy released when another neutron is absorbed, and undergo fission only by reaction with fast neutrons whose kinetic energies exceed about 1 MeV. Fission can occur after excitation by other means besides neutron capture, for instance, by gamma-ray or proton bombardment. Some nuclides are so unstable as to be capable of spontaneous fission, but they are more likely to undergo alpha decay before this takes place.

The most striking aspect of nuclear fission is the magnitude of the energy evolved. This energy is readily computed. The heavy fissionable nuclides, whose mass numbers are about 240, have binding energies of ~7.6 MeV/nucleon, while fission fragments, whose mass numbers are about 120, have binding energies of ~8.5 MeV/nucleon. Hence 0.9 MeV/nucleon is released during fission—over

FIGURE 12-19   Nuclear fission according to the liquid-drop model.

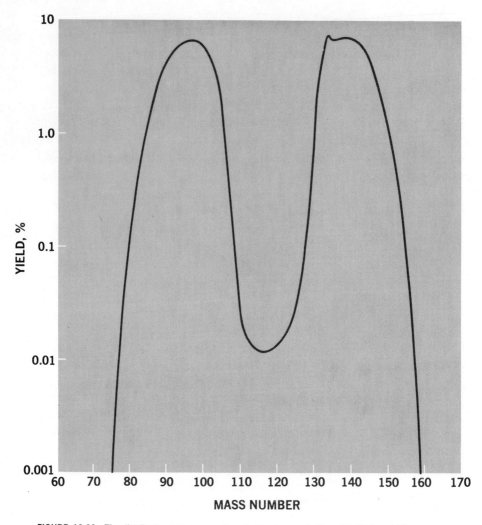

FIGURE 12-20 The distribution of mass numbers in the fragments from the fission of $^{235}_{92}$U.

200 MeV for the 240 or so nucleons involved! Ordinary chemical reactions, such as those that participate in the combustion of coal and oil, liberate only a few electron volts per individual reaction, and even nuclear reactions (other than fission) liberate no more than several million electron volts. Most of the energy that is released during fission goes into the kinetic energy of the fission fragments: the emitted neutrons, beta and gamma rays, and neutrinos carry off perhaps 20 percent of the total energy.

Almost immediately after the discovery of nuclear fission in 1939 it was recognized that, because a neutron can induce fission in a suitable nucleus with the consequent evolution of additional neutrons, a self-sustaining sequence of fissions is, in principle, possible. The condition for such a chain reaction to occur in an assembly of fissionable material is simple: at least one neutron produced during each fission must, on the average, initiate another fission. If too few neutrons initiate fissions, the reaction will slow down and stop; if precisely one neutron per fission causes another fission, energy will be released at a constant rate (which is the case in a nuclear reactor); and if the frequency of fissions increases, the energy release will be so rapid that an explosion will occur (which is the case in an atomic bomb). These situations are respectively called *subcritical*, *critical*, and *supercritical*.

12.12 TRANSURANIC ELEMENTS

Elements of atomic number greater than 92, which is that of uranium, have such short half lives that, had they been formed when the universe came into being, they would have disappeared long ago. Such *transuranic elements* may be produced in the laboratory by the bombardment of certain heavy nuclides with neutrons. Thus $^{238}_{92}U$ may absorb a neutron to become $^{239}_{92}U$, which beta-decays ($T_{1/2} = 23$ min) into $^{239}_{93}Np$, an isotope of the transuranic element *neptunium*:

$$^{238}_{92}U + {}^{1}_{0}n \rightarrow {}^{239}_{92}U$$
$$^{239}_{92}U \rightarrow {}^{239}_{93}Np + e^-$$

This neptunium isotope is itself radioactive, undergoing beta decay with a half life of 2.3 d into an isotope of the transuranic element *plutonium*:

$$^{239}_{93}Np \rightarrow {}^{239}_{94}Pu + e^-$$

Plutonium alpha-decays into $^{235}_{92}U$ with a half life of 24,000 yr:

$$^{239}_{94}Pu \rightarrow {}^{235}_{92}U + {}^{4}_{2}He$$

It is interesting to note that $^{239}_{94}Pu$, like $^{235}_{92}U$, is fissionable and can be used in nuclear reactors and weapons. Plutonium is chemically different from uranium; its separation from the remaining $^{238}_{92}U$ after neutron irradiation is more easily accomplished than the separation of $^{235}_{92}U$ from the much more abundant $^{238}_{92}U$ in natural uranium.

Transuranic elements past einsteinum ($Z = 99$) have half-lives too short for their isolation in weighable quantities, though they can be identified by chemical means. The transuranic element of the highest atomic number yet discovered has $Z = 105$.

12.13 THERMONUCLEAR ENERGY

The basic exothermic reaction in stars—and hence the source of nearly all the energy in the universe—is the fusion of hydrogen nuclei into helium nuclei. This can take place under stellar conditions in two different series of processes. In one of them, the *proton-proton cycle*, direct collisions of protons result in the formation of heavier nuclei whose collisions in turn yield helium nuclei. The other, the *carbon cycle*, is a series of steps in which carbon nuclei absorb a succession of protons until they ultimately disgorge alpha particles to become carbon nuclei once more.

The initial reaction in the proton-proton cycle is

$$\,^1_1\text{H} + \,^1_1\text{H} \rightarrow \,^2_1\text{H} + e^+ + \nu$$

the formation of deuterons by the direct combination of two protons accompanied by the emission of a positron. A deuteron may then join with a proton to form a $\,^3_2\text{He}$ nucleus:

$$\,^1_1\text{H} + \,^2_1\text{H} \rightarrow \,^3_2\text{He} + \gamma$$

Finally two $\,^3_2\text{He}$ nuclei react to produce a $\,^4_2\text{He}$ nucleus plus two protons:

$$\,^3_2\text{He} + \,^3_2\text{He} \rightarrow \,^4_2\text{He} + \,^1_1\text{H} + \,^1_1\text{H}$$

The total evolved energy is $(\Delta m)c^2$, where Δm is the difference between the mass of four protons and the mass of an alpha particle plus two positrons; it turns out to be 24.7 MeV. Figure 12-21 shows the entire sequence.

The carbon cycle proceeds in the following way:

$$
\begin{aligned}
\,^1_1\text{H} + \,^{12}_6\text{C} &\rightarrow \,^{13}_7\text{N} \\
\,^{13}_7\text{N} &\rightarrow \,^{13}_6\text{C} + e^+ + \nu \\
\,^1_1\text{H} + \,^{13}_6\text{C} &\rightarrow \,^{14}_7\text{N} + \gamma \\
\,^1_1\text{H} + \,^{14}_7\text{N} &\rightarrow \,^{15}_8\text{O} + \gamma \\
\,^{15}_8\text{O} &\rightarrow \,^{15}_7\text{N} + e^+ + \nu \\
\,^1_1\text{H} + \,^{15}_7\text{N} &\rightarrow \,^{12}_6\text{C} + \,^4_2\text{He}
\end{aligned}
$$

Carbon cycle

The net result again is the formation of an alpha particle and two positrons from four protons, with the evolution of 24.7 MeV; the initial $\,^{12}_6\text{C}$ acts as a kind of catalyst for the process, since it reappears at its end (Fig. 12-22).

Self-sustaining fusion reactions can occur only under conditions of extreme temperature and pressure, to ensure that the participating nuclei have enough energy to react despite their mutual electrostatic repulsion and that reactions occur frequently enough to counterbalance losses of energy to the surroundings. Stellar interiors meet these specifications. In the sun, whose interior temperature

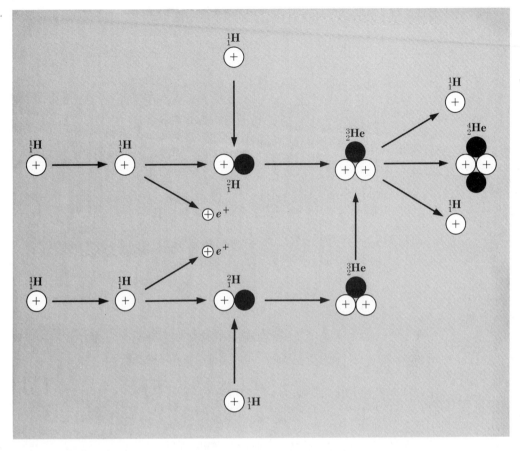

FIGURE 12-21 The proton-proton cycle. This is one of the two nuclear reaction sequences that take place in the sun and that involve the combination of four hydrogen nuclei to form a helium nucleus with the evolution of energy.

is estimated to be 2×10^6 K, the proton-proton cycle has the greater probability for occurrence. In general, the carbon cycle is more efficient at high temperatures, while the proton-proton cycle is more efficient at low temperatures. Hence stars hotter than the sun obtain their energy largely from the former cycle, while those cooler than the sun obtain the greater part of their energy from the latter cycle. The neutrinos carry away about 10 percent of the energy produced by a typical star.

The energy liberated in the fusion of light nuclei into heavier ones is often called *thermonuclear energy*, particularly when the fusion takes place under man's control. On the earth neither the proton-proton nor carbon cycle offers any hope of practical application, since their several steps require a great deal of time. Two fusion reactions that seem promising as terrestrial energy sources are the

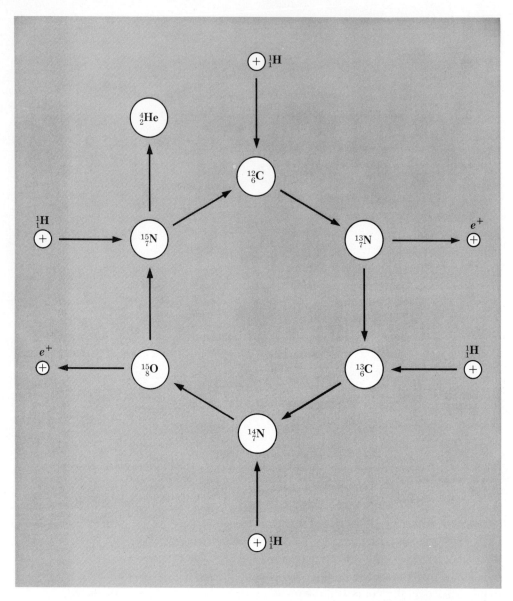

FIGURE 12-22 The carbon cycle also involves the combination of four hydrogen nuclei to form a helium nucleus with the evolution of energy. The $^{12}_{6}C$ nucleus is unchanged by the series of reactions.

direct combination of two deuterons in either of the following ways:

$$^2_1H + ^2_1H \rightarrow ^3_2He + ^1_0n + 3.3 \text{ MeV}$$
$$^2_1H + ^2_1H \rightarrow ^3_1H + ^1_1H + 4.0 \text{ MeV}$$

Another is the direct combination of a deuteron and a triton to form an alpha particle,

$$\mathstrut^3_1H + \mathstrut^2_1H \rightarrow \mathstrut^4_2He + \mathstrut^1_0n + 17.6 \text{ MeV}$$

Capitalizing upon the above reactions requires an abundant, cheap source of deuterium. Such a source is the oceans and seas of the world, which contain about 0.015 percent deuterium—a total of perhaps 10^{15} tons! In addition, a more efficient means of promoting fusion reactions than merely bombarding a target with fast particles from an accelerator is required, since the operation of an accelerator consumes far more power than can be evolved by the relatively few reactions that occur in the target. Current approaches to this problem all involve very hot plasmas (fully ionized gases) of deuterium or deuterium-tritium mixtures which are contained by strong magnetic fields. The purpose of high temperature is to ensure that the individual $\mathstrut^2_1H$ and $\mathstrut^3_1H$ nuclei have enough energy to come together and react despite their electrostatic repulsion. A magnetic field is used as a container to keep the reactive gas from contacting any other material which might cool it down or contaminate it; there is little likelihood that the wall will melt since the gas, though at a temperature of several million degrees K, actually does not have a high energy density. While nuclear-fusion reactors present more severe practical difficulties than fission reactors, there is little doubt that they will eventually become a reality.

Problems

(The masses in u of neutral atoms of nuclides mentioned below are: $\mathstrut^1_1H$, 1.007825; $\mathstrut^3_1H$, 3.016050; $\mathstrut^3_2He$, 3.016030; $\mathstrut^4_2He$, 4.002603; $\mathstrut^7_3Li$, 7.0160; $\mathstrut^7_4Be$, 7.0169; $\mathstrut^{10}_5B$, 10.0129; $\mathstrut^{12}_5B$, 12.0144; $\mathstrut^{12}_6C$, 12.0000; $\mathstrut^{13}_6C$, 13.0034; $\mathstrut^{14}_7N$, 14.0031; $\mathstrut^{16}_8O$, 15.9949; $\mathstrut^{17}_8O$, 16.9994. The neutron mass is 1.008665 u. Atomic masses of the elements are listed in Table 7.1.)

1. Tritium ($\mathstrut^3_1H$) has a half life of 12.5 yr against beta decay. What fraction of a sample of pure tritium will remain undecayed after 25 yr?

2. The half life of $\mathstrut^{24}_{11}Na$ is 15 h. How long does it take for 93.75 percent of a sample of this isotope to decay?

3. One g of radium has an activity of 1 Ci. From this fact determine the half life of radium.

4. The mass of a millicurie of $\mathstrut^{214}_{82}Pb$ is 3×10^{-14} kg. From this fact find the decay constant of $\mathstrut^{214}_{82}Pb$. (Assume atomic mass in u equal to mass number in Probs. 4 to 6.)

5. The half life of $^{238}_{92}U$ against alpha decay is 4.5×10^9 yr. How many disintegrations per second occur in 1 g of $^{238}_{92}U$?

6. The potassium isotope $^{40}_{19}K$ undergoes beta decay with a half life of 1.83×10^9 yr. Find the number of beta decays that occur per second in 1 g of pure $^{40}_{19}K$.

7. A 5.78-MeV alpha particle is emitted in the decay of radium. If the diameter of the radium nucleus is 2×10^{-14} m, how many alpha-particle de Broglie wavelengths fit inside the nucleus?

8. Calculate the maximum energy of the electrons emitted in the beta decay of $^{12}_{5}B$.

9. Why does ^7_4Be invariably decay by electron capture instead of by positron emission? Note that ^7_4Be contains one more atomic electron than does ^7_3Li.

10. Positron emission resembles electron emission in all respects except that the shapes of their respective energy spectra are different: there are many low-energy electrons emitted, but few low-energy positrons. Thus the average electron energy in beta decay is about $0.3T_{max}$, whereas the average positron energy is about $0.4T_{max}$. Can you suggest a simple reason for this difference?

11. Determine the ground and lowest excited states of the 39th proton in $^{89}_{39}Y$ with the help of Fig. 11-10. Use this information to explain the isomerism of $^{89}_{39}Y$ together with the fact, noted in Sec. 6.10, that radiative transitions between states with very different angular momenta are extremely improbable.

12. Find the minimum energy in the laboratory system that a neutron must have in order to initiate the reaction

$$^1_0n + ^{16}_8O + 2.20 \text{ MeV} \rightarrow ^{13}_6C + ^4_2He$$

13. Find the minimum energy in the laboratory system that a proton must have in order to initiate the reaction

$$p + d + 2.22 \text{ MeV} \rightarrow p + p + n$$

14. Find the minimum energy in the laboratory system that an alpha particle must have in order to initiate the reaction

$$^4_2He + ^{14}_7N + 1.18 \text{ MeV} \rightarrow ^{17}_8O + ^1_1H$$

15. The cross sections for comparable neutron- and proton-induced nuclear reactions vary with energy in approximately the manner shown in Fig. 12-23. Why does the neutron cross section decrease with increasing energy whereas the proton cross section increases?

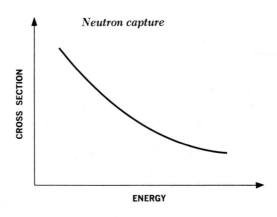

Neutron capture

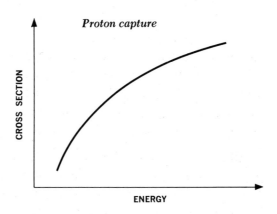

Proton capture

FIGURE 12-23 Neutron and proton capture cross sections vary differently with particle energy.

16. When a neutron is absorbed by a target nucleus, the resulting compound nucleus is usually more likely to emit a gamma ray than a proton, deuteron, or alpha particle. Why?

17. There are approximately 6×10^{28} atoms/m³ in solid aluminum. A beam of 0.5-MeV neutrons is directed at an aluminum foil 0.1 mm thick. If the capture cross section for neutrons of this energy in aluminum is 2×10^{-31} m², find the fraction of incident neutrons that are captured.

18. The density of $^{10}_{5}B$ is 2.5×10^3 kg/m³. The capture cross section of $^{10}_{5}B$ is about 4,000 b for "thermal" neutrons, that is, neutrons in thermal equilibrium with matter at room temperature. How thick a layer of $^{10}_{5}B$ is required to absorb 99 percent of an incident beam of thermal neutrons?

19. The density of iron is about 8×10^3 kg/m³. The neutron-capture cross section of iron is about 2.5 b. What fraction of an incident beam of neutrons is absorbed by a sheet of iron 1 cm thick?

20. The cross section of iron for neutron capture is 2.5 b. What is the mean free path of neutrons in iron?

21. The fission of $^{235}_{92}U$ releases approximately 200 MeV. What percentage of the original mass of $^{235}_{92}U + n$ disappears?

22. Certain stars obtain part of their energy by the fusion of three alpha particles to form a $^{12}_{6}C$ nucleus. How much energy does each such reaction evolve?

ELEMENTARY PARTICLES 13

While nuclei are apparently composed solely of protons and neutrons, several score other elementary particles have been observed to be emitted by nuclei under appropriate circumstances. These particles, christened "strange particles" soon after their discovery about two decades ago, bring the total number merely of relatively stable elementary particles to over 30. To discern order in this multiplicity of particles has not proved to be an easy task. While certain regularities in elementary-particle properties have been established, and while such particles as the electron, the neutrino, and the π meson are relatively well understood, no comprehensive theory of elementary particles has yet found wide acceptance. It is fitting to conclude our survey of modern physics with this topic, then, as a reminder that there remains much to be learned about the natural world.

13.1 ANTIPARTICLES

The electron is the only elementary particle for which a satisfactory theory is known. This theory was developed in 1928 by P. A. M. Dirac, who obtained a wave equation for a charged particle in an electromagnetic field that incorporated the results of special relativity. When the observed mass and charge of the electron are inserted in the appropriate solutions of this equation, the intrinsic angular momentum of the electron is found to be $\frac{1}{2}\hbar$ (that is, spin $\frac{1}{2}$) and its magnetic moment is found to be $e\hbar/2m$, one Bohr magneton. These predictions agree with experiment, and the agreement is strong evidence for the correctness of the Dirac theory.

An unexpected result of the Dirac theory was its prediction that positive as well as negative electrons should exist. At first it was thought that the proton was the positive counterpart of the electron despite the difference in their masses, but in 1932 a positive electron was unambiguously detected in the flux of cosmic radiation at the earth's surface. Positive electrons, as mentioned earlier, are usually called *positrons*. The materialization of an electron-positron pair from a photon of sufficient energy (>1.02 MeV) and the annihilation of an electron and a positron that come together were described in Sec. 2.6.

The positron is often spoken of as the *antiparticle* of the electron, since it is able to undergo mutual annihilation with an electron. All other known elementary particles except for the photon and the π^0 and η^0 mesons also have antiparticle counterparts: the latter constitute their own antiparticles. The antiparticle of a particle has the same mass, spin, and lifetime if unstable, but its charge (if any) has the opposite sign and the alignment or antialignment between its spin and magnetic moment is also opposite to that of the particle.

The distinction between the neutrino and the antineutrino is a particularly interesting one. The spin of the neutrino is opposite in direction to the direction of its motion; viewed from behind, as in Fig. 13-1, the neutrino spins counterclockwise. The spin of the antineutrino, on the other hand, is in the same direction as its direction of motion; viewed from behind, it spins clockwise. Thus the neutrino moves through space in the manner of a left-handed screw, while the antineutrino does so in the manner of a right-handed screw.

Prior to 1956 it had been universally assumed that neutrinos could be either left-handed or right-handed, implying that, since no difference was possible between them except one of spin direction, the neutrino and antineutrino are identical. This assumption had roots going all the way back to Leibniz, Newton's contemporary and an independent inventor of calculus. The argument may be stated as follows: if we observe an object or a physical process of some kind

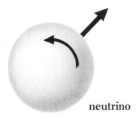

neutrino

FIGURE 13-1 Neutrinos and antineutrinos have opposite directions of spin.

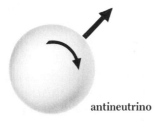

antineutrino

both directly and in a mirror, we cannot ideally distinguish which object or process is being viewed directly and which by reflection. By definition, distinctions in physical reality must be capable of discernment or they are meaningless. Now the only difference between something seen directly and the same thing seen in a mirror is the interchange of right and left, and so *all* objects and processes must occur with equal probability with right and left interchanged. This plausible doctrine is indeed experimentally valid for nuclear and electromagnetic interactions, but until 1956 its applicability to neutrinos had never been actually tested. In that year T. D. Lee and C. N. Yang suggested that several serious theoretical discrepancies would be removed if neutrinos and antineutrinos have different handedness, even though it meant that neither particle could therefore be reflected in a mirror. Experiments performed soon after their proposal showed unequivocally that neutrinos and antineutrinos are distinguishable, having left-handed and right-handed spins respectively. We might note that the absence of right-left symmetry in neutrinos can occur only if the neutrino mass is exactly zero, thereby resolving what had been the very difficult experimental problem of measuring the neutrino mass.

13.2 MESON THEORY OF NUCLEAR FORCES

If nuclear forces were exclusively attractive, a nucleus would be stable only if its size were so small (about 2 fm in radius) that each nucleon interacted with all the others. The binding energy per nucleon would then be proportional to A, the number of nucleons present. In fact, nuclear volumes are found to be proportional to A and the binding energy per nucleon is roughly the same for all nuclei: each nucleon interacts only with a small number of its nearest neighbors. Thus there must be a repulsive component in nuclear forces that keeps nuclei from collapsing, as indicated in Fig. 11-6, which means that these forces are not analogous to the "ordinary" gravitational and electrical forces.

We encountered a somewhat similar situation in Sec. 8.3, where the forces present in the H_2^+ molecular ion can be thought of as including an *exchange force* which arises because of the possibility that the electron can shift from one of the protons to the other. Depending on whether the wave function of the system is symmetric or antisymmetric for the particle exchange, the exchange force is either attractive or repulsive. It is tempting to consider the interaction between nucleons to be, at least in part, a consequence of some kind of exchange force as well. For instance, exchange forces provide an explanation for the stability of the triplet state of the deuteron, which is described by a symmetric wave function since the spins are parallel, and the instability of the singlet state, which is described by an antisymmetric wave function. Since the nucleons in a nucleus are all in different quantum states (by the exclusion principle), both attractive and repulsive exchange forces would occur, and a mixture of an "ordinary" attractive nuclear force and such exchange forces is able to account in a general way for a great many nuclear properties.

The next question is, what kinds of particles are exchanged between nearby nucleons? In 1932, Heisenberg suggested that electrons and positrons shift back and forth between nucleons; for instance, a neutron might emit an electron and become a proton, while a proton absorbing the electron would then become a neutron. However, calculations based on beta-decay data showed that the forces resulting from electron and positron exchange by nucleons are too small by the huge factor of 10^{14} to be significant in nuclear structure. Then, in 1935, the Japanese physicist Hideki Yukawa proposed that particles called *mesons*, heavier than electrons, are involved in nuclear forces, and he was able to show that the interactions they produce between nucleons are of the correct order of magnitude.

According to the meson theory of nuclear forces, all nucleons consist of identical cores surrounded by a "cloud" of one or more mesons. Mesons may be neutral or carry either charge, and the sole difference between neutrons and protons is supposed to lie in the composition of their respective meson clouds. The forces that act between one neutron and another and between one proton and another are the result of the exchange of neutral mesons (designated π^0) between them. The force between a neutron and a proton is the result of the exchange of charged mesons (π^+ and π^-) between them. Thus a neutron emits a π^- meson and is converted into a proton:

$$n \rightarrow p + \pi^-$$

while the absorption of the π^- by the proton the neutron was interacting with converts it into a neutron:

$$p + \pi^- \rightarrow n$$

In the reverse process, a proton emits a π^+ meson whose absorption by a neutron converts it into a proton:

$$p \rightarrow n + \pi^+$$
$$n + \pi^- \rightarrow p$$

While there is no simple mathematical way of demonstrating how the exchange of particles between two bodies can lead to attractive and repulsive forces, a rough analogy may make the process intuitively meaningful. Let us imagine two boys exchanging basketballs (Fig. 13-2). If they throw the balls at each other, they each move backward, and when they catch the balls thrown at them, their backward momentum increases. Thus this method of exchanging the basketballs

repulsive force due to particle exchange

FIGURE 13-2 Attractive and repulsive forces can both arise from particle exchange.

attractive force due to particle exchange

yields the same effect as a repulsive force between the boys. If the boys snatch the basketballs from each other's hands, however, the result will be equivalent to an attractive force acting between them.

A fundamental problem presents itself at this point. If nucleons constantly emit and absorb mesons, why are neutrons or protons never found with other than their usual masses? The answer is based upon the uncertainty principle. The laws of physics refer exclusively to experimentally measurable quantities, and the uncertainty principle limits the accuracy with which certain combinations of measurements can be made. The emission of a meson by a nucleon which does not change in mass—a clear violation of the law of conservation of energy—can occur provided that the nucleon absorbs a meson emitted by the neighboring nucleon it is interacting with so soon afterward that *even in principle* it is impossible to determine whether or not any mass change actually has been involved. Since the uncertainty principle may be written

13.1 $$\Delta E \, \Delta t \geqslant \hbar$$

an event in which an amount of energy ΔE is not conserved is not prohibited so long as the duration of the event does not exceed approximately $\hbar/\Delta E$.

We know that nuclear forces have a maximum range R of about 1.7 fm, so that if we assume a meson travels between nuclei at approximately the speed of light c, the time interval Δt during which it is in flight is

13.2 $$\Delta t = \frac{R}{c}$$

The emission of a meson of mass m_π represents the nonconservation of

13.3 $$\Delta E = m_\pi c^2$$

of energy. According to Eq. 13.1 this can occur if $\Delta E \, \Delta t \geqslant \hbar$; that is, if

$$(m_\pi c^2)\left(\frac{R}{c}\right) \geqslant \hbar$$

Hence the minimum meson mass is specified by

13.4 $$m_\pi \geqslant \frac{\hbar}{Rc}$$

$$\geqslant 1.9 \times 10^{-28} \text{ kg}$$

which is about $200 \, m_e$, that is, 200 electron masses.

13.3 PIONS AND MUONS

Twelve years after the meson theory was formulated, particles with the predicted properties were actually found outside nuclei. Today π mesons are usually called *pions*.

Two factors contributed to the belated discovery of the free pion. First, enough energy must be supplied to a nucleon so that its emission of a pion conserves energy. Thus at least $m_\pi c^2$ of energy, about 140 MeV, is required. To furnish a stationary nucleon with this much energy in a collision, the incident particle must have considerably more kinetic energy than $m_\pi c^2$ in order that momentum as well as energy be conserved. Particles with kinetic energies of several hundred MeV are therefore required to produce free pions, and such particles are found in nature only in the diffuse stream of cosmic radiation that bombards the earth. Hence the discovery of the pion had to await the development of sufficiently sensitive and precise methods of investigating cosmic-ray interactions. More recently high-energy accelerators were placed in operation; they yielded the necessary particle energies, and the profusion of pions that were created with their help could be studied readily.

The second reason for the lag between the prediction and experimental discovery of the pion is its instability: the half life of the charged pion is only

1.8 × 10⁻⁸ s, and that of the neutral pion is 7×10^{-17} s. The lifetime of the π^0 is so short, in fact, that its existence was not established until 1950.

Charged pions almost invariably decay into lighter mesons called μ *mesons* (or *muons*) and neutrinos:

$$\pi^+ \rightarrow \mu^+ + \nu_\mu$$
$$\pi^- \rightarrow \mu^- + \overline{\nu}_\mu$$

These neutrinos are not the same as those involved in beta decay, which is why their symbols are ν_μ and $\overline{\nu}_\mu$. The existence of two classes of neutrino was established in 1962. A metal target was bombarded with high-energy protons, and pions were created in profusion. Inverse reactions traceable to the neutrinos from the decay of these pions produced muons only, and no electrons. Hence these neutrinos must be somehow different from those associated with beta decay. The neutral pion decays into a pair of gamma rays:

$$\pi^0 \rightarrow \gamma + \gamma$$

The π^+ and π^- have rest masses of $273\ m_e$, while that of the π^0 is slightly less, $264\ m_e$. The π^- is the antiparticle of the π^+, and the π^0 is its own antiparticle, a distinction it shares only with the photon and the η^0 meson.

Whereas the existence of pions is so readily understandable that they were predicted many years before their actual discovery, muons even today represent something of a puzzle. Their physical properties are known quite accurately. Positive and negative muons have the same rest mass, $207\ m_e$, and the same spin, ½. Both decay with a half life of 1.5×10^{-6} s into electrons and neutrino-antineutrino pairs:

$$\mu^+ \rightarrow e^+ + \nu_e + \overline{\nu}_\mu$$
$$\mu^- \rightarrow e^- + \nu_\mu + \overline{\nu}_e$$

As with electrons, the positive-charge state of the muon represents the antiparticle. There is no neutral muon.

Unlike the case of pions which, as we would expect, interact strongly with nuclei, the only interaction between muons and matter is an electric one. Accordingly muons readily penetrate considerable amounts of matter before being absorbed. The majority of cosmic-ray particles at sea level are muons from the decay of pions created in nuclear collisions caused by fast primary cosmic-ray atomic nuclei, since nearly all the other particles in the cosmic-ray stream either decay or lose energy rapidly and are absorbed far above the earth's surface.

The mysterious aspect of the muon is its function—or, rather, its apparent lack of any function. Only in its mass and instability does the muon differ significantly from the electron, leading to the hypothesis that the muon is merely a kind of "heavy electron" rather than a unique entity. Other evidence, which we shall examine later in this chapter, is less unflattering to the muon, although it is still not wholly clear why, for instance, pions should preferentially decay into muons rather than directly into electrons; only about 0.01 percent of pions decay directly into electrons and neutrinos.

13.4 KAONS AND HYPERONS

Pions and muons do not exhaust the list of known particles with masses intermediate between those of the electron and proton. A third class of mesons, called *K mesons* (or *kaons*), has been discovered whose members may decay in a variety of ways. Charged kaons have rest masses of $966m_e$, spins of 0, and half lives of 8×10^{-9} s. The following decay schemes are possible for K^+ mesons, in order of relative probability:

$$\begin{aligned} K^+ &\to \mu^+ + \nu_\mu \\ &\to \pi^+ + \pi^0 \\ &\to \pi^+ + \pi^+ + \pi^- \\ &\to \pi^0 + e^+ + \nu_e \\ &\to \pi^0 + \mu^+ + \nu_\mu \\ &\to \pi^+ + \pi^0 + \pi^0 \end{aligned}$$

There are apparently two distinct varieties of neutral K mesons, the $K_1{}^0$ and $K_2{}^0$. Both have rest masses of $974 \, m_e$ and spins of 0, but the former has a half life of about 7×10^{-11} s, while that of the latter is about 4×10^{-8} s. The following decay modes are known for neutral K mesons, again in order of relative probability:

$$\begin{aligned} K_1{}^0 &\to \pi^+ + \pi^- \\ &\to \pi^0 + \pi^0 \\ K_2{}^0 &\to \pi^- + e^+ + \nu_e \\ &\to \pi^+ + e^- + \bar{\nu}_e \\ &\to \pi^- + \mu^+ + \nu_\mu \\ &\to \pi^+ + \mu^- + \bar{\nu}_\mu \\ &\to \pi^0 + \pi^0 + \pi^0 \\ &\to \pi^+ + \pi^- + \pi^0 \end{aligned}$$

In addition to their electromagnetic interaction with matter through which they pass, K mesons exhibit varying degrees of specifically nuclear interactions. The K^+ and K^0 mesons interact only weakly with nuclei, while their antiparticle counterparts are readily scattered and absorbed by nuclei in their paths.

Elementary particles heavier than protons are called *hyperons*. The known hyperons fall into four classes, Λ, Σ, Ξ, and Ω hyperons, in order of increasing mass. (Λ, Σ, Ξ, and Ω are, respectively, the Greek capital letters *lambda, sigma, xi,* and *omega*.) All are unstable with extremely brief mean lifetimes. The spin of all hyperons is $\frac{1}{2}$ except that of the Ω hyperon, which is $\frac{3}{2}$. The masses, half lives, and decay schemes of various hyperons are given in Table 13.1.

Like pions and kaons (but unlike muons), hyperons exhibit definite interactions with nuclei. The Λ^0 hyperon is even able to act as a nuclear constituent. A nucleus containing a bound Λ^0 hyperon is called a *hyperfragment;* eventually the Λ^0 decays, of course, with the resulting nucleon and π meson either reacting with the parent nucleus or emerging from it entirely.

13.5   SYSTEMATICS OF ELEMENTARY PARTICLES

Despite the multiplicity of elementary particles and the diversity of their properties, it is possible to discern an underlying order in their behavior. The fact of this order does not constitute a theory of elementary particles, however, any more than the order found in atomic spectra constitutes a theory of the atom, but it does provide hope that there may indeed be a single theoretical picture that can encompass elementary-particle phenomena in the manner that the quantum theory encompasses atomic phenomena. Thus far no such picture has emerged, although some intriguing lines of approach have been proposed. In the remainder of this chapter we shall examine the regularities observed in elementary particles and their apparent significance.

Table 13.2 is a listing in order of rest mass of the relatively stable elementary particles we have thus far mentioned plus the η meson, which we shall discuss shortly. By relatively stable is meant that the half lives of the particles all greatly exceed the time required for light to travel a distance equal to the "diameter"

Table 13.1.

HYPERON PROPERTIES.

Particle	Mass, m_e	Half life, s	Decay
Λ^0	2,184	1.7×10^{-10}	$\Lambda^0 \rightarrow p + \pi^-$
			$\rightarrow n + \pi^0$
Σ^+	2,328	0.6×10^{-10}	$\Sigma^+ \rightarrow p + \pi^0$
			$\rightarrow n + \pi^+$
Σ^-	2,342	1.1×10^{-10}	$\Sigma^- \rightarrow n + \pi^-$
Σ^0	2,334	$<10^{-14}$	$\Sigma^0 \rightarrow \Lambda + \gamma$
Ξ^-	2,585	1.2×10^{-10}	$\Xi^- \rightarrow \Lambda + \pi^-$
Ξ^0	2,573	2.0×10^{-10}	$\Xi^0 \rightarrow \Lambda + \pi^0$
Ω^-	3,276	$\sim 10^{-10}$	$\Omega^- \rightarrow \Lambda + K^-$
			$\rightarrow \Xi^0 + \pi^-$

Table 13.2.

ELEMENTARY PARTICLES STABLE AGAINST DECAY BY THE STRONG NUCLEAR INTERACTION.

Class	Name	Particle +e	Particle 0	Particle −e	Antiparticle +e	Antiparticle 0	Antiparticle −e	Spin	Rest mass m_e	Rest mass MeV	Half life s	L	M	B	S	Y	I
												(Antiparticles have opposite signs)					
PHOTON	photon		γ			(γ)		1	0	0	stable	0	0	0	0	0	0
LEPTON	e-neutrino		ν_e			$\bar\nu_e$		1/2	0	0	stable	+1	0				
	μ-neutrino		ν_μ			$\bar\nu_\mu$		1/2	0	0	stable	0	+1				
	electron			e^-	e^+			1/2	1	0.51	stable	+1	0				
	μ meson			μ^-	μ^+			1/2	207	106	1.5×10^{-6}	0	+1				
MESON	π meson		π^0			(π^0)		0	264	135	7×10^{-17}			0	0	0	1
		π^+					π^-	0	273	140	1.8×10^{-8}						
	K meson	K^+					K^-	0	966	494	8×10^{-9}			0	+1	+1	1/2
			K^0			$\bar K^0$		0	974	498	$7 \times 10^{-11}; \ 4 \times 10^{-8}$						
	η meson		η^0			(η^0)		0	1,073	548	$\sim 10^{-18}$			0	0	0	0
BARYON	nucleon proton		p			$\bar p$		1/2	1,836	938	stable			+1	0	+1	1/2
	neutron		n			$\bar n$		1/2	1,839	940	6.5×10^2			+1	0	+1	
	Λ hyperon		Λ^0			$\bar\Lambda^0$		1/2	2,184	1,116	1.7×10^{-10}			+1	−1	0	0
	Σ hyperon	Σ^+					$\bar\Sigma^-$	1/2	2,328	1,192	0.6×10^{-10}			+1	−1	0	1
			Σ^0			$\bar\Sigma^0$			2,334	1,194	$<10^{-12}$						
				Σ^-	$\bar\Sigma^+$				2,342	1,197	1.1×10^{-10}						
	Ξ hyperon		Ξ^0			$\bar\Xi^0$		1/2	2,571	1,310	2.0×10^{-10}			+1	−2	−1	1/2
				Ξ^-	$\bar\Xi^+$				2,585	1,321	1.2×10^{-10}						
	Ω hyperon			Ω^-	$\bar\Omega^+$			3/2	3,276	1,674	$\sim 10^{-10}$			+1	−3	−2	0

of an elementary particle. This diameter is probably a little over 10^{-15} m, and the characteristic time required to traverse it at the speed of light is therefore of the order of magnitude of 10^{-23} s. Thus the particles in Table 13.2 are almost all capable of traveling through space as distinct entities along paths of measurable length in such devices as bubble chambers.

A considerable body of experimental evidence also points to the existence of many different "particles" whose lifetimes against decay are only about 10^{-23} s. What can be meant by a particle which exists for so brief an interval? Indeed, how can a time of 10^{-23} s even be measured? Such particles cannot be detected by observing their formation and subsequent decay in a bubble chamber or other instrument, but instead appear as resonant states in the interaction of more stable (and hence more readily observable) particles. Resonant states occur in atoms as energy levels; in Chap. 4 we reviewed the Franck-Hertz experiment, which showed the existence of atomic energy levels through the occurrence of inelastic electron scattering from atoms at certain energies only. An atom in a specific excited state is not the same as that atom in its ground state or in another excited state, but we do not usually speak of such an excited atom as though it were a member of a special species only because the interaction that gives rise to the excited state—the electromagnetic interaction—is well understood. A rather different situation holds in the case of elementary particles, where the various interactions involved are, except for the electromagnetic one, only partially understood, and much of our information comes from the properties of the resonances.

Let us see what is involved in a resonance in the case of elementary particles. An experiment is performed, for instance the bombardment of protons by energetic π^+ mesons, and a certain reaction is studied, for instance

$$\pi^+ + p \rightarrow \pi^+ + p + \pi^+ + \pi^- + \pi^0$$

The effect of the interaction of the π^+ and the proton is the creation of three new pions. In each such reaction the new mesons have a certain total energy that consists of their rest energies plus their kinetic energies relative to their center of mass. If we plot the number of events observed versus the total energy of the new mesons in each event, we obtain a graph like that of Fig. 13-3. Evidently there is a strong tendency for the total meson energy to be 785 MeV and a somewhat weaker tendency for it to be 548 MeV. We can say that the reaction exhibits resonances at 548 and 785 MeV or, equivalently, we can say that this reaction proceeds via the creation of an intermediate particle which can be either one whose mass is 548 MeV or one whose mass is 785 MeV. From the graph we can even estimate the mean lifetimes of these intermediate particles, which are known as the η and ω mesons, respectively. According to the uncertainty principle, the uncertainty in decay time of an unstable particle—which

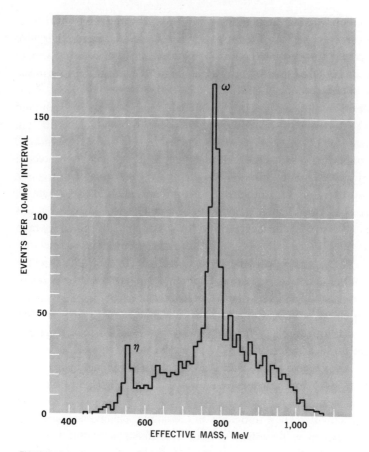

FIGURE 13-3 Resonant states in the reaction $\pi^+ + p \rightarrow \pi^+ + p + \pi^+ + \pi^- + \pi^0$ occur at effective masses of 548 and 785 MeV. By effective mass is meant the total energy, including mass energy, of the three new mesons relative to their center of mass.

is its mean lifetime τ—will give rise to an uncertainty in the determination of its energy—which is the width ΔE at half maximum of the corresponding peak in Fig. 13-3—whose relationship is

13.5 $\qquad \tau \Delta E \approx \hbar$

Hence the lifetimes of the resonances, or, just as well, the lifetimes of the η and ω mesons, can be established. The η lifetime is sufficiently long for it to be regarded as a relatively stable particle and it is included in Table 13.2, while the ω lifetime is too short by many orders of magnitude. We shall return to the resonance particles later in this chapter.

The particles in Table 13.2 seem to fall naturally into four general categories. In a class by itself is the photon, a stable particle with zero rest mass and unit spin. If there is a *graviton*, a particle that is the quantum of the gravitational field in the same sense that the photon is the quantum of the electromagnetic field or the pion the quantum of the nuclear force field, it would be another member of this class. The graviton, as yet undetected, should be massless and stable, and should have a spin of 2. Its interaction with matter would be extremely weak, and it is unlikely that present techniques are capable of verifying its existence. (The zero mass of the graviton can be inferred from the unlimited range of gravitational forces. As we saw in Sec. 13.2, the mutual forces between two bodies can be regarded as transmitted by the exchange of particles between them. If energy conservation is to be preserved, the uncertainty principle requires that the range of the force be inversely proportional to the mass of the exchanging particles, and so gravitational forces can have an infinite range only if the graviton mass is zero. A similar argument holds for the photon mass.)

After the photon in Table 13.2 come the e-neutrino and μ-neutrino, the electron, and the muon, all with spins of $\frac{1}{2}$. These particles are jointly called *leptons*. The π, K, and η mesons, all with spins of 0, are classed as *mesons*. (Despite its name, the μ meson has more in common with the other leptons than with the π, K, and η mesons.) The heaviest particles, namely the nucleons and hyperons, comprise the *baryons*.

While this grouping is reasonable on the basis of mass and spin alone, there is further evidence in its favor. Let us introduce three new quantum numbers, L, M, and B as follows. We assign the number $L = 1$ to the electron and the e-neutrino, and $L = -1$ to their antiparticles; all other particles have $L = 0$. We assign the number $M = 1$ to the μ meson and its neutrino, and $M = -1$ to their antiparticles; all other particles have $M = 0$. Finally, we assign $B = 1$ to all baryons, and $B = -1$ to all antibaryons; all other particles have $B = 0$. The significance of these numbers is that, in every process of whatever kind that involves elementary particles, the total values of L, M, and B remain constant. The classical conservation laws of energy, momentum, angular momentum, and electric charge plus the new conservation laws of L, M, and B help us to determine whether any given process is capable of taking place or not. An example is the decay of the neutron,

$$n^0 \rightarrow p^+ + e^- + \bar{\nu}_e$$

While $L = 0$ for the neutron and proton, $L = 1$ for the electron and -1 for the antineutrino, so that the total value of L before and after the decay is 0. Similarly, $B = 1$ for both neutron and proton, so that the total value of B before and after the decay is 1. The stability of the proton is a consequence of energy and baryon-number conservation: There are no baryons of smaller mass than the proton, and so the proton cannot decay.

13.6 STRANGENESS NUMBER

Despite the introduction of the quantum numbers L, M, and B, certain aspects of elementary-particle behavior still defied explanation. For instance, it was hard to see why certain heavy particles decay into lighter ones together with the emission of a gamma ray while others do not undergo apparently equally permissible decays. Thus the Σ^0 baryon decays into a Λ^0 baryon and a gamma ray,

$$\Sigma^0 \to \Lambda^0 + \gamma$$

while the Σ^+ baryon is never observed to decay into a proton and a gamma ray,

$$\Sigma^+ \nrightarrow p^+ + \gamma$$

Another peculiarity is based upon the general observation that physical processes in nature that release large amounts of energy take place more rapidly than processes that release small amounts. However, many strange particles whose decay releases considerable energy have relatively long lifetimes, well over a billion times longer than theoretical calculations predict. A third odd feature is that strange particles are never created singly, but always two or more at a time. These and still other considerations led to the introduction of a quantity called *strangeness number* S. Table 13.2 shows the values of S that are assigned to the various elementary particles. We note that L, B, and S are 0 for the photon and π^0 and η^0 mesons. Since these particles are also uncharged, there is no way to distinguish between them and their antiparticles. For this reason the photon and π^0 and η^0 mesons are regarded as their own antiparticles. Before we consider the interpretation of the strangeness number, we shall have to examine the various kinds of particle interaction.

There are apparently four types of interaction between elementary particles that, in principle, give rise to all the physical processes in the universe. The feeblest of these is the gravitational interaction. Next is the so-called weak interaction that is present between leptons and other leptons, mesons, or baryons in addition to any electromagnetic forces that may exist. The weak interaction is responsible for particle decays in which neutrinos are involved, notably beta decays. Stronger than gravitational and weak interactions are the electromagnetic interactions between all charged particles and also those with electric or magnetic moments. Finally, strongest of all are the nuclear forces (usually called simply strong forces when elementary particles are being discussed) that are found between mesons, baryons, and mesons and baryons.

The relative strengths of the strong, electromagnetic, weak, and gravitational interactions are in the ratios $1:10^{-2}:10^{-14}:10^{-40}$. Of course, the distances through which the corresponding forces act are very different. While the strong force between nearby nucleons is many powers of 10 greater than the gravitational force between them, when they are a meter apart the proportion is the other way. The structure of nuclei is determined by the properties of the strong interaction, while the structure of atoms is determined by those of the electromagnetic interaction. Matter in bulk is electrically neutral, and the strong and weak forces are severely limited in their range. Hence the gravitational interaction, utterly insignificant on a small scale, becomes the dominant one on a large scale. The role of the weak force in the structure of matter is apparently that of a minor perturbation that sees to it that nuclei with inappropriate neutron/proton ratios undergo corrective beta decays.

Let us now return to the strangeness number S. It is found that in all processes involving strong and electromagnetic interactions the strangeness number is conserved. The decay

$$\Sigma^0 \rightarrow \Lambda^0 + \gamma$$
$$S = -1 \quad -1 \quad 0$$

conserves S and is observed to occur, while the superficially similar decay

$$\Sigma^+ \nrightarrow p^+ + \gamma$$
$$S = -1 \quad 0 \quad 0$$

does not conserve S and has never been observed. Strange particles are created in high-energy nuclear collisions which involve strong interactions, and their multiple appearance results from the necessity of conserving S. The relative slowness with which all unstable elementary particles save the π^0 meson and η^0 meson decay is accounted for if we assume that weak interactions are also characteristic of mesons and baryons as well as leptons, though normally dominated by strong or electromagnetic interactions. With strong or electromagnetic processes impossible except in the above cases owing to the lack of conservation of S, only the weak interaction is available for processes in which the total value of S changes. Events governed by weak interactions take place slowly, as borne out by experiment. Even the weak interaction, however, is unable to permit S to change by more than +1 or −1 in a decay. Thus the Ξ^- hyperon does not decay directly into a neutron since

$$\Xi^- \nrightarrow n^0 + \pi^-$$
$$S = -2 \quad 0 \quad 0$$

but instead via the two steps

$$\Xi^- \rightarrow \Lambda^0 + \pi^-$$
$$S = -2 \quad -1 \quad 0$$

$$\Lambda^0 \rightarrow n^0 + \pi^0$$
$$S = -1 \quad 0 \quad 0$$

A quantity called *hypercharge*, Y, has also been found useful in characterizing particle families; it is conserved in strong interactions. Hypercharge is equal to the sum of the strangeness and baryon numbers of the particle families:

13.6 $\qquad Y = S + B$

For mesons the hypercharge is equal to the strangeness. The various hypercharge assignments are listed in Table 13.2.

13.7 ISOTOPIC SPIN

It is obvious from Table 13.2 that there are a number of particle families each of whose members has essentially the same mass and interaction properties but different charge. These families are called *multiplets,* and it is natural to think of the members of a multiplet as representing different charge states of a single fundamental entity. It has proved useful to categorize each multiplet according to the number of charge states it exhibits by a number I such that the multiplicity of the state is given by $2I + 1$. Thus the nucleon multiplet is assigned $I = \frac{1}{2}$, and its $2 \cdot \frac{1}{2} + 1 = 2$ states are the neutron and the proton. The π meson multiplet has $I = 1$, and its $2 \cdot 1 + 1 = 3$ states are the π^+, π^-, and π^0 mesons. The η meson has $I = 0$ since it occurs in only a single state and $2 \cdot 0 + 1 = 1$. There is evidently an analogy here with the splitting of an angular-momentum state of quantum number l into $2l + 1$ substates, and this has led to the somewhat misleading name of *isotopic spin quantum number* for I.

Pursuing the analogy with angular momentum, isotopic spin can be represented by a vector $\mathbf{I}$ in "isotopic spin space" whose component in any specified direction is governed by a quantum number customarily denoted I_3. The possible values of I_3 are restricted to $I, I - 1, \ldots, 0, \ldots, -(I - 1), -I$, so that I_3 is half-integral if I is half-integral and integral or zero if I is integral. The isotopic spin of the nucleon is $I = \frac{1}{2}$, which means that I_3 can be either $\frac{1}{2}$ or $-\frac{1}{2}$; the former is taken to represent the proton and the latter the neutron. In the case of the π meson, $I = 1$ and $I_3 = 1$ corresponds to the π^+ meson, $I_3 = 0$ to the π^0 meson, and $I_3 = -1$ to the π^- meson. The values of I_3 for the other mesons and baryons are assigned in a similar way.

The charge of a meson or baryon is related to its baryon number B, its strangeness number S, and the component I_3 of its isotopic spin by the formula

13.7 $\qquad q = e\left(I_3 + \dfrac{B}{2} + \dfrac{S}{2}\right)$

Each allowed orientation of the isotopic spin vector $\mathbf{I}$ hence is directly connected to the charge of the particle thus represented. In the case of the nucleon multiplet, the proton has $I_3 = \frac{1}{2}$, $B = 1$, and $S = 0$, so that $q = e$, while the neutron has $I_3 = -\frac{1}{2}$, $B = 1$, and $S = 0$, so that $q = 0$. In the case of the π meson multiplet, $B = S = 0$ and the three values of I_3 of 1, 0, and -1 respectively yield $q = e$, 0, and $-e$. Charge and baryon number B are conserved in all interactions. Thus I_3 must be conserved whenever S is conserved, namely in strong and electromagnetic interactions. Only in weak interactions does the total I_3 change.

An additional conservation law is suggested by the observed charge independence of nuclear forces, which result from the strong interaction. Such properties of a nucleus as its binding energy and pattern of energy levels change when a neutron is substituted for a proton or vice versa only by amounts that follow from purely electromagnetic considerations, implying that the strong interaction itself does not depend upon electric charge. Now the difference between a proton and a neutron in isotopic spin space lies only in the orientation of their isotopic spin vectors, and so we can say that the charge independence of the strong interaction means that this interaction is independent of orientation in isotopic spin space. Angular momentum is likewise independent of orientation in real space and it is conserved in all interactions, which might lead us to surmise that isotopic spin is conserved in strong interactions. This surmise happens to be correct; the isotopic spin quantum number I is found to be conserved in strong, but not in weak or in electromagnetic, interactions. We shall return to the relation between conservation principles and invariance with respect to symmetry operations in the next section.

We note that, although I_3 is conserved in electromagnetic interactions, I itself need not be. An example of a process in which I changes while I_3 does not is the decay of the π^0 meson into two photons:

$$\pi^0 \rightarrow \gamma + \gamma$$

A π^0 meson has $I = 1$ and $I_3 = 0$, while I is not defined for photons; there is no component I_3 of isotopic spin on either side of the equation, which is consistent with its conservation, although I has changed.

13.8 SYMMETRIES AND CONSERVATION PRINCIPLES

In the previous section the charge independence of the strong interaction was expressed in terms of the isotropy of isotopic spin space. By analogy with angular momentum, this symmetry was said to imply the conservation of isotopic spin in strong interactions. It is a remarkable fact that *all* known symmetries in the physical world lead directly to conservation laws, so that the relationship between symmetry under rotations of **I** and the conservation of **I** is wholly plausible.

Let us survey some of these symmetry-conservation relationships before continuing our discussion of elementary particles.

What is meant by a "symmetry"? Formally, if rather vaguely, we might say that a symmetry of a particular kind exists when a certain operation leaves something unchanged. A candle is symmetric about a vertical axis because it can be rotated about that axis without changing in appearance or any other feature; it is also symmetric with respect to reflection in a mirror. Table 13.3 lists the principal symmetry operations which leave the laws of physics unchanged under some or all circumstances. The simplest symmetry operation is translation in space, which means that the laws of physics do not depend upon where we choose the origin of our coordinate system to be. By more advanced methods than we are employing in this book, it is possible to show that the invariance of the description of nature to translations in space has as a consequence the conservation of linear momentum. Another simple symmetry operation is translation in time, which means that the laws of physics do not depend upon when we choose $t = 0$ to be, and this invariance has as a consequence the conservation of energy. Invariance under rotations in space, which means that the laws of physics do not depend upon the orientation of the coordinate

Table 13.3.

SOME SYMMETRY OPERATIONS AND THEIR ASSOCIATED CONSERVATION PRINCIPLES.

Symmetry operation	Conserved quantity
All interactions are dependent of:	
Translation in space	Linear momentum p
Translation in time	Energy E
Rotation in space	Angular momentum L
Electromagnetic gauge transformation	Electric charge q
Interchange of identical particles	Type of statistical behavior
Inversion of space, time, and charge	Product of charge parity, space parity, and time parity CPT
?	Baryon number B
?	Lepton number L
?	Lepton number M
The strong and electromagnetic interactions only are independent of:	
Inversion of space	Parity P
Reflection of charge	Charge parity C, isotopic spin component I_3, and strangeness S
The strong interaction only is independent of:	
Charge	Isotopic spin I

system in which they are expressed, has as a consequence the conservation of angular momentum.

Conservation of electric charge is related to gauge transformations, which are shifts in the zeros of the scalar and vector electromagnetic potentials V and $\mathbf{A}$. (As elaborated in electromagnetic theory, the electromagnetic field can be described in terms of the potentials V and $\mathbf{A}$ instead of in terms of $\mathbf{E}$ and $\mathbf{B}$, where the two descriptions are related by the vector calculus formulas $\mathbf{E} = -\nabla V$ and $\mathbf{B} = \nabla \times \mathbf{A}$.) Gauge transformations leave $\mathbf{E}$ and $\mathbf{B}$ unaffected since they are obtained by differentiating the potentials, and this invariance leads to charge conservation.

The interchange of identical particles in a system is a type of symmetry operation which leads to the preservation of the character of the wave function of a system. The wave function may be symmetric under such an interchange, in which case the particles do not obey the exclusion principle and the system follows Bose-Einstein statistics, or it may be antisymmetric, in which case the particles obey the exclusion principle and the system follows Fermi-Dirac statistics. Conservation of statistics (or, equivalently, of wave-function symmetry or antisymmetry) signifies that no process occurring within an isolated system can change the statistical behavior of that system. A system exhibiting Bose-Einstein behavior cannot spontaneously alter itself to exhibit Fermi-Dirac statistical behavior or vice versa. This conservation principle has applications in nuclear physics, where it is found that nuclei that contain an odd number of nucleons (odd mass number A) obey Fermi-Dirac statistics while those with even A obey Bose-Einstein statistics; conservation of statistics is thus a further condition a nuclear reaction must observe.

The conservations of the baryon number B and the lepton numbers L and M are alone among the principal conservation principles in having no known symmetries associated with them.

Apart from the charge independence of the strong interaction and its associated conservation of isotopic spin, which we have already mentioned, the remaining symmetry operations in Table 13.3 all involve *parities* of one kind or another. The term parity with no qualification refers to the behavior of a wave function under an inversion in space. By inversion in space is meant the reflection of spatial coordinates through the origin, with $-x$ replacing x, $-y$ replacing y, and $-z$ replacing z. If the sign of the wave function ψ does not change under such an inversion,

$$\psi(x, y, z) = \psi(-x, -y, -z)$$

and ψ is said to have *even* parity. If the sign of ψ changes,

$$\psi(x, y, z) = -\psi(-x, -y, -z)$$

and ψ is said to have *odd* parity. Thus the function $\cos x$ has even parity since $\cos x = \cos(-x)$, while $\sin x$ has odd parity since $\sin x = -\sin(-x)$.

If we write

$$\psi(x, y, z) = P\psi(-x, -y, -x)$$

we can regard P as a quantum number characterizing ψ whose possible values are $+1$ (even parity) and -1 (odd parity). Every elementary particle has a certain parity associated with it, and the parity of a system such as an atom or a nucleus is the product of the parity of the wave function that describes the coordinates of its constituent particles and the intrinsic parities of the particles themselves. Since $|\psi|^2$ is independent of P, the parity of a system is not a quantity that has an obvious physical consequence. However, it *is* found that the initial parity of an isolated system does not change during whatever events occur within it, which can be ascertained by comparing the parities of known final states of a reaction or transformation with the parities of equally plausible final states that are not observed to occur. A system of even parity retains even parity, a system of odd parity retains odd parity; this principle is known as conservation of parity.

The conservation of parity is an expression of the inversion symmetry of space, that is, of the lack of dependence of the laws of physics upon whether a left-handed or a right-handed coordinate system is used to describe phenomena. In Sec. 13.1 it was noted that the neutrino has a left-handed spin and the anti-neutrino a right-handed spin, so that there is a profound difference between the mirror image of either particle and the particle itself. This asymmetry implies that interactions in which neutrinos and antineutrinos participate—the weak interactions—need not conserve parity, and indeed parity conservation is found to hold true only in the strong and electromagnetic interactions. Historically the fact that spatial inversion is not invariably a symmetry operation was suggested by the failure of parity conservation in the decay of the K^+ meson, and was later confirmed by experiments showing the specific handedness of ν and $\bar{\nu}$.

Two other parities occur in Table 13.3, time parity T and charge parity C, which respectively describe the behavior of a wave function when t is replaced by $-t$ and when q is replaced by $-q$. The symmetry operation that corresponds to the conservation of time parity is *time reversal*. Time reversal symmetry implies that the direction of increasing time is not significant, so that the reverse of any process that can occur is also a process that can occur. In other words, if symmetry under time reversal holds, it is impossible to establish by viewing it whether a motion picture of an event is being run forwards or backwards. Although time parity T was long considered to be conserved in every interaction, it was discovered in 1964 that the $K_2{}^0$ meson can decay into a π^+ and a π^- meson, which violates the conservation of T. The symmetry of phenomena under time reversal thus has an ambiguous status at present. The symmetry operation that corresponds to the conservation of charge parity C is *charge conjugation*, which is the replacement of every particle in a system by its antiparticle. Charge parity C, like space parity P, is not conserved in weak interactions. However, despite the limited validities of the conservation of C, P, and T, there are good theoretical reasons for believing that the product CPT of the charge, space, and time parities of a system is invariably conserved. The conservation of CPT means that for every process there is an antimatter mirror-image counterpart that takes place in reverse, and this particular symmetry seems to hold even though its component symmetries sometimes fail individually.

13.9 THEORIES OF ELEMENTARY PARTICLES

In addition to the particles listed in Table 13.2 there are, as mentioned earlier, a great many "particles" of extremely brief lifetimes whose existence is revealed by resonances in interactions involving their longer-lived brethren. These resonant states are characterized by definite values of mass, charge, angular momentum, isotopic spin, parity, strangeness, and so on, and it is no more logical to disqualify them as legitimate particles because their existences are so transient than it is to consider the neutron, say, as merely an unstable state of the proton. Of course, it is possible to make out an excellent case for supposing the latter, and then to go on to generalize that *all* the various "elementary" particles are actually excited states of a very few truly elementary particles, as yet unidentified; this constitutes one line of attack toward a comprehensive theory of elementary particles. On the other hand, if we accept the particles of Table 13.2 as legitimate, then it is consistent to include the resonant states as well, and to seek a theoretical framework that embraces the entire collection of well over a hundred entities.

A recent proposal attempts to account for the various elementary particles in terms of another kind of particle called the *quark*. Three varieties of quark are postulated, plus their antiparticles, and all elementary particles are supposed to consist of combinations of quarks and antiquarks. The really revolutionary thing about quarks is that two of them should have charges of $-\frac{1}{3}e$ and the third should have a charge of $+\frac{2}{3}e$. According to this theory, each baryon is composed of three quarks, and each meson is composed of quark-antiquark pairs. Despite much effort, no experimental evidence in support of the existence of quarks has been found thus far, but the ideas that underlie their prediction are so persuasive that the hunt continues.

Several interesting and suggestive classification schemes have been devised for the strongly interacting particles based upon the abstract theory of groups. One of these schemes, the so-called *eightfold way*, collects isotopic spin multiplets into supermultiplets whose members have the same spin and parity but differ in charge and hypercharge (Figs. 13-4 and 13-5). The scheme prescribes the number of members each particular supermultiplet should have and also relates mass differences among these members. The great triumph of the eightfold way was its prediction of a previously unknown particle, the Ω^- hyperon, which was subsequently searched for and finally discovered in 1964. Other group-theoretic approaches have related the supermultiplets of the eightfold way to one another and have attempted to incorporate relativistic considerations into the comprehensive picture that is emerging.

The success of the eightfold way in organizing our knowledge of the strongly interacting particles implies that the symmetry of its mathematical structure has a counterpart in a symmetry in nature. The further we probe into nature, the more hints we receive of a profound order that underlies the complications and confusions of experience. But for all the elegance of the symmetries that have been revealed, there still remains the problem of the fundamental interactions themselves, what they signify, and how they are related to one another and to the properties of the particles through which they are manifested.

FIGURE 13-4 Supermultiplets of spin-½ baryons and spin-0 mesons stable against decay by the strong nuclear interaction. Arrows indicate possible transformations according to the eightfold way.

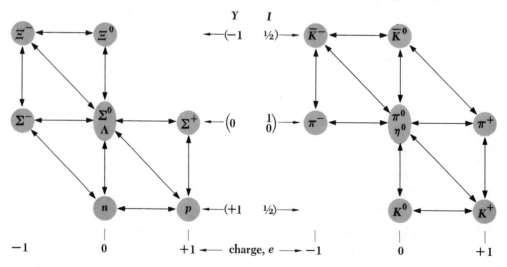

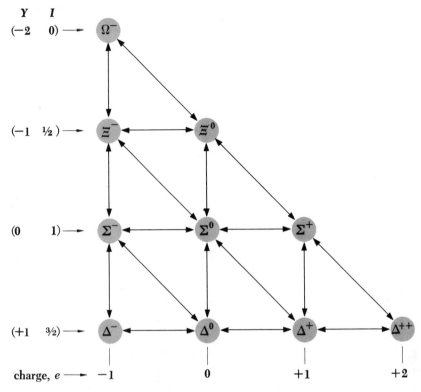

FIGURE 13-5 Baryon supermultiplet whose members have spin $\frac{3}{2}$ and (except Ω^-) are short-lived resonance particles; the Ξ and Σ particles here are heavier and have different spins from the ones in Table 13.2. Arrows indicate possible transformations according to the eightfold way. The Ω^- particle was predicted from this scheme.

Problems

1. (a) Find the maximum kinetic energy of the electron emitted in the beta decay of the free neutron. (b) What is the minimum binding energy that must be contributed by a neutron to a nucleus so that the neutron does not decay? Compare this energy with the observed binding energies per nucleon in stable nuclei.

2. Van der Waals forces are limited to very short ranges and do not have an inverse-square dependence on distance, yet nobody suggests that the exchange of a special meson-like particle is responsible for such forces. Why not?

3. What is the energy of each of the gamma rays produced when a π^0 meson decays? Must they be equal?

4. How much energy must a gamma-ray photon have if it is to materialize into a neutron-antineutron pair? Prove that such an event cannot occur in the absence of another body without violating either the conservation of energy or the conservation of linear momentum.

5. Why does a free neutron not decay into an electron and a positron? Into a proton-antiproton pair?

6. A π^0 meson whose kinetic energy is equal to its rest energy decays in flight. Find the angle between the two gamma-ray photons that are produced.

7. A proton of kinetic energy T_0 collides with a stationary proton, and a proton-antiproton pair is produced. If the momentum of the bombarding proton is shared equally by the four particles that emerge from the collision, find the minimum value of T_0.

8. Trace the decay of the Ξ^0 particle into stable particles.

9. A μ^- meson collides with a proton, and a neutron plus another particle are created. What is the other particle?

10. One theory of the evolution of the universe postulates that matter spontaneously comes into being in free space. If this matter were in the form of neutron-antineutron pairs, what conservation laws would be violated?

11. Which of the following reactions can occur? State the conservation laws violated by the others.

(a) $\qquad p + p \rightarrow n + p + \pi^+$
(b) $\qquad p + p \rightarrow p + \Lambda^0 + \Sigma^+$
(c) $\qquad e^+ + e^- \rightarrow \mu^+ + \pi^-$
(d) $\qquad \Lambda^0 \rightarrow \pi^+ + \pi^-$
(e) $\qquad \pi^- + p \rightarrow n + \pi^0$

12. The π^0 meson has neither charge nor magnetic moment, which makes it hard to understand how it can decay into a pair of electromagnetic quanta. One way to account for this process is to assume that the π^0 first becomes a "virtual" nucleon-antinucleon pair, whose members then interact electromagnetically to yield two photons whose energies total the mass energy of the π^0. How long does the uncertainty principle allow the virtual nucleon-antinucleon pair to exist? Is this long enough for the process to be observed?

13. The interaction of one photon with another can be understood by assuming that each photon can temporarily become a "virtual" electron-positron pair in free space, and the respective pairs can then interact electromagnetically.

(a) How long does the uncertainty principle allow a virtual electron-positron pair to exist if $h\nu \ll 2m_0c^2$, where m_0 is the electron rest mass? (b) If $h\nu > 2m_0c^2$, can you use the notion of virtual electron-positron pairs to explain the role of a nucleus in the production of an actual pair, apart from its function in assuring the conservation of both energy and momentum?

ANSWERS TO ODD-NUMBERED PROBLEMS

Chapter 1

1. 213 m
3. 2.6×10^8 m/s
5. 6 ft; 2.6 ft
7. 4.2×10^7 m/s
9. $0.8c$; $0.988c$; $0.9c$; $0.988c$
11. 4.2×10^7 m/s
13. 1.87×10^8 m/s; 1.64×10^8 m/s
15. 8.9×10^{-28} kg
17. 0.294 MeV
19. 2.7×10^{11} kg
21. 4.4×10^9 kg
23. $F = \dfrac{m_0 (dv/dt)}{(1 - v^2/c^2)^{3/2}}$
25. The results are different because all observers find the same value for the speed of light, whereas the speed of sound measured by an observer depends upon his own motion relative to the medium in which the sound waves propagate.
27. (a) 6, 10; (b) 10, 6, yes

Chapter 2

1. 1,800 Å
3. 5,400 Å; 3.9 eV
5. 2.83×10^{-19} J
7. 1.71×10^{30} photons/s
9. (a) 6.82×10^{-10} lb/in.^{2}
   (b) 4.24×10^{21} photons/m^2-s
   (c) 3.9×10^{26} watts; 1.2×10^{45} photons/s
   (d) 1.41×10^{13} photons/m^3

11. 1.24×10^5 V
13. 3.14 Å
15. 5×10^{18} Hz
17. 0.015 Å
19. 2.4×10^{19} Hz
21. 0.565 Å
23. (a) 2×10^{-3} eV
    (b) 2×10^{-25} eV
    (c) 3.5×10^{18} Hz; 7.7×10^3 Hz

Chapter 3

1. 6.86×10^{-12} m
3. 1.24 GeV; 616 MeV
5. $\lambda = 12.27 \left[V \left(\dfrac{eV}{2m_0c^2} + 1 \right) \right]^{-1/2}$
9. $u = w/2$
11. 6.2%
13. 1.16×10^5 m/s; 63 m/s
15. 1.05×10^{-23} kg-m/s; 1.16×10^7 m (7,200 mi!); the original narrow wave packet has spread out in 1 s to a much wider one because the phase velocities of the waves involved vary with k and a large range of wave numbers was present in the original packet.
17. Each atom in a solid is limited to a certain definite region of space—otherwise the assembly of atoms would not be a solid. The uncertainty in the position of each atom is therefore finite, and its momentum and hence energy cannot be zero. There is no restriction on the position of a mole-

457

cule in an ideal gas, and so the uncertainty in its position is effectively infinite and its momentum and hence energy can be zero.

Chapter 4

1. $10°$

3. 0.9986

7. 1.14×10^{-13} m

9. $m/m_0 = 1.002$

11. $f = \dfrac{1}{2\pi}\sqrt{\dfrac{Qe}{4\pi\varepsilon_0 mR^3}}$. For the hydro-

gen atom, $f = 6.6 \times 10^{15}$ s^{-1}, which is comparable with the highest frequencies in the hydrogen spectrum.

13. 920 Å

15. 12 V

17. 8.2×10^6 rev

19. 1.05×10^5 K

21. 2.4 Å

23. 1.9 keV

25. $(a)\ E_n = -\dfrac{m'Z^2 e^4}{8\varepsilon_0{}^2 h^2}\left(\dfrac{1}{n^2}\right)$

(b)

$(c)\ 2.28 \times 10^{-8}$ m

Chapter 5

3. 2.1 MeV

5. 2.07×10^{-15} eV

7. Classically $\bar{T} = \bar{V} = E/2$, where $\bar{T}$ and $\bar{V}$ are averages over an entire period of oscillation.

11. $b,\ d,\ f.$

Chapter 6

5. $r = (3 \pm \sqrt{5})a_0$

7. $68\%;\ 25\%$

9. $p,\ 29\%;\ d,\ 18\%;\ f,\ 13\%$

Chapter 7

1. 182

3. The alkali metals are the largest in each period of the periodic table, since their atomic structures consist of a single electron outside closed inner shells that shield the electron from all but $+e$ of nuclear charge. There is then a regular decrease in size within each period as the nuclear charge increases, which pulls the outer electrons in closer to the nucleus. At the end of each period there is a small increase in size due to the mutual repulsion of the outer electrons.

5. 1.39×10^{-4} eV

7. 0.0283 Å

9.

state	S	L	J
$^1S_0$	0	0	0
$^3P_2$	1	1	2
$^2D_{3/2}$	$\frac{1}{2}$	2	$\frac{3}{2}$
$^5F_5$	2	3	5
$^6H_{5/2}$	$\frac{5}{2}$	5	$\frac{5}{2}$

11. There are no other allowed states.

13. $^2P_{1/2}$

15. $2.58\mu_B;\ 4$

17. 18.5 T

19. Cobalt $(Z = 27)$; molybdenum $(Z = 42)$

Chapter 8

1. 3.5×10^4 K

3. $F_2{}^+;\ F_2{}^-$

5. C^{13}

7. 1.27 Å

9. 2.23 Å

11. No.

13. 1.24×10^{14} Hz

Chapter 9

3. $2/\sqrt{\pi}\ v_p$

5. 4.0×10^8 neutrons/m^3

7. $(a)\ 1.00:1.68:0.89:0.22:0.027$

(b) yes; 1533 K

9. $1.00:2.3 \times 10^{-10}:6.2 \times 10^{-12}:2.3 \times 10^{-12}$

11. 5800 K

13. A fermion gas will exert the greatest pressure because the Fermi distribution has a larger proportion of high-energy particles than the other distributions; a boson gas will exert the least pressure because the Bose distribution has a larger proportion of low-energy particles than the others.

15. 18.70 eV; 16.84 eV. The He atoms are needed to maintain an inverted energy population in the Ne atoms by transferring energy to them in collisions, which supplements the direct excitation of Ne atoms by electron impact.

Chapter 10

1. (a) The van der Waals forces increase the cohesive energy since they are attractive. (b) The zero-point oscillations decrease the cohesive energy since they represent a mode of energy possession present in a solid but not in individual atoms or ions.

3. The heat lost by the expanding gas is equal to the work done against the attractive van der Waals forces between its molecules.

5. (a) In a metal, valence electrons can find unoccupied excited energy states in the conduction band for any excitation energy, however small.

(b) The energy gap in semiconductors is small ($\leqslant 1.5$ eV), and so photons of visible light can excite valence electrons to the conduction band although photons of infrared light have insufficient energy for this purpose.

(c) The energy gap is so large that photons of visible light cannot provide enough excitation energy from electrons in the valence band to reach the conduction band.

7. p-type

9. kT is a very small fraction of ε_F, and so the electron energy distribution is not very temperature sensitive.

11. 1.9 eV; 7.6×10^{-10} m

13. 7.29 eV; $n = 9.4$

17. 3.3 eV; 2.56×10^4 K; 1.08×10^6 m/s

19. 11 eV

21. 2

23. 50 Å; the ionization energy of the electron is 0.009 eV, which is much smaller than the energy gap and not very far from the 0.025 eV value of kT at 20°C.

25. Because $m°/m = 1.01$ in copper.

Chapter 11

1. 8.83 cm

3. 19 percent B^{10}, 81 percent B^{11}

5. 34.97 u

7. 7.98 MeV

9. 15.6 MeV

11. Nuclear forces cannot be strongly charge-dependent.

13. The nucleon kinetic energy that corresponds to the momentum uncertainty implied by an uncertainty in position of 2 fm is 5.2 MeV, which is entirely consistent with a potential well 35 MeV deep.

15. $u_F = \dfrac{(3/\pi)^{4/3} h^2}{32 m R_0{}^2}$; 34 MeV

Chapter 12

1. 1/4

3. 1,620 yr

5. 1.23×10^4 s^{-1}

7. 3.37

9. The mass of ${}_4^7$Be is not sufficiently larger than that of ${}_3^7$Li to permit the creation of a positron.

11. Hint: the 39th proton in ${}_{39}^{89}$Y is normally in a $p_{1/2}$ state, and the next higher state open to this proton is a $g_{9/2}$ state.

13. 3.33 MeV

15. The neutron cross section decreases

with increasing energy because the likelihood that a neutron will be captured depends upon how much time it spends near a particular nucleus, and this is inversely proportional to its speed. The proton cross section is smaller at low energies because of the repulsive force exerted by the positive nuclear charge, which provides a potential barrier the proton must tunnel through.

17. 1.2×10^{-6}

19. 0.21

21. 0.1%

Chapter 13

1. 0.78 MeV; 1.29 MeV, which is well under the observed binding energies per nucleon of stable nuclei.

3. 68 MeV; yes, in order that momentum be conserved.

5. This decay conserves neither baryon number nor spin; this decay conserves neither baryon number, spin, nor energy.

7. 5,630 MeV

9. A neutrino.

11. (*a*) and (*e*) can occur; (*b*) violates conservation of *B* and spin; (*c*) violates conservation of *L*, *M*, and spin; (*d*) violates conservation of *B* and spin.

13. 6.4×10^{-22} s; the strong electric field of the nucleus separates the electron and positron far enough so that they cannot recombine afterward to reconstitute the photon.

INDEX